Texts in Theoretical Computer Science
An EATCS Series

T0155768

Springer
Berlin
Heidelberg
New York
Barcelona
Hong Kong
London
Milan
Paris
Singapore
Tokyo

Heribert Vollmer

Introduction to Circuit Complexity

A Uniform Approach

With 29 Figures

 Springer

Author

Dr. Heribert Vollmer
Universität Würzburg
Theoretische Informatik
Am Exerzierplatz 3
D-97072 Würzburg, Germany
vollmer@informatik.uni-wuerzburg.de

Series Editors

Prof. Dr. Wilfried Brauer
Institut für Informatik, Technische Universität München
Arcisstrasse 21, D-80333 München, Germany

Prof. Dr. Grzegorz Rozenberg
Department of Computer Science, University of Leiden
Niels Bohrweg 1, P.O.Box 9512, 2300 RA Leiden, The Netherlands

Prof. Dr. Arto Salomaa
Data City, Turku Centre for Computer Science
FIN-20520 Turku, Finland

ISBN 978-3-642-08398-3

ACM Computing Classification (1998): F.1.1, F.1.3, F.2.0, F.2.3, F.4.1, F.4.3

Library of Congress Cataloging-in-Publication Data applied for

Die Deutsche Bibliothek – CIP-Einheitsaufnahme
Vollmer, Heribert: Introduction to circuit complexity: a uniform approach /
Heribert Vollmer. – Berlin; Heidelberg; New York; Barcelona; Hong Kong;
London; Milan; Paris; Singapore; Tokyo: Springer 1999

Cover design: design & production GmbH, Heidelberg

For Siani

Omnia sub specie æternitatis.

SPINOZA

Preface

This introductory textbook presents an algorithmic and computability based approach to circuit complexity. Intertwined with the consideration of practical examples and the design of efficient circuits for these, a lot of care is spent on the formal development of the computation model of uniform circuit families and the motivation of the complexity classes defined in this model.

Boolean circuits gain much of their attractiveness from the fact that non-trivial lower bounds for the complexity of particular practical problems in this model are known. However, I feel that circuit complexity should not be reduced to the theory of lower bounds. In the past few years, the study of complexity classes defined by Boolean circuits, as well as their properties and how they relate to other computational devices, has become more and more important. Best known are probably the classes that form the NC hierarchy, which was defined to capture the notion of problems that have very fast parallel algorithms using a feasible amount of hardware. This study is thus motivated from an algorithmic point of view, and therefore an important issue is the so-called *uniformity* of circuits, because only a uniform circuit family can be regarded as an implementation of an algorithm.

This book presents some classical as well as some recent lower bound proofs; however, I restrict myself to a small and subjective selection of important examples. The reader who is interested in exploring this topic further is directed to other textbooks (an extensive list of pointers to excellent literature is given in the Bibliographic Remarks of this book). Here, instead, the theory of circuit complexity classes, mostly connected with the NC hierarchy mentioned above, is developed thoroughly.

It is difficult, in times of such rapid developments in a field as we witness today in circuit complexity, to write a textbook which is not already out of date when it is published. Therefore the material presented here is chosen from a basic core of fundamental results which underlies most current research in circuit complexity and hopefully will still be important in the future. Of course a textbook of this size cannot even touch all important and interesting developments in circuit complexity. However, I have increased the number of topics and results by covering some of them in the exercises, and introducing others in the Bibliographic Remarks sections.

The first three chapters can be used for a one-semester introduction to circuit complexity on the advanced undergraduate or first-year graduate level. (For readers familiar with the German university system, I have used Chaps. 1–3 for a Hauptstudiumsvorlesung Schaltkreiskomplexität for students with a moderate background in theoretical computer science.) The second half of the book, suited for an ongoing course or an advanced seminar, requires more experience from the reader, but is certainly still accessible to first-year graduate students.

The reader is expected to possess some knowledge and experience about formal languages and complexity of deterministic and nondeterministic Turing machines, as usually gained in a one-semester introduction to foundations of computing (refer, e. g., to [HU79]). Besides this only basic mathematics is required (however, experience in a standard programming language will be helpful). An exception from this is Chap. 6. While I have tried to keep the exposition self-contained by introducing all necessary concepts, a full comprehension of most of the results presented in this final chapter will probably only be possible for a student familiar with the basic notions and results from complexity theory.

A few notes about how to use this book: Generally, each chapter presupposes the material of the preceding chapters. It is suggested that the reader quickly browse the Appendix "Mathematical Preliminaries" (p. 233) before starting with the main text, and then later study them in more detail as necessary. Each chapter ends with a number of exercises. Most exercises can be answered immediately if the material of the text has been understood. Starred exercises require some afterthought and more time. Double-starred exercises present quite difficult or lengthy problems. For most starred and all double-starred exercises, a pointer to the literature where a solution can be found is given.

Current information on this book may be found on the Web at the following location:

`http://www-info4.informatik.uni-wuerzburg.de/CC/`

If you find errors in the book, please report them using the above URL; or send an e-mail to vollmer@informatik.uni-wuerzburg.de.

Many people have contributed to the development of this book with helpful hints, discussions, and proof reading of the manuscript. I am grateful to Eric Allender, David Mix Barrington, Christian Glaßer, Georg Gottlob, Sven Kosub, Pierre McKenzie, Ken Regan, Steffen Reith, Heinz Schmitz, Klaus W. Wagner, and, in particular, to Ulrich Hertrampf. I also thank Hans Wössner and Ingeborg Mayer from Springer-Verlag for their excellent cooperation, and the copy editor, Reginald Harris, for many very helpful remarks.

Würzburg, Spring 1999   *Heribert Vollmer*

Table of Contents

Introduction

Theoretical investigations of *switching circuits* go back to the early papers by Shannon and Lupanov [Sha38, RS42, Sha49, Lup58]. Shannon (and his co-author Riordan) used Boolean algebra to design and analyze switching circuits. Lupanov was the head of a group of Russian mathematicians, working on the question of how many gates a switching circuit must have to be able to perform certain tasks.

The theory of logical design and switching circuits then developed rapidly. The direction which *complexity theoretic research on circuits* took was heavily influenced by Savage's textbook [Sav76]. Savage (also in earlier papers) related Boolean circuits to other well-established computation models, most prominently the Turing machine. This offered the hope that the $P \overset{?}{=}$ NP-problem might be solved via circuit complexity. This hope was boosted by the development of strong lower bound techniques in the 1980s.

Around that time, a variety of concrete and abstract models for parallel computers also emerged. Boolean circuits could be related to these models (in particular, the *parallel random access machine*, PRAM), facilitating the study of the computational power of such devices.

Thus the field of circuit complexity has attracted interest from a number of different viewpoints, both practical and theoretical. The reader will see throughout this book that a lot of the questions that were asked, and a lot of the definitions and constructions made, can only be understood with this two-fold influence in mind.

In Chap. 1 we start by examining a number of problems of highly practical relevance, such as basic arithmetic operations (addition, multiplication, division), counting, sorting, etc., from a circuit complexity viewpoint. We formally develop the computation model of Boolean circuits and construct efficient circuits for these problems.

The computation model of circuits has one characteristic which makes it different from other familiar models such as the random access machine or the Turing machine: A Boolean circuit has a fixed number of input gates. Thus a single circuit can only work on inputs of one fixed length (in contrast to a machine program which works for inputs of arbitrary length). If we want to solve the problems above (which are defined for different input lengths) by circuits, we therefore have to construct one circuit for each input length. This is why the Boolean circuit model is referred to as a *non-uniform* computation model, in contrast to, e. g., the Turing machine, which is a *uniform* model. The name stems from the fact that the various circuits for different input lengths may have a completely dissimilar structure. This has the consequence, as we will see, that there are even non-computable functions which can be "computed" by small circuits.

From an algorithmic point of view, this is certainly not what we wish: An algorithm for a problem should be a finite object (as a finite text in a programming language, or a finite transition function of a Turing machine). Hence an infinite family of circuits, one for each input length, will not be considered an algorithm. However, the examples we construct will have the property that the distinct circuits in the family have a very regular structure, in particular, circuits for different input lengths do not differ too much. This makes it possible to give a finite description of the structure of the infinite family. If such a description is possible then we say the circuit family is *uniform*. We will examine several ways to make these ideas precise, leading to various so-called *uniformity conditions*.

In Chap. 2 we compare the model of Boolean circuits with a number of other computation models, such as deterministic and alternating Turing machines and parallel random access machines. We will show how these machines can simulate uniform circuit families, and, vice versa, how circuits can be constructed to simulate the computation of such machines. We will also consider so-called *non-uniform Turing machines* and relate them to non-uniform circuits.

In Chap. 3 we turn to the theory of lower bounds. We start with the examination of some abstract problems which, however, in the end lead to results about the complexity of the problems defined and examined in Chap. 1. We will thus be able to give very precise statements about the circuit complexity of problems like sorting, multiplication, etc.

All lower bounds given in this chapter will indeed hold for non-uniform circuits; e. g., we will show that multiplication cannot even be computed by non-uniform circuits of a certain complexity. In Chap. 4 we come back to our algorithmic standpoint and hence to the study of uniform circuits. We examine the class NC, a class of immense importance in the theory of parallel algorithms. This class, in a sense, captures the notion of problems with very fast parallel algorithms using a feasible amount of hardware. We will consider different aspects of the class NC and its structure, and come up with relations to diverse fields of mathematics such as group theory or finite model theory. A further computation model, the so-called *branching programs*, will turn out to be important in this chapter.

In Chap. 4 we also develop a theory of P-completeness, analogous to the well-known NP-completeness theory. P-complete problems are the "hardest" problems in the class P, and they will most likely not be in NC, i. e., they will not have efficient parallel algorithms. We will encounter a large number of examples in this book.

In Chap. 5 we turn to circuits whose basic components are not logical gates but addition and multiplication gates. This study is thus not immediately motivated from a VLSI or engineering point of view, as is the case for Boolean circuits. The importance of this model stems from the fact that it serves as a theoretical basis for the field of computer algebra. There are interesting connections between Boolean circuits and these so-called *arithmetic circuits*, as we will see.

All problems considered in Chaps. 1–5 have efficient sequential algorithms, i. e., they are members of the class P. In Chap. 6 we turn to higher classes and show how they relate to Boolean circuits. We also come back to the theory of lower bounds in this final chapter, and we examine the power of (non-uniform) small circuits, where "small" means having a number of gates bounded by a polynomial in the input length. Can, for example, NP-complete problems be solved by such circuits? We will also present lower bounds which are known to hold only for uniform circuits. We will, for instance, show that certain functions cannot be computed by so-called uniform *threshold circuits* of small size, but it is still open whether small non-uniform threshold circuits exist for these problems.

1. Complexity Measures and Reductions

1.1 An Example: Addition

Suppose we are given two binary strings, each consisting of n bits, $a = a_{n-1}a_{n-2}\cdots a_0$, and $b = b_{n-1}b_{n-2}\cdots b_0$. We want to solve the following problem: Interpret a and b as binary representations of two natural numbers and compute their sum (again in binary). We refer to this problem as ADD.

In this book we will use the following shorthand to define such problems.

> *Problem:* ADD
> *Input:* two n bit numbers $a = a_{n-1}\cdots a_0$ and $b = b_{n-1}\cdots b_0$
> *Output:* $s = s_n \cdots s_0$, $s =_{\text{def}} a + b$

This defines for every n a function $\text{ADD}^{2n} : \{0,1\}^{2n} \to \{0,1\}^{n+1}$.

How can we compute ADD^{2n}? The well-known algorithm proceeds as follows: It uses auxiliary Boolean variables $c_n, \ldots, c_0$. We set $s_0 = a_0 \oplus b_0$, $c_0 = 0$, and then define inductively $c_i = (a_{i-1} \wedge b_{i-1}) \vee (a_{i-1} \wedge c_{i-1}) \vee (b_{i-1} \wedge c_{i-1})$ (i.e., c_i is on if at least two of $a_{i-1}, b_{i-1}, c_{i-1}$ are on), $s_i = a_i \oplus b_i \oplus c_i$ for $1 \le i < n$, and $s_n = c_n$. (Boolean connectives such as $\oplus$ are defined in Appendix A5.)

This algorithm adds two n bit numbers in $O(n)$ steps. It is essentially an implementation of the school method for addition in the binary number system. The bit c_i is the *carry* that ripples from position $i - 1$ into position i.

Let us now address the question whether we can do better in parallel. The above procedure is sequential, but is there another way?

The answer is "yes," and it is based on the following simple idea: There is a carry that ripples into position i if and only if there is some position $j < i$ to the right where this carry is *generated*, and all positions in between *propagate* (i.e., do not eliminate) this carry. A carry is generated at position i if and only if both input bits a_i and b_i are on, and a carry is eliminated at position i, if and only if both input bits a_i and b_i are off. This leads to the following definitions:

For $0 \le i < n$, let

$$g_i = a_i \wedge b_i \qquad \text{(position } i \text{ generates a carry);}$$
$$p_i = a_i \vee b_i \qquad \text{(position } i \text{ propagates a carry that ripples into it).}$$

Now the following holds: A carry ripples into position i if and only if there is a position $j < i$ where a carry is generated and all positions in between propagate it. Formally:

$$c_i = \bigvee_{j=0}^{i-1} \left(g_j \wedge \bigwedge_{k=j+1}^{i-1} p_k \right) \qquad \text{for } 1 \leq i \leq n. \qquad (1.1)$$

Once we have computed the carry, the bits of the sum are computed as before: $s_0 = a_0 \oplus b_0$, $s_i = a_i \oplus b_i \oplus c_i$ for $1 \leq i < n$, and $s_n = c_n$. Observe that these formulae allow all the c_i to be computed in parallel, since they only depend on the inputs a and b, and once we have computed the carry bits, all the bits of the sum s can be computed in parallel. This procedure is called the *carry look-ahead adder*. In an idealized parallel computer, where we do not care about the number of processors used and moreover assume a perfect interconnection network where every processor can read the input and is connected to all other processors, we may compute all the g_i and p_i in one time step, then the c_i in a second step, and finally all of the output bits c_i. In such a way the function ADD can be computed in parallel in constant time. We come back to parallel machines and develop the above ideas precisely in Sect. 2.7.

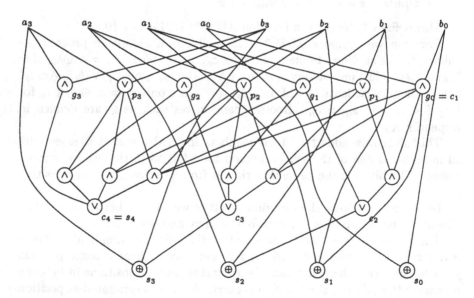

Fig. 1.1. Carry look-ahead adder for 3 bit numbers

These formulae for addition can be visualized as in Fig. 1.1. (Observe that the value p_0 is actually never needed.) A Boolean circuit now is essentially nothing else than a labeled directed acyclic graph as in Fig. 1.1. It has nodes with in-degree 0 corresponding to the inputs, nodes with out-degree 0 corresponding to the outputs, and inner nodes corresponding to the computation of a Boolean function. We define these ideas more precisely in the next section.

1.2 Circuit Size and Depth

A Boolean circuit is a graphical representation of a (sequence of) propositional formulas where common sub-formulas may be re-used by allowing what we will call *fan-out* greater than one. Let us develop these ideas in a formal way. In Fig. 1.1 we had nodes corresponding to inputs and nodes where some computation takes place. Some of the latter nodes are designated as output nodes. The computation taking place in non-input nodes is specified by associating the node with a *Boolean function*.

Definition 1.1. A *Boolean function* is a function $f: \{0,1\}^m \to \{0,1\}$ for some $m \in \mathbb{N}$.

Example 1.2. The functions $\wedge, \vee, \oplus: \{0,1\}^2 \to \{0,1\}$ (mostly used in infix notation, e. g., $x \wedge y$ instead of $\wedge(x,y)$), and $\neg: \{0,1\} \to \{0,1\}$ are well known (see also Appendix A5). In the following we will consider generalizations of the first three functions to arbitrary arities, i. e., we define

$$\wedge^m: \{0,1\}^m \to \{0,1\}, \quad \wedge^m(x_1,\ldots,x_m) =_{\text{def}} x_1 \wedge \cdots \wedge x_m$$
$$\vee^m: \{0,1\}^m \to \{0,1\}, \quad \vee^m(x_1,\ldots,x_m) =_{\text{def}} x_1 \vee \cdots \vee x_m$$
$$\oplus^m: \{0,1\}^m \to \{0,1\}, \quad \oplus^m(x_1,\ldots,x_m) =_{\text{def}} x_1 \oplus \cdots \oplus x_m$$

The consideration of sequences of Boolean functions with one function for every possible number of inputs leads us to the following definition.

Definition 1.3. A *family of Boolean functions* is a sequence $f = (f^n)_{n \in \mathbb{N}}$, where f^n is an n-ary Boolean function.

Example 1.4. We set $\wedge =_{\text{def}} (\wedge^n)_{n \in \mathbb{N}}$, $\vee =_{\text{def}} (\vee^n)_{n \in \mathbb{N}}$, and $\oplus =_{\text{def}} (\oplus^n)_{n \in \mathbb{N}}$. In cases where no confusion can arise, we will simply write, e. g., $\wedge(a_1,\ldots,a_n)$ for $\wedge^n(a_1,\ldots,a_n)$ (as we already have in Example 1.2).

The set of functions allowed in a Boolean circuit is given by a *basis*.

Definition 1.5. A *basis* is a finite set consisting of Boolean functions and families of Boolean functions.

Observe that a basis always has a finite number of elements, but some of its elements may be infinite objects (families of Boolean functions). A Boolean circuit (or circuit for short) is defined now as follows:

Definition 1.6. Let B be a basis. A *Boolean circuit* over B with n inputs and m outputs is a tuple

$$C = (V, E, \alpha, \beta, \omega),$$

where (V, E) is a finite directed acyclic graph, $\alpha \colon E \to \mathbb{N}$ is an injective function, $\beta \colon V \to B \cup \{x_1, \ldots, x_n\}$, and $\omega \colon V \to \{y_1, \ldots, y_m\} \cup \{*\}$, such that the following conditions hold:

1. If $v \in V$ has in-degree 0, then $\beta(v) \in \{x_1, \ldots, x_n\}$ or $\beta(v)$ is a 0-ary Boolean function (i. e. a Boolean constant) from B.
2. If $v \in V$ has in-degree $k > 0$, then $\beta(v)$ is a k-ary Boolean function from B or a family of Boolean functions from B.
3. For every i, $1 \leq i \leq n$, there is at most one node $v \in V$ such that $\beta(v) = x_i$.
4. For every i, $1 \leq i \leq m$, there is exactly one node $v \in V$ such that $\omega(v) = y_i$.

If $v \in V$ has in-degree k_0 and out-degree k_1, then we say: v is a *gate* in C with *fan-in* k_0 and *fan-out* k_1. If v is a gate in C then we also write $v \in C$ instead of $v \in V$. If $e = (u, v) \in E$ then we say: e is a *wire* in C; more specifically we say that e is an input wire of v and an output wire of u; and that u is a *predecessor gate* of v. If $\beta(v) = x_i$ for some i then v is an *input gate* or *input node*. If $\omega(v) \neq *$ then v is an *output gate* or *output node*.

$x_1, \ldots, x_n, y_1, \ldots, y_m, *$ are special symbols that are (besides elements from B) attached to certain nodes $v \in V$. The idea behind the definition is that the β function gives the *type* of a gate v, i. e., either input node or computation node. The ω function designates certain nodes as output nodes. If $\omega(v) = y_i$ then v yields the ith bit of the output, see below. If $\omega(v) = *$ then v is not an output node. Note that output nodes are usual computation nodes that can have fan-out greater 0, which then means that the particular output bit is needed for further computation. The importance of α will shortly become clear. At the moment, let us just remark that it defines an ordering on the edges of C.

The labeled directed graph in Fig. 1.1 is a circuit in this sense. The basis is $\{\wedge, \vee, \oplus^2, \oplus^3\}$. The nodes that are labeled $a_0, \ldots, a_3, b_0, \ldots, b_3$ are the nodes with β-values $x_1, \ldots, x_8$. The nodes labeled $s_0, \ldots, s_4$ are the nodes with ω-values $y_1, \ldots, y_5$.

A circuit $C = (V, E, \alpha, \beta, \omega)$ with n inputs and m outputs computes a function

$$f_C \colon \{0, 1\}^n \to \{0, 1\}^m.$$

This is formalized in the following definition.

Definition 1.7. Let $C = (V, E, \alpha, \beta, \omega)$ be a circuit over B with n inputs and m outputs. First, we define inductively for every $v \in V$ a function $\mathrm{val}_v \colon \{0, 1\}^n \to \{0, 1\}$ as follows. Let $a_1, \ldots, a_n$ be arbitrary binary values.

1. If $v \in V$ has fan-in 0 and if $\beta(v) = x_i$ for some i, $1 \leq i \leq n$, then $\mathrm{val}_v(a_1, \ldots, a_n) =_{\mathrm{def}} a_i$. If $v \in V$ has fan-in 0 and if $\beta(v) = b$ is a 0-ary function from B, then $\mathrm{val}_v(a_1, \ldots, a_n) =_{\mathrm{def}} b$.

2. Let $v \in V$ have fan-in $k > 0$, and let $v_1, \ldots, v_k$ be the gates that are predecessors of v ordered in such a way that $\alpha(v_1) < \cdots < \alpha(v_k)$. Let $\beta(v) = f \in B$. If f is a k-ary function, then let

$$\mathrm{val}_v(a_1, \ldots, a_n) =_{\mathrm{def}} f\left(\mathrm{val}_{v_1}(a_1, \ldots, a_n), \ldots, \mathrm{val}_{v_k}(a_1, \ldots, a_n)\right).$$

Otherwise f must be a family of Boolean functions, $f = (f^n)_{n \in \mathbb{N}}$. In this case, we define

$$\mathrm{val}_v(a_1, \ldots, a_n) =_{\mathrm{def}} f^k\left(\mathrm{val}_{v_1}(a_1, \ldots, a_n), \ldots, \mathrm{val}_{v_k}(a_1, \ldots, a_n)\right).$$

For $1 \leq i \leq m$ let v_i be the unique gate $v_i \in V$ with $\omega(v_i) = y_i$. Then the function computed by C, $f_C \colon \{0,1\}^n \to \{0,1\}^m$, is given for all $a_1, \ldots, a_n \in \{0,1\}$ by

$$f_C(a_1, \ldots, a_n) =_{\mathrm{def}} \left(\mathrm{val}_{v_1}(a_1, \ldots, a_n), \ldots, \mathrm{val}_{v_m}(a_1, \ldots, a_n)\right).$$

Thus we see that the function α is needed to specify the order of the predecessor gates to a given gate. This is not important for the functions $\wedge$, $\vee$, and $\oplus$, since permuting the input bits here does not change the output value. However, bases B are conceivable which contain functions for which the order of the arguments is important. Generally, α is obsolete if B consists only of *symmetric functions*.

Definition 1.8. A Boolean function f is *symmetric*, if for any input x, the value $f(x)$ depends only on the number of ones in x; in other words: $f(x) = f(x')$ whenever x and x' have the same number of bits that are equal to 1.

If there are only symmetric functions in B, then we denote a circuit over B simply by $C = (V, E, \beta, \omega)$.

Complexity theorists usually pay special attention to 0-1-valued functions, because they can be interpreted as characteristic functions of some set of words.

Definition 1.9. If $C = (V, E, \alpha, \beta, \omega)$ is a circuit with one output gate, i.e., C computes the characteristic function of some language A, then we say: C *accepts* A. If $x \in \{0,1\}^*$ such that $f_C(x) = 1$ then we also say: C accepts x.

A single circuit computes only a *finite function*, i.e., a function with finite domain and range. As an example, for every $n \in \mathbb{N}$ the function ADD^{2n} from Sect. 1.1 is finite. If a circuit accepts a language A, then A must be finite ($A \subseteq \{0,1\}^n$ for some $n \in \mathbb{N}$). However, generally we are interested in infinite functions; e.g., we want to add numbers of arbitrary length or we want to build acceptors for arbitrary, possibly infinite sets. A circuit however by definition has a fixed finite number of inputs and outputs. Thus we will have to consider different circuits for different input lengths. Therefore we consider circuit families.

Definition 1.10. Let B be a basis. A *circuit family* over B is a sequence $C = (C_0, C_1, C_2 \ldots)$, where for every $n \in \mathbb{N}$, C_n is a circuit over B with n inputs. Let f^n be the function computed by C_n. Then we say that C computes the function $f \colon \{0,1\}^* \to \{0,1\}^*$, defined for every $w \in \{0,1\}^*$ by

$$f(w) =_{\text{def}} f^{|w|}(w).$$

We write $f = (f^n)_{n \in \mathbb{N}}$ and $C = (C_n)_{n \in \mathbb{N}}$. We say that C accepts $A \subseteq \{0,1\}^*$ if and only if C computes c_A. In this context we also use the notation $A = (A^n)_{n \in \mathbb{N}}$ (where $A^n =_{\text{def}} A \cap \{0,1\}^n$) and $c_A = (c_{A^n})_{n \in \mathbb{N}}$. If C is a circuit family then we use the notation f_C for the function computed by C.

Now it makes sense to consider complexity measures in the length of the input.

Definition 1.11. Let $C = (V, E, \alpha, \beta, \omega)$ be a circuit over B. The *size* of C is defined to be the number of non-input gates in V, i. e., $|\{\, v \in V \mid \beta(v) \in B \,\}|$, and the *depth* of C is defined to be the length of a longest directed path in the graph (V, E) (see page 237).

Let $C = (C_n)_{n \in \mathbb{N}}$ be a circuit family, and let $s, d \colon \mathbb{N} \to \mathbb{N}$. C has size s and depth d if for every n, C_n has size $s(n)$ and depth $d(n)$.

In the following we will consider some particular bases.

Definition 1.12. $B_0 =_{\text{def}} \{\neg, \wedge^2, \vee^2\}$ is the *standard bounded fan-in basis*. $B_1 =_{\text{def}} \{\neg, (\wedge^n)_{n \in \mathbb{N}}, (\vee^n)_{n \in \mathbb{N}}\}$ is the *standard unbounded fan-in basis*.

When we refer to *bounded fan-in circuits* we always mean circuits over a basis B consisting only of Boolean functions, i. e., B does not contain a family of Boolean functions (cf. Def. 1.5). In this case B is called a *bounded fan-in basis*. If on the other hand B contains at least one family of Boolean functions, then B is called an *unbounded fan-in basis*, and we will use *unbounded fan-in circuit* as a shorthand for a circuit over an unbounded fan-in basis.

Definition 1.13. Let B be a basis, and let $s, d \colon \mathbb{N} \to \mathbb{N}$. We define the following complexity classes:

1. $\text{SIZE}_B(s)$ is the class of all sets $A \subseteq \{0,1\}^*$ for which there is a circuit family C over basis B of size $O(s)$ that accepts A.
2. $\text{DEPTH}_B(d)$ is the class of all sets $A \subseteq \{0,1\}^*$ for which there is a circuit family C over basis B of depth $O(d)$ that accepts A.
3. $\text{SIZE-DEPTH}_B(s, d)$ is the class of all sets $A \subseteq \{0,1\}^*$ for which there is a circuit family C over basis B of size $O(s)$ and depth $O(d)$ that accepts A.
4. $\text{FSIZE}_B(s)$ is the class of all functions $f \colon \{0,1\}^* \to \{0,1\}^*$ for which there is a circuit family C over basis B of size $O(s)$ that computes f.
5. $\text{FDEPTH}_B(d)$ is the class of all functions $f \colon \{0,1\}^* \to \{0,1\}^*$ for which there is a circuit family C over basis B of depth $O(d)$ that computes f.

6. FSIZE-DEPTH$_B(s, d)$ is the class of all functions $f: \{0,1\}^* \to \{0,1\}^*$
   for which there is a circuit family $\mathcal{C}$ over basis B of size $O(s)$ and depth
   $O(d)$ that computes f.

In the case where we have the basis $B = \mathcal{B}_0$ we simply omit the index,
that is SIZE(s) $=_{\mathrm{def}}$ SIZE$_{\mathcal{B}_0}(s)$, etc. In the case where the basis B is
$\mathcal{B}_1$ we omit the index but we prefix the class with "Unb", e.g., we write
UnbSIZE-DEPTH(s, d) for SIZE-DEPTH$_{\mathcal{B}_1}(s, d)$.

Sometimes it will be tedious, or even impossible, to bound the size or
depth of a circuit family exactly by an explicit function. We will often be
satisfied by knowing that the size is bounded by *some* polynomial. In this case
we use the notation SIZE$(n^{O(1)})$. Generally, if $\mathcal{F}$ is a class of functions then we
let SIZE$(\mathcal{F})$ $=_{\mathrm{def}} \bigcup_{f \in \mathcal{F}}$ SIZE(f), and analogously for the other complexity
measures we defined.

By considering circuit families we can now talk about circuits computing
not necessarily finite functions. For every input length, we have one circuit
in our family. However, this one circuit will have a fixed number of output
bits. Thus all functions computed by circuits have a property which we call
length-respecting.

Definition 1.14. Let $f: \{0,1\}^* \to \{0,1\}^*$. f is *length-respecting* if whenever
$|x| = |y|$ then also $|f(x)| = |f(y)|$.

When we talk about number theoretic functions, i.e., functions $f: \mathrm{N} \to \mathrm{N}$,
and we use binary encoding, this is no restriction. If $|x| = |y|$ but $|f(x)| <
|f(y)|$ our circuit can fill up the representation of $f(x)$ with leading zeroes.

In the example from Sect. 1.1 we constructed for every n a circuit computing ADD^{2n} of size $O(n^2)$ and depth 4. As a basis we used unbounded
fan-in $\lor$ and $\land$ gates and fan-in 2 and 3 $\oplus$ gates. However, the $\oplus$ gates can
be replaced by small circuits of constant size over basis $\mathcal{B}_0$ simulating their
behavior. This shows:

Theorem 1.15. ADD $\in$ FUnbSIZE-DEPTH$(n^2, 1)$.

A word on the notation we used: We will generally specify problems or
functions as on p. 5 in Sect. 1.1. This usually defines for certain designated
input lengths a function f; in the example we defined the functions ADD^m
for every integer m which is a multiple of 2. We then only construct circuits
for inputs of these designated lengths, and we do not care about other input lengths. The function ADD is not defined by this specification for input
lengths $2n+1$. However to be mathematically precise in such a case we implicitly assume that ADD takes the value 0 for all inputs of such a length. Thus,
the specification given at the beginning of Sect. 1.1 should be taken as the definition of the function ADD $= (\mathrm{ADD}^m)_{m \in \mathrm{N}}$, where ADD$^m(a_1, \ldots, a_m) = 0$
if $m = 2n+1$, and ADD$^m(a_1, \ldots, a_m)$ gives the result of adding $a_1 \cdots a_n$ and
$a_{n+1} \cdots a_m$ in binary, if $m = 2n$.

A circuit family is an infinite object. It is not immediate that one should accept this as a manifestation or implementation of an algorithm, which we always require to be finite. However most circuit families we construct will be very regular and thus will have a *finite description*. This is what we will call *uniformity of the circuit family*. There will be a lot to say with respect to this in the upcoming chapters. For the moment we do not restrict the definition of a circuit family any further.

For later use we note the following immediate properties of our definitions.

Proposition 1.16. *Let B be a bounded fan-in basis and let $s, d \colon \mathbb{N} \to \mathbb{N}$. Then the following holds:*

1. $\text{SIZE-DEPTH}_{B \cup \mathcal{B}_0}(s, d) = \text{SIZE-DEPTH}(s, d)$.
2. $\text{SIZE-DEPTH}_{B \cup \mathcal{B}_1}(s, d) = \text{UnbSIZE-DEPTH}(s, d)$.

Proof. For every function from B we construct a $\mathcal{B}_0$ circuit computing it. Then replacing every B gate in a circuit over $B \cup \mathcal{B}_0$ by such a small subcircuit gives an equivalent circuit over $\mathcal{B}_0$, and the size and depth only grow by a constant factor. The proof for the second statement is analogous. $\square$

Proposition 1.17. *Let $s, d \colon \mathbb{N} \to \mathbb{N}$. Then*

$$\text{UnbSIZE-DEPTH}(s, d) \subseteq \text{SIZE-DEPTH}(s^2, d \cdot \log s).$$

Proof. We simulate every $\vee$ and $\wedge$ gate of fan-in greater than 2 by a small tree consisting only of $\vee^2$ ($\wedge^2$, respectively) gates. Every gate in the original circuit can have at most s inputs, thus the depth of this construction is $\log s$ and the size is $O(s)$. $\square$

Corollary 1.18. $\text{ADD} \in \text{FSIZE-DEPTH}(n^{O(1)}, \log n)$.

Of most practical interest are those functions for which the length of the output is not much longer than the length of the input.

Definition 1.19. Let $f \colon \{0,1\}^* \to \{0,1\}^*$. f is *polynomially length-bounded* if there is a polynomial p such that for every $x \in \{0,1\}^*$, $|f(x)| \le p(|x|)$.

Of course, polynomial size circuits (i.e. circuits of size $n^{O(1)}$) can only compute polynomially length-bounded functions.

If f is computed by circuit family $\mathcal{C} = (C_n)_{n \in \mathbb{N}}$ of logarithmic depth over the standard bounded fan-in basis, this means that every output bit can be computed in logarithmic depth by a polynomial size circuit. (The length of a path in $\mathcal{C}$ of an output gate to the inputs is logarithmic, which in the worst case can give rise only to a polynomial size tree which computes this one output.) If now f is polynomially length-bounded, then the number of output gates in C_n is polynomial in n. Thus if $f \in \text{FDEPTH}(\log n)$, then we know that $f \in \text{FSIZE-DEPTH}(n^{O(1)}, \log n)$. Therefore the corollary just given could equally well be stated simply as $\text{ADD} \in \text{FDEPTH}(\log n)$.

1.3 More Examples

In this section we will construct more examples for small circuits. As we will see, the examples will fall in one of the classes SIZE-DEPTH$(n^{O(1)}, \log n)$ or UnbSIZE-DEPTH$(n^{O(1)}, 1)$. Hence the most important arithmetical functions can be located in these low classes.

1.3.1 Iterated Addition

Consider the following problem:

> *Problem:* ITADD
> *Input:* n numbers in binary with n bits each
> *Output:* the sum of the input numbers in binary

This only defines the function ITADD $= (\text{ITADD}^n)_{n \in \mathbb{N}}$ for square numbers n. We appeal to the discussion from Sect. 1.2 (p. 11) and recall that for non-square input lengths, ITADD is the constant zero function.

A straightforward implementation of the sequential algorithm would give a circuit of linear depth. A divide-and-conquer algorithm adding the given numbers in a tree-like fashion (because of associativity of addition) using the circuits for ADD as components would give a circuit of logarithmic depth, but unbounded fan-in (i. e., ITADD $\in$ FUnbDEPTH$(\log n)$). We want to do better. We want to design a logarithmic depth *bounded fan-in* circuit family.

Suppose we are given three numbers in binary, $\text{bin}(a) = a_{n-1} \cdots a_0$, $\text{bin}(b) = b_{n-1} \cdots b_0$, and $\text{bin}(c) = c_{n-1} \cdots c_0$. We want to reduce the addition of a, b, c to the addition of two numbers d and e, which of course then can possibly have $n + 1$ bits. This can be achieved by defining

$$
\begin{aligned}
e_0 &= 0 \\
e_i &= (a_{i-1} \wedge b_{i-1}) \vee (a_{i-1} \wedge c_{i-1}) \vee (b_{i-1} \wedge c_{i-1}) \quad \text{for } 1 \le i \le n \\
d_i &= a_i \oplus b_i \oplus c_i \quad \text{for } 0 \le i < n \\
d_n &= 0
\end{aligned}
$$

Look at one single bit position i. We have to add a_i, b_i and c_i. The result is given by the two bit number $e_{i+1} d_i$. Bit e_{i+1} is on if at least two of the bits a_i, b_i and c_i are on, and d_i is on if an odd number of a_i, b_i and c_i is on.

Thus we conclude that $a + b + c = d + e$.

If we are given m numbers with r bits each, we group them into three-element sets (plus one set with only one or two numbers, if m is not a multiple of 3), and then compute for each set as just explained two numbers whose sum is equal to the sum of all three numbers from the set. In this way we end up with m' numbers of $r + 1$ bits each, where

$$m' = \begin{cases} \frac{2}{3}m & \text{if } m \equiv 0 \pmod 3, \\ \frac{2}{3}(m-1)+1 & \text{if } m \equiv 1 \pmod 3, \\ \frac{2}{3}(m-2)+2 & \text{if } m \equiv 2 \pmod 3. \end{cases}$$

Observe that if $m > 2$ then $m' \le \frac{4}{5}m$. Thus given n input numbers with n bits each, iterating the above procedure $O(\log n)$ times gives two numbers with $n + O(\log n)$ bits. These numbers are then added as in Sect. 1.1.

What is the complexity of the resulting circuit family? In every iteration the numbers d and e can be computed in constant depth by just hardwiring their defining formulas presented above. Thus the $O(\log n)$ iterations of reducing the number of terms take depth $O(\log n)$. By Corollary 1.18, adding two numbers of $n + O(\log n)$ bits can be done in depth $O(\log n)$. The number of output bits is $O(n)$, each output bit can be computed by an $O(\log n)$ depth circuit which cannot have more than polynomially many gates. Thus the overall size is polynomial, and we have proved:

Theorem 1.20. ITADD $\in$ FSIZE-DEPTH$(n^{O(1)}, \log n)$.

Let us now consider the problem where we have only a small number of terms to add, say, $\log n$ numbers with n bits.

Problem: LOGITADD
Input: $\log n$ numbers with n bits each
Output: the sum of the input numbers in binary

The technique above of reducing the number of terms to add leads to a number of iterations of order $\log\log n$. Combining this with Theorem 1.15 shows that LOGITADD $\subseteq$ FUnbSIZE-DEPTH$(n^{O(1)}, \log\log n)$. We want to show that constant depth is sufficient.

Suppose we are given $\ell(n)$ numbers $a_1, \ldots, a_{\ell(n)}$; $\text{bin}(a_i) = a_{i,n-1} \cdots a_{i,0}$ for $1 \le i \le \ell(n)$. (For a definition of the function ℓ, see Appendix A2.) We start by computing for every input position k, $0 \le k < n$, the sum

$$s_k =_{\text{def}} \sum_{i=1}^{\ell(n)} a_{i,k}.$$

Every s_k can be written in binary using no more than $\ell^{(2)}(n) = \ell(\ell(n))$ bits: $s_k = s_{k,\ell^{(2)}(n)-1} \cdots s_{k,0}$. Therefore,

$$\sum_{i=1}^{\ell(n)} a_{i,k} = \sum_{j=0}^{\ell^{(2)}(n)-1} s_{k,j} \cdot 2^j,$$

which implies

$$\sum_{i=1}^{\ell(n)} a_i = \sum_{i=1}^{\ell(n)} \sum_{k=0}^{n-1} a_{i,k} \cdot 2^k$$

$$= \sum_{j=0}^{\ell^{(2)}(n)-1} \sum_{k=0}^{n-1} s_{k,j} \cdot 2^j \cdot 2^k.$$

Thus we have reduced the problem of adding $\ell(n)$ numbers $a_1, \ldots, a_{\ell(n)}$ to that of adding the $\ell^{(2)}(n)$ numbers

$$\sum_{k=0}^{n-1} s_{k,j} \cdot 2^{k+j}$$

for $0 \le j < \ell^{(2)}(n)$. Each of these numbers has no more than $n + \ell^{(2)}(n)$ bits.

Next, we iterate the above procedure. After the next iteration we have $\ell^{(3)}(n)$ numbers with $n + \ell^{(2)}(n) + \ell^{(3)}(n)$ bits; after three iterations we have $\ell^{(4)}(n)$ numbers, and so on. We perform k iterations, where k is the smallest number such that $\ell^{(k+1)}(n) \le 2$.

Thus we end up with 2 numbers of $n + \ell^{(2)}(n) + \ell^{(3)}(n) + \cdots + \ell^{(k+1)}(n)$ bits each, which we add using the circuit from Sect. 1.1.

For the complexity of this circuit family, first observe that every function $f : \{0,1\}^m \to \{0,1\}$ can be computed over basis B_0 in constant depth with $O(2^m)$ gates (see Exercise 1.1). Thus step 1 can be implemented in constant depth and polynomial size, since every s_k depends only on a logarithmic number of bits. For iterations 2 to k the following holds: Every bit computed in iteration i depends only on $\ell^{(i)}(n)$ many bits of the previous iteration. Thus, every bit computed in iteration k depends on $t(n) =_{\text{def}} \ell^{(2)}(n) \cdot \ell^{(3)}(n) \cdots \ell^{(k)}(n)$ many bits computed in iteration 1. By Exercise 1.3(2), $t(n) = O(\log n)$; thus altogether iterations 2 to k can be implemented by a constant depth circuit of polynomial size (see Exercise 1.1). The two numbers that have to be summed up in the end have $O(n)$ bits each (Exercise 1.3(1)) and thus can be added in constant depth and polynomial size. All in all we have proved:

Theorem 1.21. LOGITADD $\in$ FUnbSIZE-DEPTH($n^{O(1)}, 1$).

A special case we will consider in the next section is the addition of n bits.

> *Problem:* BCOUNT
> *Input:* n bits $a_{n-1}, \ldots, a_0$
> *Output:* $\sum_{i=0}^{n-1} a_i$

BCOUNT stands for *binary count*.
Immediately from Theorem 1.20 we have

Corollary 1.22. BCOUNT $\in$ FDEPTH($\log n$).

1.3.2 Multiplication

Next we consider the problem of multiplying two numbers.

> *Problem:* MULT
> *Input:* two n bit numbers
> *Output:* the product of the input numbers in binary

Our circuit will be a direct implementation of the school method for multiplication.

Given a, b, $\text{bin}(a) = a_{n-1} \cdots a_0$, $\text{bin}(b) = b_{n-1} \cdots b_0$, i.e. $b = \sum_{i=0}^{n-1} b_i \cdot 2^i$, then obviously $a \cdot b = \sum_{i=0}^{n-1} a \cdot b_i \cdot 2^i$. Define for $0 \le i < n$ the strings

$$c_i =_{\text{def}} \begin{cases} 0^{n-i-1} a_{n-1} \cdots a_1 a_0 0^i & \text{if } b_i = 1, \\ 0^{2n-1} & \text{otherwise;} \end{cases}$$

observe that interpreted as a number, $c_i = a \cdot 2^i$ if $b_i = 1$, and $c_i = 0$ else. Thus we have $a \cdot b = \sum_{i=0}^{n-1} c_i$. Each of the c_i can be computed in constant depth. Computing the sum of the c_i can be done in logarithmic depth by Theorem 1.20. Thus:

Theorem 1.23. MULT $\in$ FDEPTH($\log n$).

1.3.3 Majority

Of utmost importance for the examination of small depth circuit classes will be the problem of determining whether the majority of a given number of bits is on.

> *Problem:* MAJ
> *Input:* n bits $a_{n-1}, \ldots, a_0$
> *Question:* Are at least half of the a_i on?

Theorem 1.24. MAJ $\in$ DEPTH($\log n$).

Proof. We first compute the sum of the input bits as in Corollary 1.22. Then we compare this number with the number $\frac{n}{2}$ which we hardwire into the circuit. (See also Exercise 1.4.)   □

1.3.4 Transitive Closure

We now design depth-efficient circuits for Boolean matrix multiplication and transitive closure.

Boolean matrix multiplication is matrix multiplication where instead of addition we use logical disjunction and instead of multiplication we use logical conjunction.

Definition 1.25. Let $\vec{x} = (x_1, \ldots, x_n)$ and $\vec{y} = (y_1, \ldots, y_n)$ be Boolean vectors with n components. The *inner product* of $\vec{x}$ and $\vec{y}$ is defined as

$$\vec{x} \cdot \vec{y} =_{\text{def}} \bigvee_{i=1}^{n} (x_i \wedge y_i).$$

Lemma 1.26. *Let $\vec{x}$ and $\vec{y}$ be Boolean vectors. Then their inner product $\vec{x} \cdot \vec{y}$ can be computed in constant depth by unbounded fan-in circuits.*

Proof. Hardwire the given defining formula into a circuit. ❑

Definition 1.27. Let $A = (a_{i,j})_{1 \leq i,j \leq n}$ and $B = (b_{i,j})_{1 \leq i,j \leq n}$ be square Boolean matrices. Their product $C = A \cdot B$ is defined as follows: Let $C = (c_{i,j})_{1 \leq i,j \leq n}$. Then

$$c_{i,j} = (a_{i,1}, \ldots, a_{i,n}) \cdot (b_{1,j}, \ldots, b_{n,j}).$$

Lemma 1.28. *Let A and B be $n \times n$ Boolean matrices. Then their product $A \cdot B$ can be computed in constant depth by unbounded fan-in circuits.*

Proof. Let $C = A \cdot B$, $C = (c_{i,j})_{1 \leq i,j \leq n}$. Then all entries $c_{i,j}$ can be computed in parallel using Lemma 1.26. ❑

Definition 1.29. Let $A = (a_{i,j})_{1 \leq i,j \leq n}$ be a Boolean matrix. The *transitive closure* of A is defined to be

$$A^* =_{\text{def}} \bigvee_{i \geq 0} A^i,$$

i.e., if we set $A^* = (a_{i,j}^*)_{1 \leq i,j \leq n}$, then for every i,j the element $a_{i,j}^*$ is the logical disjunction of the elements at row i, column j in the matrices A^0, A^1, $A^2, \ldots$ (A^0 is the $n \times n$ identity matrix.)

Remark 1.30. The name *transitive closure* stems from the following, easily proved fact (see Exercise 1.6). Let A and A^* be as above. Then for all $1 \leq i,j \leq n$, we have $a_{i,j}^* = 1$ iff $i = j$ or there is a sequence $k_1, k_2, \ldots, k_l \in \{1, \ldots, n\}$, $1 \leq l \leq n - 1$, such that $a_{i,k_1} = a_{k_1,k_2} = a_{k_2,k_3} = \cdots = a_{k_l,j} = 1$. This is equivalent to saying: The directed graph defined by A (with an edge from node μ to node ν if and only if $a_{\mu,\nu} = 1$) has a path from node i to node j.

Theorem 1.31. *Let $A = (a_{i,j})_{1 \leq i,j \leq n}$ be a Boolean matrix. Then its transitive closure A^* can be computed in depth $O\big((\log n)^2\big)$ and polynomial size by bounded fan-in circuits.*

Proof. Let I be the $n \times n$ identity matrix. Let $(A \vee I)$ denote the $n \times n$ matrix obtained by element-wise disjunction of A and I. Then $A^* = (A \vee I)^{n-1}$ (see Exercise 1.7). This product of $n - 1$ matrices can be evaluated in a tree-wise manner, making use of the associativity of matrix multiplication. This gives a tree of depth $\lceil \log n \rceil$, where in every node one single matrix multiplication is performed. By Lemma 1.28 we need bounded fan-in circuits of logarithmic depth for each such multiplication. All in all this results in a Boolean circuit of depth $O((\log n)^2)$ over basis $\mathcal{B}_0$. The size is clearly polynomial. $\square$

1.4 Constant-Depth Reductions

In the previous sections, two classes turned out to be of interest when examining the circuit complexity of elementary arithmetical operations: the class of polynomial size unbounded fan-in circuits of constant depth, and the class of polynomial size bounded fan-in circuits of logarithmic depth: FUnbSIZE-DEPTH($n^{O(1)}$, 1) and FSIZE-DEPTH($n^{O(1)}$, $\log n$). To show containment of particular problems in those classes we used, besides constructing a circuit family from scratch, also the following method: We showed how the computation of a function can be conveniently carried out making use of circuits we have already constructed for other functions. That is, our circuits used subcircuits for other problems. For example, the circuit family for majority is very easy once we constructed the circuit family for iterated addition. We say that majority *reduces* to iterated addition. We want to make this concept precise in this section.

Definition 1.32. Let $f \colon \{0,1\}^* \to \{0,1\}^*$, $f = (f^n)_{n \in \mathbb{N}}$, be length-respecting. Let for every n and $|x| = n$ the length of $f^n(x)$ be $r(n)$. Let $f_i^n \colon \{0,1\}^n \to \{0,1\}$ be the Boolean function that computes the i-th bit of f^n, i.e., if $f^n(x) = a_1 a_2 \cdots a_{r(n)}$ then $f_i^n(x) = a_i$, where $1 \le i \le r(n)$. Then bits(f) denotes the class of all those functions, i.e., bits(f) $=_{\text{def}} \{ f_i^n \mid n, i \in \mathbb{N}, i \le r(n) \}$.

Definition 1.33. Let $f, g \colon \{0,1\}^* \to \{0,1\}^*$ be length-respecting. We say that f is *constant-depth reducible* to g; in symbols: $f \le_{\text{cd}} g$, if $f \in$ FSIZE-DEPTH$_{\mathcal{B}_1 \cup \text{bits}(g)}(n^{O(1)}, 1)$. Here, a gate v for a function from bits(g) contributes to the size of its circuit with k, where k is the fan-in of v. We write $f \equiv_{\text{cd}} g$ if $f \le_{\text{cd}} g$ and $g \le_{\text{cd}} f$.

For f to be reducible to g there must be a polynomial size constant depth circuit family computing f where we allow, besides the standard unbounded fan-in basis, gates for g also. (These gates are sometimes called *oracle gates* for g, see also Sect. 4.4.) The requirement that a gate v computing the function g of fan-in k contributes with the number k to the size of a circuit reflects the fact that for such a gate it may be reasonable (depending on g) to have more than one input wire from the same predecessor gate (for gates from the standard unbounded fan-in basis, this is not reasonable). This should be taken into consideration when defining the size of the circuit—otherwise we could have, e. g., polynomial size circuits with an exponential number of wires, which does not seem natural. We choose the way above to handle this; in the literature this issue is sometimes treated differently by defining circuit size via number of wires in the circuit, but in all cases which are of interest in this book this leads to equivalent results. (See also Exercise 1.8.) Note that here we allow, deviating from our previous definitions, a basis consisting of infinitely many Boolean functions. (FSIZE-DEPTH$_{B_1 \cup \text{bits}(g)}(s,d)$ is formally defined completely analogously to the definition of FSIZE-DEPTH$_B(s,d)$ for finite B, given in Sect. 1.2.) These infinitely many functions however all originate from the one function $g \colon \{0,1\}^* \to \{0,1\}^*$. We want to stress that only here, in the context of reductions, are infinite bases allowed in this book.

Example 1.34. MAJ $\leq_{cd}$ BCOUNT $\leq_{cd}$ ITADD: The proof of Theorem 1.24 shows MAJ $\leq_{cd}$ BCOUNT: The reduction is given by the circuit family that first, using ITADD gates, computes the sum of the input bits and then compares this to the number of input bits divided by 2. BCOUNT $\leq_{cd}$ ITADD, since BCOUNT is just a special case of ITADD; thus the circuit family witnessing the reduction is trivial.

Example 1.35. MULT $\leq_{cd}$ ITADD: This is established by the proof of Theorem 1.23. All we did there is construct the numbers c_i in constant depth and then add. This gives a constant depth circuit with ITADD gates.

The following properties of the above definition are easy to see (see Exercise 1.10).

Lemma 1.36. *1. If f is length-respecting and polynomially length-bounded then $f \leq_{cd} f$. Thus $\leq_{cd}$ is reflexive on the set of all length-respecting polynomially length-bounded functions.*

2. $\leq_{cd}$ is transitive.

3. If $f \leq_{cd} g$ and $g \in$ FUnbSIZE-DEPTH$(n^{O(1)}, 1)$, then we also have $f \in$ FUnbSIZE-DEPTH$(n^{O(1)}, 1)$.

4. If $f \leq_{cd} g$ and $g \in$ FSIZE-DEPTH$(n^{O(1)}, \log n)$, then we also have $f \in$ FSIZE-DEPTH$(n^{O(1)}, \log n)$.

5. If $f \in$ FUnbSIZE-DEPTH$(n^{O(1)}, 1)$ and g is arbitrary then $f \leq_{cd} g$.

Proof sketch. 1 is trivial. 2 follows by simply composing the two reductions, 3 and 4 follow by composing the reducing circuit with the circuit computing g. For 5 observe that to compute f we do not actually need the g gates. ☐

Statements 3 and 4 of Lemma 1.36 allow us to use reductions to prove containments of functions in the classes FUnbSIZE-DEPTH$(n^{O(1)}, 1)$ or FSIZE-DEPTH$(n^{O(1)}, \log n)$. From Example 1.34 and Theorem 1.20 we can thus conclude (as we already did in Sect. 1.3) that MAJ and BCOUNT are in FSIZE-DEPTH$(n^{O(1)}, \log n)$.

However, another possibly even more important use of reductions is to *compare* the complexity of problems. Containment of problems in complexity classes just gives upper bounds on their complexity. It does not say anything about the relative complexity of different problems. But if $f \leq_{cd} g$ this intuitively says that f is not more complex than g. The function f can essentially be computed with the same resources as g (plus those needed for the reducing circuit, but they are negligible). If $f \equiv_{cd} g$ then f and g share the same complexity. From statement 5 of Lemma 1.36 we see that reductions cannot be used to compare functions in FUnbSIZE-DEPTH$(n^{O(1)}, 1)$ since they reduce to every other function. But they become useful for larger classes, e. g. FSIZE-DEPTH$(n^{O(1)}, \log n)$. For instance, we know that both ITADD and MULT are in FSIZE-DEPTH$(n^{O(1)}, \log n)$, and that MULT $\leq_{cd}$ ITADD, thus multiplication is not more complicated than iterated addition. But is iterated addition really more complicated? To answer questions like these we use reductions. We give a number of examples in the upcoming subsections.

1.4.1 Iterated Addition and Multiplication

We first show that the answer to the question just posed is no.

Theorem 1.37. ITADD $\leq_{cd}$ BCOUNT $\leq_{cd}$ MAJ.

Proof. Let $a_i = a_{i,n-1} \cdots a_{i,0}$ for $1 \leq i \leq n$ be given. First we compute $s_k =_{\text{def}} \sum_{i=1}^{n} a_{i,k}$ for $0 \leq k < n$ using BCOUNT gates. Next we proceed as in the proof of Theorem 1.21. Thus our new task is to add $\ell(n)$ numbers with $n + \ell(n)$ bits each. This however is the problem LOGITADD which we know can be done in FUnbSIZE-DEPTH$(n^{O(1)}, 1)$. Thus we have proved ITADD $\leq_{cd}$ BCOUNT.

Let now MAJ $= (\text{maj}^n)_{n \in \mathbb{N}}$, $\text{maj}^n(a_1, \ldots, a_n) = 1 \iff \sum_{i=1}^{n} a_i \geq \frac{n}{2}$. We define the following generalizations of these functions, known as *threshold functions*:

$$T_m^n(a_1, \ldots, a_n) = 1 \iff_{\text{def}} \sum_{i=1}^{n} a_i \geq m.$$

The following relations hold:

1. $T_m^n(a_1, \ldots, a_n) = \text{maj}^{2n-2m}(a_1, \ldots, a_n, \underbrace{1, \ldots, 1}_{n-2m})$, if $m < \frac{n}{2}$.
2. $T_m^n(a_1, \ldots, a_n) = \text{maj}^{2m}(a_1, \ldots, a_n, \underbrace{0, \ldots, 0}_{2m-n})$, if $m \geq \frac{n}{2}$.

This shows that for every $m \in \mathbb{N}$, $(T_m^n)_{n \in \mathbb{N}} \leq_{cd}$ MAJ.

We next reduce BCOUNT to threshold computations. Let $\ell = \ell(n)$, $\sum_{i=1}^{n} a_i = s_{\ell-1} \cdots s_0$ in binary. Let $0 \leq j < \ell$. Let R_j be the set of those numbers $r \in \{0, \ldots, n\}$ whose j-th bit is on (where the 0th bit is the rightmost bit and the $(\ell-1)$st bit is the leftmost bit). Then obviously,

$$s_j = \bigvee_{r \in R_j} \left[\left[\sum_{i=1}^{n} a_i = r \right] \right].$$

Certainly $[x = r] = [x \geq r] \wedge \neg[x \geq r + 1]$, thus

$$s_j = \bigvee_{r \in R_j} \left(T_r^n(a_1, \ldots, a_n) \wedge \neg T_{r+1}^n(a_1, \ldots, a_n) \right).$$

Observe that the sets R_j do not depend on the input but only on the input length n and thus can be hardwired into the circuit. Thus we have a circuit for BCOUNT which uses $(T_r^n)_{n \in \mathbb{N}}$ and $(T_{r+1}^n)_{n \in \mathbb{N}}$ gates. By the above these can be reduced to MAJ which all in all yields a reduction of BCOUNT to MAJ.   □

In the proof just given, we constructed a circuit for BCOUNT which made use of both T_r^n and T_{r+1}^n gates. This phenomenon is examined in a more abstract way in Exercise 1.12.

Theorem 1.38. MAJ $\leq_{cd}$ MULT.

Proof. Let $a = a_{n-1} \cdots a_0$ and $\ell = \ell(n)$. Define $A = \sum_{i=0}^{n-1} a_i \cdot 2^{\ell i}$ and $B = \sum_{i=0}^{n-1} 2^{\ell i}$. Then the binary representation of A consists of n blocks of length ℓ each, where the i-th block starts with $\ell - 1$ 0-bits followed by bit a_i. The binary representation of B consists of n blocks of length ℓ each, where each block starts with $\ell - 1$ 0-bits followed by a 1-bit. Let $C = A \cdot B$. Now partition the binary representation of C into blocks of length ℓ each, i.e. write C as $C = \sum_{i=0}^{2n-1} c_i \cdot 2^{\ell i}$, where each c_i is a binary word of length ℓ. If we now recall the school method for multiplication then we see that in the middle block (i.e., c_{n-1}) we have exactly the computation of the sum of the a_i. No conflicts with neighboring blocks can occur since we started with block length ℓ which is large enough to hold this sum. Hence, $c_{n-1} = \sum_{i=0}^{n-1} a_i$.

Our circuit thus looks as follows: First A and B are constructed in constant depth and fed into multiplication gates yielding C. Finally we just have to compare c_{n-1} with $\frac{n}{2}$, which can be done in constant depth (Exercise 1.4).   □

This answers the question above. Iterated addition is of the same complexity as multiplication.

Corollary 1.39. MAJ $\equiv_{cd}$ BCOUNT $\equiv_{cd}$ ITADD $\equiv_{cd}$ MULT.

1.4.2 Iterated Multiplication

Next we want to consider the problem of multiplying a given sequence of numbers, i.e., iterated multiplication.

Problem: ITMULT

Input: n numbers in binary with n bits each
Output: the product of the input numbers in binary

Certainly MULT $\leq_{cd}$ ITMULT; we will show that the converse also holds.

Theorem 1.40. ITMULT $\leq_{cd}$ MAJ.

Proof. Our proof proceeds by constructing a constant-depth circuit with both ITADD and MAJ gates. Since both problems reduce to MAJ our claim follows (see Exercise 1.12).

Let $a_i = a_{i,n-1} \cdots a_{i,0}$ for $1 \leq i \leq n$ be the numbers which we want to multiply up. Certainly $a =_{\text{def}} a_1 \cdot a_2 \cdots a_n < (2^n)^n = 2^{n^2} < p_1 \cdot p_2 \cdots p_{n^2}$; where p_k denotes the k-th prime number, see Appendix A7. It will be our goal to compute the product a of the input numbers modulo $p =_{\text{def}} p_1 \cdot p_2 \cdots p_{n^2}$ by using the Chinese Remainder Theorem (Appendix A7, p. 238). Therefore it will be our first goal to compute the product $a_1 \cdot a_2 \cdots a_n$ modulo all the primes $p_1, p_2, \ldots, p_{n^2}$. The Chinese Remainder Theorem then tells us that if we take all these products and compute the sum

$$\sum_{j=1}^{n^2} (a_1 \cdot a_2 \cdots a_n \bmod p_j) \cdot r_j \cdot s_j,$$

where r_j and s_j are as on p. 238, we can recover the actual product $a_1 \cdot a_2 \cdots a_n$. Of course one might ask why we choose this way; instead of solving one iterated multiplication problem we now have to solve one such problem for each of the first n^2 primes. The reason for doing so is that—as we will see—we will be able to use finite field arithmetic to reduce the problem of iterated multiplication to a simple iterated sum.

Formally we proceed as follows: First we compute

$$a'_{i,j} =_{\text{def}} \sum_{k=1}^{n-1} a_{i,k} \cdot (2^k \bmod p_j) \tag{1.2}$$

for $1 \leq i \leq n$, $1 \leq j \leq n^2$. This gives n^3 numbers, all of which can be computed in constant depth. Note that ITADD and MULT gates are allowed, and the values $(2^k \bmod p_j)$ can be hardwired into the circuit since they do not depend on the input but only the input length.

Next we compute

$$a_i \bmod p_j = a'_{i,j} \bmod p_j \tag{1.3}$$

for $1 \leq i \leq n$, $1 \leq j \leq n^2$. By the Prime Number Theorem (see Appendix A7) $p_j = O(n^2 \log n)$ for every j; therefore all the numbers from (1.2) consist of $O(\log n)$ bits only, and we use Exercise 1.1 to conclude that the computations in (1.3) can be performed in constant depth.

For $1 \leq j \leq n^2$, let g_j be a generator of the multiplicative group of the finite field $\mathbb{Z}_{p_j}$ (see Appendix A8). For $k \in \{1, \ldots, p_j - 1\}$, let $m_j(k) \in \{0, 1, \ldots, p_j - 2\}$ be such that $g_j^{m_j(k)} = k$; we set $m_j(0) =_{\mathrm{def}} p_j - 1$. Now we compute

$$m_j(a_i \bmod p_j) \tag{1.4}$$

for $1 \leq i \leq n$, $1 \leq j \leq n^2$. Since the numbers $a_i \bmod p_j$ consist of logarithmically many bits, these computations can again be done in constant depth.

Next we compute

$$m_j(a_1 \cdot a_2 \cdots a_n \bmod p_j) = \left(\sum_{i=1}^{n} m_j(a_i \bmod p_j) \right) \bmod p_j - 1 \tag{1.5}$$

for $1 \leq j \leq n^2$. This can be achieved in constant depth using ITADD gates. The particular case that one of the $a_i \bmod p_j$ is equal to 0 can easily be taken care of. Again these n^2 numbers have logarithmically many bits. Thus we can compute the values

$$a_1 \cdot a_2 \cdots a_n \bmod p_j \tag{1.6}$$

for $1 \leq j \leq n^2$ in constant depth. Finally we compute

$$a' =_{\mathrm{def}} \sum_{j=1}^{n^2} (a_1 \cdot a_2 \cdots a_n \bmod p_j) \cdot r_j \cdot s_j \tag{1.7}$$

in constant depth with ITADD and MULT gates. The values r_j, s_j can be hardwired into the circuit, since again they do not depend on the input but only on n.

The Chinese Remainder Theorem now tells us that $a = a' \bmod p$, that is, for $q =_{\mathrm{def}} \lfloor \frac{a'}{p} \rfloor$ we have $a = a' - q \cdot p$. But since $r_j = \frac{p}{p_j}$ and $s_j \leq p_j$ we have $r_j \cdot s_j \leq p$. Hence $a' \leq n^2 \cdot p_{n^2} \cdot p$ which implies $q \leq n^2 \cdot p_{n^2}$. Thus the value of q is polynomial in n and can be determined as follows: We test in parallel for all $i \leq n^2 \cdot p_{n^2}$ if $i \cdot p \leq a' < (i+1) \cdot p$. This can be done in constant depth (see Exercise 1.4). q is the one value i for which the answer is "yes." The result of the iterated multiplication then is $a = a' - q \cdot p$. This is again a constant depth computation (subtraction can be realized similarly to addition, see Exercise 1.5). □

It is essential in the proof above that we have, given an input of length n, the first n^2 prime numbers available (as well as the numbers $2^k \bmod p_j$, r_j, s_j for $1 \leq k < n$, $1 \leq j \leq n^2$). Similarly, in the proof of Theorem 1.37 we needed the sets R_j, depending on n. This poses no problem since we have different circuits for different values of n; hence we may "hardwire" the required values as constants into C_n, the circuit responsible for input length n. This is a peculiarity of the non-uniform model of Boolean circuits. We come back to this point later.

1.4.3 Counting and Sorting

As a last problem in this section we want to consider the problem of sorting numbers.

> *Problem:* SORT
> *Input:* n numbers with n bits each
> *Output:* the sequence of the input numbers in non-decreasing order

We want to show that SORT is equivalent to the other problems considered in this section, i.e., SORT $\equiv_{cd}$ MAJ. For this we need as an auxiliary problem the unary counting problem. The *k-bit unary representation* of a number $n \le k$ is the string $\mathrm{un}_k(n) =_{\mathrm{def}} 1^n 0^{k-n}$. Unary representations are words in the regular language $1^* 0^*$.

> *Problem:* UCOUNT
> *Input:* $a_1, \ldots, a_n \in \{0, 1\}$
> *Output:* $\mathrm{un}_n\left(\sum_{i=1}^n a_i\right)$

We first want to show that counting in unary or binary is of the same complexity.

Lemma 1.41. BCOUNT $\le_{cd}$ UCOUNT.

Proof. Given $a_1, \ldots, a_n$, let $b_1 \cdots b_n = \mathrm{un}_n(\sum_{i=1}^n a_i)$. Furthermore let $b_0 = 1$ and $b_{n+1} = 0$. This means that $b_0 b_1 \cdots b_n b_{n+1} \in 1^+ 0^+$. Now set $d_j = b_j \wedge \neg b_{j+1}$ for $0 \le j \le n$. Observe that $d_j = 1$ iff $\sum_{i=1}^n a_i = j$. Define as in the proof of Theorem 1.37 the sets R_j to consist of those numbers $r \in \{0, \ldots, n\}$ whose j-th bit is on. Then BCOUNT$(a_1, \ldots, a_n) = c_{\ell(n)-1} \cdots c_0$ is given by

$$c_j = \bigvee_{i \in R_j} d_i$$

for $0 \le j < \ell(n)$. □

Corollary 1.42. BCOUNT $\equiv_{cd}$ UCOUNT.

Proof. UCOUNT $\le_{cd}$ BCOUNT is immediate, see Exercise 1.15. □

Observe that the UCOUNT problem is essentially nothing more than the sorting problem restricted to input values 0 and 1. This leads to the following lemma:

Lemma 1.43. UCOUNT $\le_{cd}$ SORT.

Proof. Given $a_1, \ldots, a_n$, define $A_i =_{\mathrm{def}} a_i \underbrace{0 \cdots 0}_{n-1}$ for $1 \le i \le n$. Sort these numbers. The sequence of most significant bits in the ordered sequence is $\mathrm{un}_n(\sum_{i=1}^n a_i)$ reversed. □

We now prove the converse.

Theorem 1.44. SORT $\leq_{cd}$ UCOUNT.

Proof. Given are $a_i = a_{i,n-1} \cdots a_{i,0}$, $1 \leq i \leq n$. For $1 \leq i, j \leq n$ define

$$c_{i,j} \Longleftrightarrow_{\text{def}} (a_i < a_j) \vee (a_i = a_j \wedge i \leq j).$$

For fixed j, $1 \leq j \leq n$, determine $c_j =_{\text{def}}$ UCOUNT$(c_{1,j}, \ldots, c_{n,j})$. Then c_j is the n-bit unary representation of the position of a_j in the ordered output sequence. Let $a'_1, \ldots, a'_n$ be this ordered sequence, i.e., $a'_{c_j} = a_j$. Let $a'_i = a'_{i,n-1} \cdots a'_{i,0}$ for $1 \leq i \leq n$. Then $a'_{i,k}$ is on if and only if the i-th element in the ordered sequence is a_j and the k-th bit of a_j is on, i.e.,

$$a'_{i,k} = \bigvee_{1 \leq j \leq n} \left([c_j = 1^i 0^{n-i}] \wedge a_{j,k} \right).$$

$\square$

Thus we have seen that all the problems we considered in this section are equivalent under constant-depth reducibility:

Corollary 1.45. MAJ $\equiv_{cd}$ UCOUNT $\equiv_{cd}$ BCOUNT $\equiv_{cd}$ ITADD $\equiv_{cd}$ MULT $\equiv_{cd}$ ITMULT $\equiv_{cd}$ SORT.

By Lemma 1.36 this shows that they can all be computed in logarithmic depth.

Corollary 1.46. *All the above problems are in* FDEPTH$(\log n)$.

To summarize: So far we have considered the classes

$$\text{FUnbSIZE-DEPTH}(n^{O(1)}, 1) \subseteq \text{FSIZE-DEPTH}(n^{O(1)}, \log n)$$
$$\subseteq \text{FSIZE-DEPTH}(n^{O(1)}, (\log n)^2).$$

We identified a number of problems in these classes, particularly in the lower two, see Fig. 1.2. However all problems from FSIZE-DEPTH$(n^{O(1)}, \log n)$ turned out to be reducible to MAJ. It will be shown in Chap. 3 that majority—and therefore all problems equivalent to MAJ under constant-depth reducibility—is not contained in the class FUnbSIZE-DEPTH$(n^{O(1)}, 1)$. Therefore the above inclusion is strict. The question whether there are problems in FSIZE-DEPTH$(n^{O(1)}, \log n)$ which are *not* reducible to majority is still open. We will come back to this important point later.

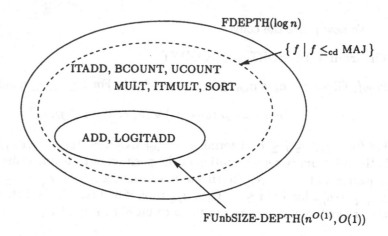

Fig. 1.2. Small circuit classes

1.5 Asymptotic Complexity

In this section we consider the complexity of Boolean functions with respect to bounded fan-in circuits. We prove two theorems which together lead to an astonishing result. First, we will show that for large n most n-ary Boolean functions require circuits of size $\frac{2^n}{n}$ for their computation (this result is known as Shannon's Theorem). Second, we show that all n-ary Boolean functions can be computed with a slightly greater number of gates (Lupanov's Theorem). That is, when n is large, we can determine exactly the complexity of most n-ary Boolean functions: they have circuit size $\Theta\left(\frac{2^n}{n}\right)$.

1.5.1 A Lower Bound

Let $\mathbb{B}^n =_{\text{def}} \{0,1\}^{\{0,1\}^n}$ denote the set of all n-ary Boolean functions. Define $\widehat{B} =_{\text{def}} \mathbb{B}^0 \cup \mathbb{B}^1 \cup \mathbb{B}^2$.

For a circuit C, $S(C)$ denotes the size of C.

Theorem 1.47 (Shannon). *Let $\varepsilon > 0$. The ratio of all n-ary Boolean functions that can be computed by circuits over $\widehat{B}$ with $(1-\varepsilon)\frac{2^n}{n}$ gates approaches 0 as $n \to \infty$.*

Proof. We start by introducing some notation. If f is a Boolean function then $S(f) =_{\text{def}} \min\{ S(C) \mid C$ is a circuit over $\widehat{B}$ that computes $f \}$. For $q, n \in \mathbb{N}$ define $N_{q,n} =_{\text{def}} |\{ f \in \mathbb{B}^n \mid S(f) \leq q \}|$. Since $|\mathbb{B}^n| = 2^{2^n}$ we have to show that $\frac{N_{q,n}}{2^{2^n}}$ approaches 0 in the limit, if q is $(1-\varepsilon)\frac{2^n}{n}$ for some $\varepsilon > 0$.

If C is a circuit over $\widehat{B}$ that computes f then there is a circuit C' over $\mathbb{B}^2$ that computes f such that $S(C') \leq S(C)$. Thus we have to consider only circuits over $\mathbb{B}^2$ in the following.

Let C be a circuit over $\mathbb{B}^2$ of size q with n inputs. Fix an ordering on the gates of C, and based on this identify the gates of C with the set $\{x_1, \ldots, x_n\} \cup \{1, \ldots, q\}$ (n inputs and q non-input gates). C can be encoded as a map

$$\tau \colon \{1, \ldots, q\} \to \left(\{1, \ldots, q\} \cup \{x_1, \ldots, x_n\}\right)^2 \times \{1, \ldots, 16\}.$$

Here $\tau(i) = (j, k, l)$ means that the i-th non-input gate in C has predecessors j and k and computes the l-th functions of $\mathbb{B}^2$ (recall $|\mathbb{B}^2| = 16$). We do not fix the output gate yet; that is, every such circuit can compute a possible $q + n$ functions.

Let $T_{q,n}$ be the set of all such maps τ,

$$|T_{q,n}| = \left(16 \cdot (n+q)^2\right)^q. \tag{1.8}$$

We want to use this to estimate the number $N_{q,n}$. For this, two points have to be considered. First, many maps $\tau \in T_{q,n}$ do not correspond to a valid circuit (for example, there might be "cycles" in τ, but the graph underlying a circuit must be acyclic). Since we are interested in an upper bound for $N_{q,n}$ it is safe to ignore this possibility. Second, many elements from $T_{q,n}$ are encodings of circuits that compute the same function. We have to take this into account, and we proceed as follows.

Let $\sigma \in S_q$ be a permutation on q elements. For $\tau \in T_{q,n}$ define $\sigma\tau$ as follows: If $\tau(i) = (j, k, l)$ then $\sigma\tau(i) = (j', k', l)$, where

$$j' =_{\text{def}} \begin{cases} x_m, & \text{if } j = x_m, \\ \sigma(j), & \text{if } j \in \{1, \ldots, q\}, \end{cases}$$

and analogously for k'. Let C and C_σ be circuits whose encodings are τ and $\sigma\tau$. Then the following holds for all $i \in \{1, \ldots, q\}$: The i-th non-input gate in C computes the same function as gate $\sigma(i)$ in circuit C_σ, i.e., the functions val_i with respect to C and $\text{val}_{\sigma(i)}$ with respect to C_σ are identical.

Say a circuit C is *reduced* if for all non-input gates $v \neq v'$ in C we have $\text{val}_v \neq \text{val}_{v'}$. It follows that if τ encodes a reduced circuit and $\sigma\tau = \tau$, then σ must be the identity permutation. (Otherwise there exists a v such that v and $\sigma(v)$ compute the same function, contradicting the fact that C is reduced.) Thus all the maps $\sigma\tau$ and $\sigma'\tau$ for $\sigma \neq \sigma'$ are different elements of $T_{q,n}$, though both C and C_σ compute the same set of functions. Since every function that can be computed with at most q gates can be computed by a reduced circuit with exactly q gates we see that

$$N_{q,n} \leq q \cdot \frac{|T_{q,n}|}{q!} = \frac{|T_{q,n}|}{(q-1)!}.$$

The extra factor q stems from the fact that given a map τ and a corresponding circuit C we still have q possibilities to fix the output gate. (As mentioned above, a circuit with q gates can possibly compute $q + n$ functions, thus we actually get $N_{q,n} \leq q \cdot \frac{|T_{q,n}|}{q!} + n$. However the extra term n refers to the n projections $(x_1, \ldots, x_n) \mapsto x_i$ for $1 \leq i \leq n$, but since $\mathbb{B}^2$ contains the identity functions, these projections are certainly also among the $q \cdot \frac{|T_{q,n}|}{q!}$ functions we obtain when we allow only non-inputs as output gates.)

By Stirling's formula, $q! \geq (2\pi q) \left(\frac{q}{e}\right)^q \geq \frac{q^q}{e^q}$, where e denotes Euler's constant, $e = 2.71828\ldots$. We now argue as follows:

$$
\begin{aligned}
N_{q,n} &\leq q \cdot \frac{(16(n+q)^2)^q}{q!} && \text{by (1.8)} \\
&\leq q \cdot d^q \cdot \frac{(n+q)^{2q}}{q^q} && \text{for } d = 16e \\
&\leq q \cdot d^q \cdot \frac{4^q \cdot q^{2q}}{q^q} && \text{for } q \geq n, \text{ i.e., } n + q \leq 2q \\
&= q \cdot (cq)^q && \text{for } c = 4d \\
&\leq (cq)^{q+1}.
\end{aligned}
$$

This gives for $q_\varepsilon = (1 - \varepsilon)\frac{2^n}{n}$:

$$
\begin{aligned}
N_{q_\varepsilon, n} &\leq \left(c(1-\varepsilon)\frac{2^n}{n}\right)^{(1-\varepsilon)2^n/n+1} \\
&\leq (2^n)^{(1-\varepsilon)2^n/n+1} && \text{for } n \geq c(1-\varepsilon) \\
&= 2^{(1-\varepsilon)2^n+n}
\end{aligned}
$$

Consequently,

$$
\frac{N_{q_\varepsilon, n}}{|\mathbb{B}^n|} \leq \frac{2^{(1-\varepsilon)2^n+n}}{2^{2^n}} = 2^{n-\varepsilon \cdot 2^n},
$$

which approaches 0 as $n \to \infty$. $\qquad\square$

1.5.2 An Upper Bound

In this section, we want to show that the lower bound just given is met by a corresponding upper bound. That is, we want to construct circuits of size roughly $\frac{2^n}{n}$ for *every* Boolean function. These circuits rely on representations of Boolean functions as so-called *ring sum expansions*.

Theorem 1.48. *Let* $f \in \mathbb{B}^n$. *Then there is a length* 2^n *binary string* $(\alpha_a)_{a \in \{0,1\}^n}$ *such that*

$$
f(x_1, \ldots, x_n) = \bigoplus_{a_1 \cdots a_n \in \{0,1\}^n} \alpha_{a_1 \cdots a_n} \wedge x_1^{a_1} \wedge \cdots \wedge x_n^{a_n}, \tag{1.9}
$$

where we set $x_i^{a_i} =_{\text{def}} \begin{cases} x_i, & \text{if } a_i = 1, \\ 1, & \text{if } a_i = 0. \end{cases}$

Here, the vector $(\alpha_a)_{a \in \{0,1\}^n}$ *is uniquely determined by* f. *Equation (1.9) is called the* ring sum expansion *of* f.

Proof. Given f we first construct the conjunctive normal form for f. Next we replace disjunctions by conjunctions and negations making use of de Morgan's laws, and then we replace negations applying the identity $\neg x = x \oplus 1$. Finally we bring the resulting expression into the required form using distributive laws. Recalling that for every term t we have $t \oplus t = 0$ we define $(\alpha_a)_{a \in \{0,1\}^n}$ in such a way that $\alpha_{a_1 \cdots a_n} = 1$ if the term $x_1^{a_1} \wedge \cdots \wedge x_n^{a_n}$ occurs an odd number of times. This proves that for every $f \in \mathbb{B}^n$ there is an $(\alpha_a)_{a \in \{0,1\}^n}$ with the stated properties. Certainly, different functions will need different such vectors. Since there are $2^{2^n} = |\mathbb{B}^n|$ such vectors, it follows that there is one unique vector $(\alpha_a)_{a \in \{0,1\}^n}$ for every $f \in \mathbb{B}^n$. $\qquad \square$

To construct efficient circuits for arbitrary $f \in \mathbb{B}^n$ we first try to build, by relying on a sort of two-level ring sum expansion, small circuits which realize simultaneously many sub-functions of f.

Let F be a finite set of Boolean functions. Define $S(F)$ to be the size of the smallest circuit $C = (V, E, \alpha, \beta, \omega)$ over $\widehat{B}$ such that $F \subseteq \{ \operatorname{val}_v \mid v \in V \}$. Thus C simultaneously computes all functions in F. Define $S(m,t) =_{\text{def}} \max\{ S(F) \mid F \subseteq \mathbb{B}^m \wedge |F| = t \}$.

Lemma 1.49. *Let $m, t \in \mathbb{N}$. For every $p \geq 1$ we have*

$$S(m,t) \leq 2^m + \left\lceil \frac{2^m}{p} \right\rceil \cdot 2^p + \frac{t \cdot 2^m}{p}.$$

Proof. Let $F \subseteq \mathbb{B}^m$, $|F| = t$. We construct a circuit C computing all functions $f \in F$.

First we compute the terms

$$x_1^{a_1} \wedge \cdots \wedge x_m^{a_m} \tag{1.10}$$

for every $a_1, \ldots, a_m \in \{0,1\}$. Say that a conjunction of the form (1.10) is of *length* l if $|\{ a_i \mid 1 \leq i \leq m, a_i = 1 \}| = l$. If we compute all conjunctions of length l before we compute any conjunction of length $l + 1$, and we reuse shorter conjunctions whenever possible, it can be deduced that we need just one $\wedge$ gate per conjunction. Since we have 2^m conjunctions we see that the same number of gates is sufficient.

Second, we partition the set of all these 2^m conjunctions into $q =_{\text{def}} \left\lceil \frac{2^m}{p} \right\rceil$ sets $C_1, \ldots, C_q$ in such a way that $|C_1| = \cdots = |C_{q-1}| = p$ and $|C_q| \leq p$. Within each C_i ($1 \leq i \leq q$) we compute all possible $\oplus$-sums of conjunctions

$$\bigoplus_{f \in C_i} \alpha_f \wedge f \quad \text{for all } \alpha_f \in \{0,1\}. \tag{1.11}$$

Let $D_1, \ldots, D_q$ be the sets of sums of conjunctions thus obtained. Then $|D_i| \leq 2^p$ for all $1 \leq i \leq q$. Define the length of a sum analogously to the length of a conjunction as before, and compute all sums in nondecreasing order of length, reusing shorter sums when possible. This shows that for every particular sum, one $\oplus$ gate is sufficient. Thus for this step, $q \cdot 2^p = \left\lceil \frac{2^m}{p} \right\rceil \cdot 2^p$ gates are needed.

Finally every $f \in F$ can be computed as

$$\bigoplus_{i=1}^{q} \alpha_i \wedge g_i \tag{1.12}$$

for certain $\alpha_i \in \{0,1\}$ and $g_i \in D_i$ for $1 \leq i \leq q$. This can be done with at most $q-1$ $\oplus$ gates for each function f. Thus this step needs $t \cdot (q-1) \leq t \cdot \frac{2^m}{p}$ gates.

We see that the overall size of C is not more than

$$S(m,t) \leq 2^m + \left\lceil \frac{2^m}{p} \right\rceil \cdot 2^p + \frac{t \cdot 2^m}{p}.$$

$\square$

Lemma 1.50. *Let $n \in \mathbb{N}$. For every $p \geq 1$, $0 \leq m \leq n$ we have:*

$$S(n,1) \leq 3 \cdot 2^{n-m} + 2^m + \left\lceil \frac{2^m}{p} \right\rceil \cdot 2^p + \frac{2^n}{p}.$$

Proof. Let $f \in \mathbb{B}^n$ be arbitrary. We construct a circuit C computing f.

As a first step, we compute the terms

$$x_{m+1}^{a_{m+1}} \wedge \cdots \wedge x_n^{a_n} \tag{1.13}$$

for every $a_{m+1}, \ldots, a_n \in \{0,1\}$. By the same argument as in Lemma 1.49 this can be done with 2^{n-m} $\wedge$ gates.

Now there are functions $h_{a_{m+1} \cdots a_n} : \{0,1\}^m \to \{0,1\}$ (for all $a_{m+1}, \ldots, a_n \in \{0,1\}$) such that

$$f(x_1, \ldots, x_n) = \bigoplus_{a_{m+1}, \ldots, a_n \in \{0,1\}} h_{a_{m+1} \cdots a_n}(x_1, \ldots, x_m) \wedge x_{m+1}^{a_{m+1}} \wedge \cdots \wedge x_n^{a_n}. \tag{1.14}$$

As a second step we compute the 2^{n-m} functions $h_{a_{m+1} \cdots a_n}$ as we have shown in Lemma 1.49. For this we need $2^m + \left\lceil \frac{2^m}{p} \right\rceil \cdot 2^p + \frac{2^{n-m} \cdot 2^m}{p} = 2^m + \left\lceil \frac{2^m}{p} \right\rceil \cdot 2^p + \frac{2^n}{p}$ many gates.

Finally we compute f as in (1.14), which needs 2^{n-m} $\wedge$ gates and $2^{n-m}-1$ $\oplus$ gates.

The overall size of C then is not greater than $S(n,1) \leq 3 \cdot 2^{n-m} + 2^m + \left\lceil \frac{2^m}{p} \right\rceil \cdot 2^p + \frac{2^n}{p}$.

$\square$

Theorem 1.51 (Lupanov). *Every $f \in \mathbb{B}^n$ can be computed by circuits with* $\frac{2^n}{n} + o\left(\frac{2^n}{n}\right)$ *gates over the basis* $\{0, 1, \oplus, \wedge\}$.

Proof. We choose $m = \lceil \sqrt{n} \rceil$ and $p = n - m - \lceil \log n \rceil$ in Lemma 1.50. Let us consider the different terms in the sum of Lemma 1.50.

$$3 \cdot 2^{n-m} \ \leq \ 3 \cdot \frac{2^n}{2^{\sqrt{n}}} = o\left(\frac{2^n}{n}\right).$$

$$2^m \ \leq \ 2^{\sqrt{n}+1} = o\left(\frac{2^n}{n}\right).$$

$$\left\lceil \frac{2^m}{p} \right\rceil \cdot 2^p \ \leq \ \left\lceil \frac{2^m}{n-m-\lceil \log n \rceil} \right\rceil \cdot 2^{n-m-\lceil \log n \rceil}$$

$$\leq \ \left(\frac{2^m}{n-m-\lceil \log n \rceil} + 1 \right) \cdot 2^{n-m-\lceil \log n \rceil}$$

$$\leq \ \frac{2^{n-\lceil \log n \rceil}}{n-\sqrt{n}-\lceil \log n \rceil} + 2^{n-\sqrt{n}-\lceil \log n \rceil} = o\left(\frac{2^n}{n}\right).$$

$$\frac{2^n}{p} \ = \ \frac{2^n}{n} \cdot \left(1 + \frac{n-p}{p}\right) = \frac{2^n}{n} \cdot \left(1 + \frac{\lceil \sqrt{n} \rceil + \lceil \log n \rceil}{n - \lceil \sqrt{n} \rceil - \lceil \log n \rceil}\right)$$

$$= \ \frac{2^n}{n} + o\left(\frac{2^n}{n}\right).$$

Summing up these terms shows that all in all we need no more than $\frac{2^n}{n} + o\left(\frac{2^n}{n}\right)$ gates to compute f. □

Bibliographic Remarks

The development of the computation model of Boolean circuits goes back to the early papers [Sha49] and [Lup58] (though some previous papers of Shannon already mention "switching circuits"). An early textbook which heavily influenced the development of the field is [Sav76]. The survey [Fis74] also has to be mentioned. Many of our examples above were considered in these references. The presentation of the examples in this chapter follows [Str94, Chap. VIII] and [CSV84]. The reader will find detailed constructions of circuits for arithmetic and other problems (leading to sometimes more efficient results than presented in this chapter) in [Weg87, Weg89].

Constant-depth reductions were introduced in [CSV84], and all reducibility results presented in this chapter were first established there. Stricter reducibility notions and a comparison of the problems considered here under these can be found in [Weg87]. Theorem 1.47 is from [Sha49]. Theorem 1.51 is from [Lup58]. Our presentation of the proofs was inspired by [Fis74] and [Sch86].

Exercises

1.1. Let $f: \{0,1\}^m \to \{0,1\}$. Show that f can be computed over the basis B_0 in depth 3 with $O(2^m)$ gates. (*Hint:* Use conjunctive or disjunctive normal form.)

1.2. Prove:

(1) For every $k \in \mathbb{N}$, $(\log \log n)^k = o(\log n)$.
(2) $\log^* n = o(\log \log n)$.

1.3. Let n be given, and let k be minimal such that $\ell^{(k)}(n) = 1$. Prove:

(1) $\ell^{(2)}(n) + \ell^{(3)}(n) + \cdots + \ell^{(k)}(n) = O(\log n)$.
(2) $\ell^{(2)}(n) \cdot \ell^{(3)}(n) \cdots \ell^{(k)}(n) = O(\log n)$.

1.4. Consider the following problem of comparing two numbers:

> *Problem:*   LEQ
> *Input:*   two n bit numbers a and b
> *Question:* Is $a \leq b$?

Show that LEQ $\in$ UnbSIZE-DEPTH($n^2, 1$).

1.5. Let $m, n \in \mathbb{N}$. Define $m \mathbin{\dot{-}} n =_{\text{def}} \max\{0, m-n\}$.

> *Problem:*   SUB
> *Input:*   two n bit numbers a and b
> *Output:* $a \mathbin{\dot{-}} b$

Show that SUB $\in$ FSIZE-DEPTH($n^{O(1)}, 1$).

1.6. Let $A = (a_{i,j})_{1 \leq i,j \leq n}$ be a Boolean matrix, and let $A^* = (a^*_{i,j})_{1 \leq i,j \leq n}$. Prove: For $1 \leq i,j \leq n$, $a^*_{i,j} = 1$ iff $i = j$ or there is a sequence $k_1, k_2, \ldots, k_l \in \{1, \ldots, n\}$, $1 \leq l \leq n-1$, such that $a_{i,k_1} = a_{k_1,k_2} = a_{k_2,k_3} = \cdots = a_{k_l,j} = 1$.

1.7. Conclude from Exercise 1.6 that $A^* = (A \vee I)^{n-1}$.

1.8. Let B be a bounded fan-in basis, and let $C = (V, E, \alpha, \beta, \omega)$ be a circuit over B of size $\Omega(n)$ with one output gate. Show that $|E| = \Theta(|V|)$. Hence to evaluate the size of C it is sufficient to determine $|E|$. What can be said in the case of an unbounded fan-in basis?

1.9. Let $s, d: \mathbb{N} \to \mathbb{N}$. Let $f: \{0,1\}^* \to \{0,1\}^*$ be polynomially length-bounded. Prove:

$$f \in \text{FSIZE-DEPTH}(s^{O(1)}, d) \iff \text{bits}(f) \subseteq \text{SIZE-DEPTH}(s^{O(1)}, d).$$

1.10. Give a detailed proof of Lemma 1.36.

1.11. Prove:

(1) If $g \in$ FSIZE-DEPTH(s, d) and $f \leq_{cd} g$, then

$$f \in \text{FSIZE-DEPTH}(s^{O(1)}, d + \log n).$$

(2) If $g \in$ FUnbSIZE-DEPTH(s, d) and $f \leq_{cd} g$, then

$$f \in \text{FUnbSIZE-DEPTH}(s^{O(1)}, d).$$

1.12. Let $A, D_1, D_2 \subseteq \{0,1\}^*$. Prove:

(1) $D_1 \uplus D_2 \leq_{cd} A \iff D_1 \leq_{cd} A$ and $D_2 \leq_{cd} A$.
(2) $A \leq_{cd} D_1 \uplus D_2 \iff A \in \text{SIZE-DEPTH}_{B_1 \cup \{c_{D_1}, c_{D_2}\}}(n^{O(1)}, 1)$.

1.13. For a set $A \subseteq \{0,1\}^*$ let $[A]_{\equiv_{cd}} =_{\text{def}} \{ B \subseteq \{0,1\}^* \mid B \equiv_{cd} A \}$. Let $\mathbf{D} =_{\text{def}} \{ [A]_{\equiv_{cd}} \mid A \subseteq \{0,1\}^* \}$ be the set of *constant-depth degrees*. For $\mathbf{d_1}, \mathbf{d_2} \in \mathbf{D}$, define $\mathbf{d_1} \leq \mathbf{d_2}$ iff there are $A_1 \in \mathbf{d_1}$ and $A_2 \in \mathbf{d_2}$ such that $A_1 \leq_{cd} A_2$. Show that $(\mathbf{D}, \leq)$ is an *upper semi-lattice*, that is, $(\mathbf{D}, \leq)$ is a partially ordered set (i.e., $\leq$ is a reflexive, anti-symmetric, and transitive relation) in which each pair of elements has a least upper bound. Show that the least upper bound of $[A]_{\equiv_{cd}}$ and $[B]_{\equiv_{cd}}$ is $[A \uplus B]_{\equiv_{cd}}$.

1.14. Let $B \subseteq \{0,1\}^*$. Define the set A as follows:

$$A =_{\text{def}} \{ x \in \{0,1\}^* \mid \exists y (|y| = \ell(|x|) \wedge xy \in B) \}.$$

Here xy means the concatenation of the words x and y. Show that $A \leq_{cd} B$.

1.15. Prove: UCOUNT $\leq_{cd}$ BCOUNT.

1.16. Let $\oplus = (\oplus^n)_{n \in \mathbf{N}}$. Prove: $\oplus \leq_{cd}$ MAJ.

1.17.* Consider the following problem:

    Problem: ZERO-MOD-2^c
    Input: n bits $x_1, \ldots, x_n$
    Question: Is $\sum_{i=1}^{n} x_i \equiv 0 \pmod{2^c}$?

Prove: ZERO-MOD-$2^c \leq_{cd} \oplus$.

1.18. Consider the following problem:

    Problem: COMPARISON
    Input: $2n$ bits $x_1, \ldots, x_n, y_1, \ldots, y_n$
    Question: Is $\sum_{i=1}^{n} x_i \geq \sum_{i=1}^{n} y_i$?

Prove that COMPARISON $\equiv_{cd}$ MAJ.

1.19.** Show that the following problem is in FDEPTH$(\log n)$:

    Problem: DIVISION
    Input: two n bit numbers a and b

Output: $\left\lceil \frac{a}{b} \right\rceil$

Hint: Establish a series of reductions as follows:

(a) Reduce division to the problem of computing polynomially many bits of the binary representation of $\frac{1}{b}$ for given b.
(b) Reduce the problem in (a) to the problem of computing $\frac{1}{z}$ for $\frac{1}{2} \leq z \leq 1$.
(c) Use the equality $\frac{1}{z} = \sum_{i \geq 0}(1-z)^i$ to reduce the problem in (b) to iterated multiplication.

1.20. Let $A \subseteq \{0,1\}^*$ be a regular language. Show that $A \in \text{DEPTH}(\log n)$.

1.21.* Let $N_{q,n}$ be defined as in the proof of Theorem 1.47. Show that for $q = \frac{2^n}{n} - 1$ we have $\frac{N_{q,n}}{|\mathbb{B}^n|} \to 0$ as $n \to \infty$. Thus if we say that *almost all n-ary functions have property P* if

$$\lim_{n \to \infty} \frac{|\{ f \in \mathbb{B}^n \mid f \text{ has property P} \}|}{|\mathbb{B}^n|} = 1,$$

we have proved that almost all n-ary Boolean functions need circuits of size at least $\frac{2^n}{n}$ over $\widehat{\mathcal{B}}$ for their computation.

Notes on the Exercises

1.17. See [CSV84].

1.19. This result is from [BCH86]. A thorough presentation of the construction can be found in [Weg87, Sect. 3.3].

2. Relations to Other Computation Models

The results of the previous chapter show that there is a circuit complexity class which contains *every* length-respecting function $f\colon \{0,1\}^* \to \{0,1\}^*$, computable or not:

Corollary 2.1. *For every length-respecting function $f\colon \{0,1\}^* \to \{0,1\}^*$ we have: $f \in \mathrm{FSIZE}\left(\frac{2^n}{n}\right)$.*

Proof. Let $f = (f^n)_{n\in\mathbb{N}}$. By Theorem 1.51, every f^n can be computed with $O\left(\frac{2^n}{n}\right)$ gates. $\qquad\Box$

An analogous result for Turing machine classes defined via time or space restrictions cannot hold, simply because all these classes contain only computable functions; hence they cannot contain *every* function f as above.

In fact there is a much smaller circuit class which contains non-recursive functions:

Lemma 2.2. *Let $A \subseteq \{0,1\}^*$ be such that it contains only words consisting of 1s only (i. e., A is a subset of the regular set 1^*). Then $A \in \mathrm{SIZE}(1) \cap \mathrm{DEPTH}(1)$.*

Proof. For every word length n there can be at most one string of length n in A. Design a circuit family $\mathcal{C} = (C_n)_{n\in\mathbb{N}}$ such that for every n, C_n accepts 1^n if $1^n \in A$, and C_n rejects all other words. $\qquad\Box$

Since there are non-recursive sets $A \subseteq 1^*$ we observe:

Observation 2.3. There are non-recursive sets in $\mathrm{SIZE}(1) \cap \mathrm{DEPTH}(1)$.

But even if we restrict our attention to decidable sets, a result for Turing machines similar to Corollary 2.1 cannot hold. It is well known (see, e. g., [HU79]) that for every recursive bound t, decidable sets can be constructed by diagonalization which are neither in $\mathrm{DTIME}(t)$ nor in $\mathrm{DSPACE}(t)$. This shows that recursive functions can be constructed with arbitrary large gaps between circuit complexity and Turing machine complexity. In other words, a result of the form $\mathrm{SIZE}(t) \subseteq \mathrm{DTIME}\big(E(t)\big)$ where $E(t)$ is any expression in t cannot hold. Is there a way to bridge this fundamental gap between computation models?

And what about the converse? Given a set $A \in \text{DTIME}(t)$, is there an upper bound (depending on t) for the number of gates needed to accept A?

We will address questions of this kind, relating circuit complexity to Turing machine complexity, in this chapter.

2.1 Time on Turing Machines

In this section we want to give an upper bound for the circuit size complexity of a function f given that we know an upper bound on the time for f on deterministic Turing machines. Of course we have to presuppose that f is length-respecting, since only such functions can be computed by circuits. Our Turing machine model is described in Appendix A4.

We will find it convenient to normalize our machines in a certain sense before simulating them by circuits.

Definition 2.4. A k-tape Turing machine M is called *oblivious* if

1. all heads never move to the left of the position they hold at the beginning of the computation;
2. when M reaches a final state then any further computation steps do not change the tape contents;
3. for all $n, s \in \mathbb{N}$ and $x \in \{0,1\}^n$ the positions of M's heads after working on input x for s steps are given by $K_0(n,s), \ldots, K_{k+1}(n,s)$, where $K_0, \ldots, K_{k+1} \colon \mathbb{N} \times \mathbb{N} \to \mathbb{N}$. (Here, K_0 describes the moves of the input head, $K_1, \ldots, K_k$ describe the moves of the work-tape heads, and K_{k+1} describes the moves of the head on the output tape.)

Intuitively condition 3 says that the head positions do not depend on the input but only on the input length. For all inputs of the same length the machine will make the same head movements.

Functions computed by oblivious Turing machines can easily be computed by circuits:

Lemma 2.5. *Let* $f \colon \{0,1\}^* \to \{0,1\}^*$ *be length-respecting, and let* $t \colon \mathbb{N} \to \mathbb{N}$. *If* f *can be computed by an oblivious Turing machine in time* t *then* $f \in \text{FSIZE}(t)$.

Proof. Fix an input length n. For $s = 0, 1, 2, \ldots$, let $C_s(x)$ be the configuration of M working on input x, $|x| = n$, after s steps, i.e., $C_0(x)$ is M's initial configuration. Since M is oblivious the head positions in every $C_s(x)$ are determined by n and s. Thus to encode such a configuration we just have to encode M's state and the contents of M's work tapes. Since M is time-bounded by t, it is space-bounded by the same function; hence every $C_s(x)$ can be encoded in binary by a string of length $O(t)$. Now look at two configurations $C_s(x)$ and $C_{s+1}(x)$. When moving from $C_s(x)$ to $C_{s+1}(x)$ the machine changes its state and for every work tape one symbol according to the δ function. Hence the encodings of these configurations can only differ at a constant number of positions, and these positions are independent of x; they only depend on n. Given a configuration $C_s(x)$, every bit at a position in $C_{s+1}(x)$ different from the corresponding bit in $C_s(x)$ can be computed by a constant depth circuit, since it is determined by a finite function: M's transition function δ.

Our circuit to simulate the work of M now looks as follows: First the encoding of the initial configuration is constructed—this needs $O(t)$ gates. Then t steps of M are simulated, and in every step we just have constant size circuits computing those bits that change during the execution of that step. That is, in every time step we compute only the bits that change. For all the other bits nothing has to be done. Observe that since M is oblivious, we know in advance which bits will possibly change at a particular time step. Therefore the simulation of each step of M needs a constant number of gates, and for t steps again $O(t)$ gates are required. Finally the computed value can be found in the last of M's configurations; it is the value on the output tape. $\square$

Thus it remains to show that every Turing machine computation can be simulated in an oblivious manner without too large time overhead.

Lemma 2.6. *Let $f: \{0,1\}^* \to \{0,1\}^*$ be length-respecting, and let $t: \mathrm{N} \to \mathrm{N}$ be time-constructible, $t(n) \geq n$. If $f \in$ FDTIME(t) then there is an oblivious Turing machine computing f in time $O(t \cdot \log t)$.*

Proof. Let M compute f in time t. First we simulate M by a machine M' which has the property that the moves of the heads on the input and output tape are oblivious. This can easily be achieved: M' has three more work tapes than M. At the beginning of its work M' simply copies the input to an extra work tape. Then M' simulates M for exactly $t(n)$ steps, where n is the length of the input. This can be done by running in parallel the simulation of M and—on a separate tape—a machine M^t witnessing the time-constructibility of t. As soon as M^t halts, M' stops the simulation. During the simulation, M' uses one of its work tapes instead of M's output tape. Finally M' copies the output from this particular work tape to the output tape. Obviously the heads on the input and output tape during this simulation move in an oblivious way.

Now we simulate M' by a machine M'' which is oblivious (for all heads). First every work tape of M' can be simulated by two pushdown stores (see Exercise 2.2). A pushdown store is a one-way infinite tape, where the head is always on the rightmost non-blank symbol and which can only be used in two ways:

1. Move one cell to the right and print the symbol a; we denote this operation by "push(a)".
2. Print the blank symbol b and move one cell left, denote this by "pop".

M'' now has two one-way infinite work tapes. The cells on every tape are numbered $1, 2, 3, \ldots$. The first tape, called the *master tape*, is divided into *channels;* for each of the pushdown stores described above there will be one channel, and every channel will be divided into three *tracks*. Each cell can hold one symbol of M' per track or the symbol ♯. Suppose M' was a k-tape machine; then M''s first work tape consists of $2k$ channels and $6k$ tracks $t_{1,1}, t_{1,2}, t_{1,3}, t_{2,1}, \ldots, t_{2k,3}$ (in other words, if M' works over alphabet Σ, then M'' has alphabet $(\Sigma \cup \{♯\})^{6k}$). The master tape is divided into a sequence of segments; segment number i will consist of tape cells 2^i up to $2^{i+1} - 1$ for all $i \geq 0$. A *block* is the portion of a track that lies within one segment. The simulation will be such that a block in segment i will consist of either 2^i symbols from Σ or 2^i symbols ♯. In the first case we say the block is *filled*, in the second case it is *empty*. The contents of a channel is the concatenation of the filled blocks, ordered first by segment number then by track number. Thus channel 1 in Fig. 2.1 holds the string $abcdefghijkl\cdots$, channel 2 holds $zyxwvutsr\cdots$.

cell no.	1	2	3	4	5	6	7	8	9
$t_{1,1}$	a	c	d	e	f	g	h	$\cdots$	
$t_{1,2}$	b	♯	♯	i	j	k	l	$\cdots$	
$t_{1,3}$	♯	♯	♯	♯	♯	♯	♯	$\cdots$	
$t_{2,1}$	z	y	x	u	t	s	r	$\cdots$	
$t_{2,2}$	♯	w	v	♯	♯	♯	♯	$\cdots$	
$t_{2,3}$	♯	♯	♯	♯	♯	♯	♯	$\cdots$	
$t_{3,1}$	·	·	·	·	·	·	·	$\cdots$	
$t_{3,2}$	·	·	·	·	·	·	·	$\cdots$	
$\vdots$	$\vdots$	$\vdots$	$\vdots$	$\vdots$	$\vdots$	$\vdots$	$\vdots$		$\ddots$
segment	0	1		2					

Fig. 2.1. Master tape of oblivious machine M''

The contents of a channel in reversed order are the contents of the corresponding pushdown store of M', i. e., in our example a and z will be the rightmost symbols where the pushdown heads will be found.

We show how to simulate a single pushdown by its corresponding channel. All channels will then be handled in parallel in the same way. Our simulation will pursue the goal that in each segment each channel will have either 1 or 2 filled blocks. In this case we call that segment *clean*. To make a segment clean we use the procedure which can be found in Fig. 2.2. The actual simulation is performed by procedure sim found in Fig. 2.3.

At the beginning of the simulation we assume that track one is filled entirely with blanks, all other tracks are empty. Thus all segments are clean. Procedure $\text{sim}(n)$ simulates 2^n steps of M'. It is clear that during the simula-

```
procedure clean(k: segment);
{ clean segment k }
begin if segment k has 3 filled blocks
          then concatenate blocks in track 2 and 3 and move
                                   them into segment k + 1;
          if segment k has no filled blocks
          then divide a block from segment k + 1 into two and move
                                   the results into tracks 1 and 2
end;
```

Fig. 2.2. Procedure clean(k)

```
procedure sim(n: integer);
{ simulate 2ⁿ steps of M' }
begin if n = 0 then simulate one step of M by updating cell 1 where:
               push(a) ≡ begin
                             track 3 := track 2;
                             track 2 := track 1;
                             track 1 := a
                         end;
               pop ≡ begin
                         track 1 := track 2;
                         track 2 := track 3;
                         track 3 := ♮
                     end;
      else begin
               sim(n − 1);
               clean(n − 1);
               sim(n − 1);
               clean(n − 1)
           end
end;
```

Fig. 2.3. Procedure sim(n)

tion every block is always either empty or filled. For each execution of sim(n) the following properties can be shown by an easy induction:

1. The head of M'' does not move beyond segment number n.
2. At the end segments $0, \ldots, n-1$ will be clean.
3. The number of filled blocks in segment n will change by at most 1.

Procedure clean(k) cleans segment k, but it might possibly leave segment $k+1$ unclean. However property 3 ensures that this does not cause any problems. For each call of sim(n), if in the beginning segment n is clean then it contains 1 or 2 filled blocks. Thus it can never happen that we try either two times to move a new block from segment $n-1$ to segment n or two times to remove a block from segment n. Thus after sim(n) segment n is not necessarily clean but the executions of clean($n-1$) ensure that segment $n-1$, and—by induction—all segments to the left, are clean.

The main program is now simply a sequence of calls sim(0), sim(1), sim(2), $\ldots$, until machine M' stops. It remains to show that sim and clean can be implemented obliviously. This can easily be seen for sim by an inductive argument (assuming that clean is oblivious). The situation for clean is more difficult since we have branches because of the if-statements. However we can handle this by designing M'' such that, independent of the if-conditions, we always make all head moves necessary to perform consecutively all statements in *both* the then- and the else-branch. M'' needs its second work tape just as auxiliary tape for copying and bookkeeping during this clean procedure.

Let us now turn to the time complexity of the simulation. The time needed for sim(k) is at most twice the time for sim($k-1$) plus $O(2^k)$ for the clean-calls, which together is $O(k \cdot 2^k)$, see Exercise 2.3. The sequence of calls to sim with growing arguments will stop at the latest with the call sim($\lceil \log t(|x|) \rceil$). Thus the overall time is $O(t \cdot \log t)$. □

Combining both lemmas shows:

Theorem 2.7. *Let $f \colon \{0,1\}^* \to \{0,1\}^*$ be length-respecting, and let $t(n) \geq n$ be time-constructible. If $f \in \mathrm{FDTIME}(t)$ then $f \in \mathrm{FSIZE}(t \cdot \log t)$.*

For the case of acceptance of sets we get:

Theorem 2.8. *Let $t(n) \geq n$. Then*

$$\mathrm{DTIME}(t) \subseteq \mathrm{SIZE}(t \cdot \log t).$$

Proof. This follows from Theorem 2.7 by restricting our attention to 0-1-valued functions. Observe that the requirement that f is time-constructible was only needed before to make the head moves on the output tape oblivious. This is no longer needed here. □

2.2 Space on Turing Machines

In this section we will give a depth-efficient circuit simulation of Turing machines that are space-bounded. There is a close connection between space-bounded computation and transitive closure of certain matrices, see Sect. 1.3.4. This connection will also be used here.

Theorem 2.9. *Let $s(n) \geq \log n$. Then*

$$\text{NSPACE}(s) \subseteq \text{DEPTH}(s^2).$$

Proof. Let $A \in \text{NSPACE}(s)$ be accepted by Turing machine M. Without loss of generality (see Appendix A4) let M be a single-tape machine.

We describe configurations of M by specifying

- M's state,
- the position of M's input head,
- the position of M's work tape head, and
- the contents of M's work tape.

The number of such configurations, given an input of length n, is thus bounded from above by

$$N(n) =_{\text{def}} c \cdot (n+1) \cdot s(n) \cdot d^{s(n)},$$

where c is the number of states of M and d is the number of symbols in M's alphabet.

Furthermore we assume that M has a unique accepting configuration K_a, and that if M reaches K_a then any further computation steps do not leave K_a.

Fix an input length n and a numbering of M's configurations while working on inputs of length n. Let i_0 be the number of M's initial configuration and let i_a be the number of M's accepting configuration in this numbering.

For every input x of length n we define an $N(n) \times N(n)$ matrix $T_x = (t_{i,j})$, the so-called *transition matrix* of M, as follows:

$$t_{i,j} = \begin{cases} 1 & \text{if the configuration numbered } j \text{ can be reached from} \\ & \text{the configuration numbered } i \text{ in one step,} \\ 0 & \text{otherwise.} \end{cases} \qquad (2.1)$$

Given $x = x_1 \cdots x_n$, the matrix T_x can be constructed in constant depth. Every entry $t_{i,j}$ as a function of x is either constant 0, constant 1, an identity x_k or a negation $\neg x_k$ for some $k \in \{1, \ldots, n\}$ (since the transition from the ith configuration to the jth configuration depends on only one bit on the input tape).

By definition of T_x and Remark 1.30, the input x is accepted if in the transitive closure of T_x the element at row i_0 and column i_a has entry 1, i. e.,

$x \in A \iff t^*_{i_0, i_a} = 1$. By Theorem 1.31, to compute $(T_x)^*$ we need circuits of depth $O\big((\log N(n))^2\big)$, which for $s(n) \geq \log n$ is not more than $O\big(s(n)^2\big)$. This is also the overall depth of a circuit family for A. $\square$

Corollary 2.10. *Let $f \colon \{0,1\}^* \to \{0,1\}^*$ be length-respecting, and let $s(n) \geq \log n$. If $f \in \text{FDSPACE}(s)$ then $f \in \text{FDEPTH}(s^2)$.*

Proof. If $f \in \text{FDSPACE}(s)$ then clearly bits$(f) \subseteq \text{DSPACE}(s)$. Here we interpret a function in bits(f) as a characteristic function of a set. By Theorem 2.9 there are circuits of depth $O\big(s(n)^2\big)$ for all sets whose characteristic functions are in bits(f). By combining these circuits we get a circuit of the same depth for the function f. $\square$

2.3 Non-uniform Turing machines

In the preceding two sections we simulated Turing machines by circuits. We proved inclusions of the type

$$\text{DTIME}(t) \subseteq \text{SIZE}(t \cdot \log t) \quad \text{and} \quad \text{NSPACE}(s) \subseteq \text{DEPTH}(s^2).$$

However as pointed out at the beginning of this chapter, any inclusion in the other direction cannot hold. This is due to the fact that Turing machines are *uniform* in the sense that they are a manifestation of an algorithm working for *all* inputs of arbitrary lengths. On the other hand, circuits are a *non-uniform* computation model, i. e., for every input length n, we have a single device working on inputs of that length. This fact is central in the proof of Theorem 1.51 and Observation 2.3.

There are two ways of dealing with this discrepancy between circuits and machines:

1. Make Turing machines non-uniform by supplying them with additional information depending on the length of their input.
2. Make circuit families uniform by postulating that there is a finite description of the infinite family, i. e., by requiring that the circuits for different input lengths have something in common concerning their structure.

In this section we will turn our attention to possibility 1 while in the next section we consider uniform circuit families.

Definition 2.11. Let $\mathcal{K} \subseteq \mathcal{P}(\{0,1\}^*)$ be a class of sets. Let $\mathcal{F}$ be a class of functions from $\{0,1\}^* \to \{0,1\}^*$. The class $\mathcal{K}/\mathcal{F}$ is defined to consist of all sets $A \subseteq \{0,1\}^*$ for which there exist a set $B \in \mathcal{K}$ and a function $f \in \mathcal{F}$ such that for all x,

$$x \in A \iff \langle x, f(1^{|x|}) \rangle \in B.$$

$\mathcal{K}/\mathcal{F}$ is a so-called *non-uniform complexity class*. In the notation in the definition, the function f gives, depending on the length of the actual input x, additional information to the set B (or a machine deciding B). Therefore, f is called an *advice function*, and $f(1^n)$ is the *advice* for inputs of length n. $\mathcal{K}/\mathcal{F}$ is also called an *advice class*.

In the following we mainly consider classes of advice functions given by a length restriction.

Definition 2.12. Let $f: \mathbb{N} \to \mathbb{N}$. Then $\mathrm{F}(f) =_{\mathrm{def}} \{h: \{0,1\}^* \to \{0,1\}^* \mid |h(x)| \le f(|x|)$ for all $x\}$. If $\mathcal{F}$ is a class of functions then $\mathrm{F}(\mathcal{F}) =_{\mathrm{def}} \bigcup_{f \in \mathcal{F}} \mathrm{F}(f)$. Set Poly $=_{\mathrm{def}} \mathrm{F}(n^{O(1)})$, i.e., the set of all polynomially length-bounded functions.

For non-uniform Turing machines, we now give an analogue of Observation 2.3:

Observation 2.13. There are non-recursive languages in $\mathrm{DTIME}(n)/\mathrm{F}(1)$.

The (easy) proof is left as Exercise 2.6.

We are now able to simulate non-uniform circuits by non-uniform Turing machines and give, in a sense, inverses of Theorems 2.8 and 2.9. The idea in both simulations will be that we use the advice function to encode the different circuits of the given family. The Turing machine will then, given an input and an encoding of the relevant circuit, just compute the outcome of the circuit. We now fix an encoding scheme for circuits.

Definition 2.14. Let $C = (C_n)_{n \in \mathbb{N}}$ be a circuit family over B. An *admissible encoding scheme* of C is given as follows: First fix an arbitrary numbering of the elements of B. Second, for every n, fix a numbering of the gates of C_n with the following properties:

- In C_n the input gates are numbered $0, \dots, n-1$.
- If C_n has m output gates, then these are numbered $n, \dots, n+m-1$.
- Let $s(n)$ be the size of C_n. Then there is a polynomial p such that, for every n, the highest number of a gate in C_n is bounded by $p(s(n))$. (This ensures that the encoding of a gate in binary just requires a number of bits which is logarithmic in the size of C_n.)

The encoding of a gate v in C_n is now given by a tuple

$$\langle g, b, g_1, \dots, g_k \rangle,$$

where g is the number of v, b is the number of the type of v, and $g_1, \dots, g_k$ are the numbers of the predecessor gates of v in the order of the edge numbering of C_n.

Fix an arbitrary order of the gates of C_n. Let $v_1, \dots, v_s$ be the gates of C_n in that order. Let $\overline{v_1}, \dots, \overline{v_s}$ be their encodings. The encoding of C_n is then given by

$$\langle \overline{v_1}, \ldots, \overline{v_s} \rangle.$$

If an admissible encoding scheme of $\mathcal{C}$ is fixed, then $\overline{C_n}$ denotes the encoding of C_n under this scheme.

Theorem 2.15. *Let $s(n) \geq n$. Then*

$$\mathrm{SIZE}(s) \subseteq \mathrm{DTIME}(s^2)/\mathrm{F}\left(O(s \log s)\right).$$

Proof. Let $A \in \mathrm{SIZE}(s)$ via circuit family $\mathcal{C} = (C_n)_{n \in \mathbb{N}}$. Fix an admissible encoding scheme of $\mathcal{C}$, which has the property that for all gates v in a circuit from $\mathcal{C}$ the number of v is larger than the numbers of all predecessor gates of v. Such an ordering is called a *topological order* of the gates. Define $f \colon \{0,1\}^* \to \{0,1\}^*$ as $f(x) = \overline{C_{|x|}}$. The encoding of a single gate in C_n needs no more than $O(\log s(n))$ bits (recall that we have bounded fan-in here); hence $|f(x)| = O(s(|x|) \cdot \log s(|x|))$.

Now define a Turing machine M as follows: On input $\langle x, \overline{C_n} \rangle$ where $n = |x|$, M simulates C_n gate by gate in the ordering induced by the gate numbering. Store the values of the gates on the work tape. Because of the particular choice of the ordering, when we want to compute the value of a gate v all the values of v's predecessor gates are already computed. All we have to do to evaluate v is to find the values of v's predecessors and apply the function associated with v to them. This may require that we perform a complete scan over all gate values computed so far. Since C_n has at most $s(n)$ gates, at most that many values have to be stored and therefore such a scan takes time $O(s(n))$. Since $s(n)$ gates have to be evaluated, the overall time is $O(s(n)^2)$. Therefore, $A \in \mathrm{DTIME}(s^2)/\mathrm{F}\left(O(s \log s)\right)$. □

Corollary 2.16. *Let $t(n) \geq n$. Then*

$$\mathrm{SIZE}(t^{O(1)}) = \mathrm{DTIME}(t^{O(1)})/\mathrm{F}\left(t^{O(1)}\right).$$

Proof. ($\subseteq$): By Theorem 2.15.

($\supseteq$): Given $A \in \mathrm{DTIME}(t^{O(1)})/\mathrm{F}\left(t^{O(1)}\right)$ via Turing machine M and advice function f, we construct a circuit family $\mathcal{C} = (C_n)_{n \in \mathbb{N}}$ for A simulating M as in the proof of Theorem 2.8, but with the following modification: We hardwire the value of $f(n)$ into circuit C_n. Thus C_n on input x simulates M working on input $\langle x, f(|x|) \rangle$. Thus $\mathcal{C}$ accepts A. □

Corollary 2.17. $\mathrm{SIZE}(n^{O(1)}) = \mathrm{P}/\mathrm{Poly}$.

We now come to a simulation of depth-bounded circuits by non-uniform space-bounded Turing machines.

It will be convenient if we assume that the graph underlying our circuit is actually a tree. An arbitrary circuit can be unwound into a tree as follows:

Let $C = (V, E, \alpha, \beta, \omega)$ be a Boolean circuit with n inputs and one output. Let d be the depth of C. We construct a tree $\mathrm{T}(C)$ in d stages as follows:

After the completion of stage m we will have ensured that all nodes v at depth m (i.e., the length of the longest path from v to the inputs is m) have fan-out at most one. We start with $m = 0$.

Stage m: Let v be a node in C at depth m with fan-out $k > 1$. Let $u_1, \ldots, u_k \in C$ be the nodes that have predecessor v. Replace v by k gates $v^1, \ldots, v^k$. Replace the subtree T_v with root k by k copies $T_v^1, \ldots, T_v^k$, where v^i is the root of T_v^i for $1 \le i \le k$. Replace the wire (v, u_i) by (v^i, u_i) for $1 \le i \le k$.

Note that this construction does not increase the depth of the circuit considered.

Theorem 2.18. *Let $d(n) \ge \log n$. Then*

$$\text{DEPTH}(d) \subseteq \text{DSPACE}(d)/\text{F}\left(2^{O(d)}\right).$$

Proof. Let $A \in \text{DEPTH}(d)$ via circuit family $\mathcal{C} = (C_n)_{n \in \mathbb{N}}$. We construct a circuit family $\mathcal{C}' = (C'_n)_{n \in \mathbb{N}}$ which is defined as follows: Given C_n, first construct $\text{T}(C_n)$. Formally, $\text{T}(C_n)$ is not a circuit since input gates x_i may appear more than once, contradicting Def. 1.6. Therefore we obtain C'_n from $\text{T}(C_n)$ by merging for all i, $1 \le i \le n$, all nodes with label x_i. In C'_n all nodes except the output and the inputs have out-degree one. Since the depth of C_n is at most $d(n)$ and we have only bounded fan-in gates, the size of C'_n is $2^{O(d)}$. The depth of C'_n is the same as the depth of C_n. As in Theorem 2.15 we use an advice function f to encode $\mathcal{C}'$. Because the size of C'_n is $2^{O(d)}$, we have $f \in \text{F}\left(2^{O(d)}\right)$.

In C'_n every gate except the output has fan-out 1. This means that the value computed at a gate is needed exactly once in another computation. If we evaluate the gates of C_n in postorder then we may use a stack to store the results computed so far. More exactly we evaluate the gates of C_n by the recursive procedure given in Fig. 2.4.

```
procedure eval(v: gate);
{ evaluate gate v }
begin if v is an input gate
        then push the corresponding input bit
        else begin let v1, ..., vk be the predecessors of v in this order;
                   eval(v1); ... ; eval(vk);
                   let f be the Boolean function associated with v;
                   pop k elements from the stack;
                   apply f to them and push the result
             end
end
```

Fig. 2.4. Procedure eval(v)

The simulation of C_n now consists of a call of eval where the argument is C_n's output gate. This simulation needs a stack whose height is given by the depth of the circuit, i.e., $O(d(n))$. To store a single computed value we need just one bit; therefore the space requirement of the above procedure is also $O(d(n))$.

This shows that $A \in \mathrm{DSPACE}(d)/\mathrm{F}\left(2^{O(d)}\right)$. $\Box$

Corollary 2.19.

$$\mathrm{DEPTH}\left((\log n)^{O(1)}\right) = \mathrm{DSPACE}\left((\log n)^{O(1)}\right)/\mathrm{F}\left(2^{(\log n)^{O(1)}}\right).$$

Proof. ($\subseteq$): By Theorem 2.18.

($\supseteq$): Given $A \in \mathrm{DSPACE}\left((\log n)^{O(1)}\right)/\mathrm{F}\left(2^{(\log n)^{O(1)}}\right)$ via Turing machine M and advice function f, we construct a circuit family $C = (C_n)_{n \in \mathbb{N}}$ for A simulating M as in the proof of Theorem 2.9. As in the proof of Corollary 2.16 we hardwire the advice into the circuit.

Given now an input x of length n, the length of $\langle x, f(n) \rangle$ is $n + 2^{(\log n)^{O(1)}}$. Thus machine M needs space $(\log n)^{O(1)}$, and therefore C has depth $(\log n)^{O(1)}$. $\Box$

Corollary 2.20. $\mathrm{DEPTH}(\log n) \subseteq \mathrm{L}/\mathrm{Poly}$.

2.4 Uniform Circuit Families

We now turn to the second possibility outlined in Sect. 2.3 to bridge the gap between the circuit model and the machine model: We want to consider uniform circuit families.

Circuit families are infinite objects, but an algorithm has to be finite. A program written in any programming language has a finite text. A Turing machine has a finite number of states and therefore a finite transition function. A random access machine (RAM) has a finite number of instructions. Therefore a *uniform circuit family* should be a circuit family with a finite description.

There are several ways to make this idea precise. The possibility we want to consider first is that we require that the function $n \mapsto \overline{C_n}$ is easy, i.e., computable. The machine computing this function is then in a sense a finite description of the circuit family. Let us observe that, if we do so, we obtain in contrast to Observation 2.3:

Observation 2.21. (1) Let $C = (C_n)_{n \in \mathbb{N}}$ be a circuit family. If there is an admissible encoding scheme such that the function $n \mapsto \overline{C_n}$ is computable, then f_C is computable.

(2) Let f be length-respecting and recursive. Then there is a circuit family $C = (C_n)_{n \in \mathbb{N}}$ of size $O(\frac{2^n}{n})$ and an admissible encoding scheme for C such that $f = f_C$ and $n \mapsto \overline{C_n}$ is computable.

Proof. (1) is obvious. For (2) let $f = (f^n)_{n\in\mathbb{N}}$. For every n we construct C_n as in Theorem 1.51. Given n we can first compute a complete table of f^n. Now the proofs of Theorems 1.48 and 1.51 are constructive, hence C_n can be constructed algorithmically.   □

Depending on the complexity of the function $n \mapsto \overline{C_n}$ we will distinguish different *uniformity conditions*. Certainly it is most desirable and natural to measure the complexity needed to compute $\overline{C_n}$ in n, the number of inputs of the circuit under consideration. Since it is customary in complexity theory to measure complexity in the length of the input, we will therefore not use the function $n \mapsto \overline{C_n}$ but instead consider the function

$$1^n \mapsto \overline{C_n}$$

in the formal definitions below.

Definition 2.22. A circuit family $C = (C_n)_{n\in\mathbb{N}}$ of size s is

- L-uniform (logspace-uniform), if there is an admissible encoding scheme such that the map $1^n \mapsto \overline{C_n}$ is in FDSPACE($\log s$);
- P-uniform (ptime-uniform), if there is an admissible encoding scheme such that the map $1^n \mapsto \overline{C_n}$ is in FDTIME($s^{O(1)}$).

We will denote the corresponding circuit classes by prefixing their name with U_L- for logspace-uniformity or U_P- for ptime-uniformity. For instance, we will write U_L-UnbSIZE-DEPTH($n^{O(1)}, 1$) or U_P-FDEPTH($\log n$).

To show that a certain circuit family is uniform under one of the two above notions, one has to fix an admissible encoding scheme and then present a required machine that computes the circuit encodings. Observe that the definition of admissible encoding (Def. 2.14) leaves a lot of freedom when fixing the numbering of gates. In concrete cases this is often central. In fact, the main difficulty when proving uniformity is often to derive a suitable gate numbering.

Let us show as an example that the circuits designed for addition in Sect. 1.1 are logspace-uniform.

Example 2.23. We claim that ADD $\in U_L$-FUnbSIZE-DEPTH($n^{O(1)}, 1$) over basis $B = \{0^1, \oplus^2, \oplus^3, \wedge, \vee\}$, where we added to the basis used in Sect. 1.1 for convenience the 1-ary constant 0 function. Let us number B as follows:

0^1	$\oplus^2$	$\oplus^3$	$\wedge$	$\vee$
0	1	2	3	4

For convenience we recall the defining formulae, where the inputs are labeled $a_{n-1}, \ldots, a_0, b_{n-1}, \ldots, b_0$.

$$g_i = a_i \wedge b_i \qquad\qquad \text{for } 0 \le i \le n-1$$
$$p_i = a_i \vee b_i \qquad\qquad \text{for } 0 \le i \le n-1$$
$$c_i = \bigvee_{j=0}^{i-1}\left(g_j \wedge \bigwedge_{k=j+1}^{i-1} p_k\right) \qquad \text{for } 1 \le i \le n$$
$$s_0 = a_0 \oplus b_0$$
$$s_i = a_i \oplus b_i \oplus c_i \qquad\qquad \text{for } 1 \le i \le n-1$$
$$s_n = c_n$$

The definition of the machine witnessing the uniformity of the construction is given in Fig. 2.5.

---

input 1^m:
(1)   **if** m is odd **then** print $\langle m, 0, 0\rangle$
       else begin
              $n := m/2$;
(2)   **for** $i := 0$ **to** $n - 1$ **do** print $\langle m + n + 1 + i, 3, i, n + i\rangle$;
(3)   **for** $i := 0$ **to** $n - 1$ **do** print $\langle m + 2n + 1 + i, 4, i, n + i\rangle$;
              for $i := 1$ **to** n **do**
                   for $j := 0$ **to** $i - 1$ **do**
(4)   print $\langle m + 3n + 1 + n(i - 1) + j, 3, m + 2n + 1 + j + 1, \ldots,$
                          $m + 2n + 1 + i - 1\rangle$;
              for $i := 1$ **to** n **do**
                   for $j := 0$ **to** $i - 1$ **do**
(5)   print $\langle m + 3n + 1 + n^2 + n(i - 1) + j, 3,$
                          $m + n + j + 1, m + 3n + 1 + n(i - 1) + j\rangle$;
              for $i := 1$ **to** $n - 1$ **do**
(6)   print $\langle m + 3n + 2n^2 + i, 4, m + 3n + 1 + n^2 + n(i - 1), \ldots,$
                     $m + 3n + 1 + n^2 + n(i - 1) + i - 1\rangle$;
(6')   print $\langle m + n, 4, m + 3n + 1 + n^2 + n(n - 1), \ldots$
                     $m + 3n + 1 + n^2 + n(n - 1) + n - 1\rangle$;
(7)   print $\langle m, 1, 0, n\rangle$;
(8)   **for** $i := 1$ **to** $n - 1$ **do**
                         print $\langle m + i, 2, i, n + i, m + 3n + 2n^2 + i\rangle$;
       end.

Fig. 2.5. Uniformity machine for addition circuits

We suppose the inputs $a_0, \ldots, a_{n-1}, b_0, \ldots, b_{n-1}$ are numbered (in this order) by $0, \ldots, m-1$. If the input length is not divisible by 2, then the circuit consists only of a constant 0 gate with number m, see line (1). Otherwise, in line (2), the gates computing g_i with numbers $m + n + 1 + i$ (for $0 \le i \le n-1$) are printed. Observe that the gate for a_i has number i and the gate for b_i has number $n + i$. In line (3) the gates for p_i with numbers $m + 2n + 1 + i$ (for $0 \le i \le n-1$) are printed. In line (4) gates that compute all expressions $\bigwedge_{k=j+1}^{i-1} p_k$ are printed; these carry gate numbers $m + 3n + 1 + n(i - 1) + j$ (for $1 \le i \le n, 0 \le j \le i - 1$). In line (5) gates that compute all expressions $g_j \wedge \bigwedge_{k=j+1}^{i-1} p_k$ are printed; these carry gate numbers $m+3n+1+n^2+n(i-1)+j$ (for $1 \le i \le n, 0 \le j \le i-1$). In line (6) the gates computing c_i with numbers

$m + 3n + 2n^2 + i$ (for $1 \le i \le n - 1$) are printed. The gate for c_n is also the highest-order output bit s_n, therefore it has to get number $m + n$ (line (6')). In line (7) the gate computing the first output bit s_0 is printed; it has number m. Finally, in line (8), the gates computing outputs $s_1, \ldots, s_{n-1}$ are printed at numbers $m + 1, \ldots, m + n - 1$.

This algorithm is clearly logarithmic space-bounded, since all occurring variables are polynomially bounded in m, and all arithmetical operations can be performed in the same space. Thus we see that the circuit family for ADD is logspace-uniform.

The uniformity conditions above were defined in terms of *computation of a function*. It turns out that for space-bounded uniformity restrictions there is an equivalent, but often easier to use, characterization in terms of *acceptance of a set*.

Definition 2.24. Let $C = (C_n)_{n \in \mathbb{N}}$ be a circuit family over B. Fix an admissible encoding scheme for C. The *direct connection language* of C (with respect to the fixed encoding), $L_{DC}(C)$, is the set of all tuples

$$\langle y, g, p, b \rangle,$$

where for $n = |y|$

- g is a number of a gate v in C_n;
- $p \in \{0, 1\}^*$;
- if $p = \epsilon$ then b is the number of the function from B computed at v;
- if $p = \text{bin}(k)$ then b is the number of the k-th predecessor gate to v (with respect to the order of edges in C_n).

Thus $L_{DC}(C)$ is a language encoding the family C. If $L_{DC}(C)$ is recursive then the machine accepting $L_{DC}(C)$ is a finite description of the infinite object C.

Lemma 2.25. *Let $C = (C_n)_{n \in \mathbb{N}}$ be a circuit family of size s over a bounded fan-in basis B. Let $f : \mathbb{N} \to \mathbb{N}$, $f(n) \ge \log s(n)$. For all admissible encoding schemes the following holds: The map $1^n \mapsto \overline{C_n}$ is computable in space $O(f)$ if and only if $L_{DC}(C)$ is decidable by a Turing machine which needs space $O(f(n))$ on inputs $\langle y, \cdot, \cdot, \cdot \rangle$, where $|y| = n$.*

Proof. ($\Rightarrow$): Let $1^n \mapsto \overline{C_n} \in \text{FDSPACE}(f)$ via Turing machine M. Then $L_{DC}(C)$ is accepted by a machine working as follows: To decide if a tuple $\langle y, g, p, b \rangle$ belongs to $L_{DC}(C)$ we simulate M on input 1^n, until it outputs the encoding of the gate with number g. Then we accept if and only if this encoding is consistent with $\langle y, g, p, b \rangle$. The space needed for this procedure is clearly $O(f(n))$.

($\Leftarrow$): Let $L_{DC}(C)$ be decidable by machine M working in space $O(f)$ on inputs $\langle y, \cdot, \cdot, \cdot \rangle$ where $|y| = n$. Below we will for convenience freely identify circuit gates with their numbers according to the encoding scheme.

We now construct a machine N computing $1^n \mapsto \overline{C_n}$. First we describe three particular subroutines of N:

(1) Given a gate (number) g, the type of g is the unique element $b \in B$ for which $\langle 1^n, g, \epsilon, b \rangle \in L_{DC}(\mathcal{C})$. Thus N can determine the type of g by testing for all possible $b \in B$ if the above containment holds. For this test N simulates M as a subroutine. However M's input $\langle 1^n, g, \epsilon, b \rangle$ cannot be constructed completely on a work tape of N, since this might use too much space. Instead, N just writes down $\langle g, \epsilon, b \rangle$ and then uses this to simulate the behavior of M on the virtual input $\langle 1^n, g, \epsilon, b \rangle$. Technically, N keeps track of the input head position of M. If it is not more than n, N simulates M with virtual input symbol 1. If it is larger than n, N simulates M with the input bit taken from the string $\langle g, \epsilon, b \rangle$. In this way, N behaves as M on input $\langle 1^n, g, \epsilon, b \rangle$.

Observe that once we have determined the type of g we also know how many predecessor gates g has.

(2) Given a number g, N can determine if g is a gate of C_n. This is achieved in a similar way to the above by just testing if there is any b such that $\langle 1^n, g, \epsilon, b \rangle \in L_{DC}(\mathcal{C})$. If so, g is a gate in C_n (the value b is discarded since it is no longer needed). If no such b exists, g cannot be a gate in C_n.

(3) Given a gate g with k predecessor gates, N can determine these predecessor gates of g, again by calling M as a subroutine. The i-th predecessor of g, $i \le k$, is the unique b such that $\langle 1^n, g, \mathrm{bin}(i), b \rangle \in L_{DC}(\mathcal{C})$. Thus by simulating M successively for $b = 0, 1, 2, \ldots$ we will find the predecessor gates. The actual simulation of M is performed as described above.

With these three routines in mind we now construct N as in Fig. 2.6. The idea of variable m is that it is the number of the highest gate we have

```
input 1^n:
    m := n;
    while there is a gate m + 1 in C_n do
        m := m + 1;
    g := n;
    repeat
        if g is a gate in C_n with k predecessor gates then
        begin
            let g_1, ..., g_k be the predecessor gates of g;
            let b be the function computed at g;
            print ⟨g, b, g_1, ..., g_k⟩;
            m := max{m, g_1, ..., g_k}
        end;
        g := g + 1
    until g > m.
```

Fig. 2.6. Machine computing $1^n \mapsto \overline{C_n}$

encountered so far. The **while**-loop makes sure that m is at least the number of the highest output gate of C_n. (Recall that if there are k outputs then these are numbered $n, n+1, \ldots, n+k-1$.) The **repeat**-loop constructs encodings of all gates with number between n and m. If we encounter gate numbers exceeding m then m is updated accordingly. Thus when the loop ends, we have considered all gates up to m and all their predecessors in C_n, and since during this procedure no higher gate numbers occurred we know that we have finished.

The space needed by N is given by the space needed for the simulation of M, i.e., $O(f(n))$, plus the space for the local variables which is $\log s(n)$, since they only hold gate numbers in C_n.   □

Remark 2.26. In the previous lemma, we considered a condition on the complexity of $L_{DC}(C)$: We require that

there is a machine which on inputs $\langle y, \cdot, \cdot, \cdot \rangle$ needs space $O(f(n))$ (2.2)

(where $|y| = n$). This does not necessarily mean that

$$L_{DC}(C) \in \mathrm{DSPACE}(f),   (2.3)$$

since in (2.2) n is not the length of the input $\langle y, \cdot, \cdot, \cdot \rangle$; the actual length of the input is $n + O(\log s(n))$.

However there is a special case where the distinction between (2.2) and (2.3) is of no importance. If the circuit family C is of polynomial size then the length of a tuple $\langle y, \cdot, \cdot, \cdot \rangle$ and the number n are linearly related. Hence, C is logspace-uniform iff $1^n \mapsto \overline{C_n} \in \mathrm{FDSPACE}(\log n)$ iff $L_{DC}(C) \in \mathrm{DSPACE}(\log n)$.

Now we can state that the simulations given in the preceding sections also hold in the uniform setting. First let us consider the relation between size and time.

Theorem 2.27. *Let $t(n) \geq n$ be time-constructible, and suppose the function $1^n \mapsto t(n)$ is computable in space $O(\log t(n))$. Then*

$$\mathrm{DTIME}(t) \subseteq \mathrm{U_L\text{-}SIZE}(t \log t).$$

Proof. This follows from the proof of Theorem 2.8 with the following additional observation: The simulation of an arbitrary machine by an oblivious machine in Lemma 2.6 leads to very regular right/left scans (resulting from calls to procedure sim) of the work-tape heads, such that the functions $1^n \mapsto K_i(n, s)$ (i.e., head position on tape i after s steps on input length n, see Def. 2.4) are computable in space $O(\log t(n))$. This implies that the circuit constructed in Lemma 2.5 can be constructed using space $O(\log t)$.   □

Theorem 2.28. *Let* $t(n) \geq n$. *Then*

$$U_L\text{-SIZE}(t) \subseteq \text{DTIME}(t^{O(1)}).$$

Proof. As the proof of Theorem 2.15, but this time we do not check the advice to obtain information about the structure of the circuit family; instead we use a subroutine to produce the encoding of the relevant circuit. This subroutine uses space $O(\log t)$ which can be simulated in time $t^{O(1)}$. □

Corollary 2.29. *Let* $t(n) \geq n$ *be time-constructible, and suppose the function* $1^n \mapsto t(n)$ *is computable in space* $O(\log t(n))$. *Then*

$$U_L\text{-SIZE}(t^{O(1)}) = \text{DTIME}(t^{O(1)}).$$

Corollary 2.30. $U_L\text{-SIZE}(n^{O(1)}) = P$.

Next we turn to the relation between depth and space.

Theorem 2.31. *Let* $s(n) \geq \log n$ *be space-constructible. Then*

$$\text{NSPACE}(s) \subseteq U_L\text{-DEPTH}(s^2).$$

Proof. As in the proof of Theorem 2.9. The entries of the transition matrix can be constructed in space $O(\log s)$; according to (2.1) it only has to be checked whether a given configuration is a successor of another one, which can be done in space proportional to the length of the configurations. The circuits we construct have size $2^{O(s)}$, thus the space needed is logarithmic in the size.

The circuit family for transitive closure given in Lemma 1.31 is logspace-uniform (see Exercise 2.5). □

Theorem 2.32. *Let* $s(n) \geq \log n$. *Then*

$$U_L\text{-DEPTH}(s) \subseteq \text{DSPACE}(s).$$

Proof. As in Theorem 2.18, but using a subroutine for the direct connection language instead of an advice. □

Corollary 2.33. *Let* $s(n) \geq \log n$ *be space-constructible. Then*

$$U_L\text{-DEPTH}(s^{O(1)}) = \text{DSPACE}(s^{O(1)}).$$

Combining Theorems 2.31 and 2.32 gives a result known as Savitch's Theorem [Sav70]:

Corollary 2.34 (Savitch). *Let* $s(n) \geq \log n$ *be space-constructible. Then* $\text{NSPACE}(s) \subseteq U_L\text{-DEPTH}(s^2) \subseteq \text{DSPACE}(s^2)$.

2.5 Alternating Turing Machines

In the previous sections we related circuit size to Turing machine time and circuit depth to Turing machine space. However this does not address the feeling one has that circuits are some form of parallel computation model. Remember the carry look-ahead adder designed in Sect. 1.1: it is a hardware implementation of an optimal parallel algorithm for addition.

In the following sections we will relate circuits to parallel machine models. We start by considering the alternating Turing machine, an abstract model suitable for theoretical investigations, and later turn to a more realistic parallel computer in Sect. 2.7.

2.5.1 Computation Model and Complexity Measures

In this section we formally define the computation model of alternating Turing machines and the complexity classes defined via this model which are of interest for us.

An *alternating k-tape Turing machine* is given by a 5-tuple

$$M = (Q, \Sigma, \delta, q_0, g),$$

where Q is a finite set, the so-called *state set*, Σ is an alphabet (Σ always contains the blank symbol b), $\delta: Q \times \Sigma^{k+1} \to \mathcal{P}(Q \times \Sigma^k \times \{-1, 0, +1\}^{k+1})$ is the *transition function*, $q_0 \in Q$ is the *initial state*, and finally $g: Q \to \{\wedge, \vee, 0, 1\}$ is the *state type function*.

If for a state $q \in Q$, $g(q) = 0$ then we say that q is *rejecting*, if $g(q) = 1$ then q is *accepting*, if $g(q) = \wedge$ then q is *universal*, and if $g(q) = \vee$ then q is *existential*. Rejecting and accepting states are subsumed under the name *final states*, and we assume that for a state q, $\delta(q, a_0, \ldots, a_k) = \emptyset$ for all $a_0, \ldots, a_k \in \Sigma$ if and only if q is final. Thus alternating machines are similar to nondeterministic machines in the sense that the transition function δ is of the same type, however, besides rejecting and accepting states nondeterministic machines have only regular, i. e., nondeterministic, states, but here we have two types of non-final states. At every time step, M will have the possibility to move on with its work according to the different entries in the transition function, as for nondeterministic machines. The difference lies in the way the computation is evaluated after the machine stops. Let us make this precise.

Initially we assume the input to M is given on the input tape with the input head on the first position, the k work tapes are empty (i. e., all cells contain the blank symbol), and the current state is the initial state. As we will see the machine will work in such a way that at all times only a finite portion of every work tape contains non-blank symbols. The smallest coherent part of a work tape containing all non-blank tape cells and the cell scanned by the tape head is simply referred to as the contents of that tape. A *configuration* of an alternating machine M as above is given by specifying its state, the contents of its work tapes, the position of the input head, and the position of the work-tape heads, i. e., a configuration is an element of

$$Q \times (\Sigma^*)^k \times \mathbb{N}^{k+1}.$$

Thus the configuration $(q, w_1, \ldots, w_k, n_0, n_1, \ldots, n_k)$ refers to the situation where M is in state q, the input head scans the n_0th input symbol, work tape i contains the string w_i with its head on the n_ith position in w_i (for $1 \leq i \leq k$). The *initial configuration* of M is

$$(q_0, \underbrace{b, \ldots, b}_{k \text{ times}}, \underbrace{1, \ldots, 1}_{k+1 \text{ times}}).$$

We extend g to a type function $\hat{g}$ for configurations by letting $\hat{g}(q, w_1, \ldots, w_k, n_0, n_1, \ldots, n_k) =_{\text{def}} g(q)$, and we say that, e. g., a configuration α is existential, universal, and so on.

Suppose that M is working on input $x = x_1 \cdots x_n$ and that at a certain point, M is in configuration $\alpha = (q, w_1, \ldots, w_k, n_0, n_1, \ldots, n_k)$, where $w_i = w_{i,1} \cdots w_{i,\ell_i}$ for $1 \leq i \leq k$. Suppose that $\delta(q, x_{n_0}, w_{1,n_1}, \cdots, w_{k,n_k})$ contains (possibly among others) the tuple $(q', a_1, \ldots, a_k, X_0, \ldots, X_k)$, where $a_1, \ldots, a_k \in \Sigma$ and $X_0, \ldots, X_k \in \{-1, 0, +1\}$. Then $\beta = (q', w_1', \ldots, w_k', n_0', n_1', \ldots, n_k')$ is a *successor configuration* of α, where for $1 \leq i \leq k$, w_i' is obtained from w_i as follows: replace the n_ith letter by a_i, and to cover the case where the head moves to the left or right out of the word by prefixing or suffixing a blank symbol; and for $0 \leq i \leq k$, let $n_i' = n_i + X_i$, if $n_i + X_i \geq 0$, otherwise we leave $n_i' = n_i$. That β is such a successor configuration of α is denoted by $\alpha \vdash_{M(x)} \beta$, or if M and the input x are understood simply by $\alpha \vdash \beta$. Observe that a final configuration has no successor.

Given a machine M and an input x, the *computation tree* $T_M(x)$ of M on x is a directed acyclic tree defined as follows:

1. The root of $T_M(x)$ is the initial configuration of M.
2. If $\alpha \in T_M(x)$ and $\alpha \vdash \beta$, then $\beta \in T_M(x)$ and there is an edge in $T_M(x)$ from α to β.

If $\alpha, \beta \in T_M(x)$ and $\alpha \vdash \beta$ then we say that node β is a *successor* of α. If there is a directed path from α to β in $T_M(x)$ then β is a *descendant* of α.

If a configuration β can be reached from two different configurations α and α', then there will be two copies of β in $T_M(x)$, i.e., we really want $T_M(x)$ to be a tree. Observe that the leaves of $T_M(x)$ must be final configurations, and no non-leaf can be final. However $T_M(x)$ can be infinite, since M may have non-halting computation branches.

We now define a labeling $l_M : T_M(x) \to \{0, 1\}$. The idea is that we want to label accepting configurations by 1, rejecting configurations by 0, universal configuration by 1 if all their successors are labeled by 1, and existential configurations by 1 if at least one successor is labeled by 1. But the problem with this approach is that it only works for finite computation trees. However, even in an infinite tree we want an existential configuration to be labeled 1 if one of its successors is labeled 1, even if the other successors lead to infinite branches. Similarly, a universal configuration should be labeled 0 if one of its successors is labeled 0, even if the other successors lead to infinite branches.

We overcome this problem by first defining an infinite sequence $l_0, l_1, l_2, \ldots$ of labelings. Here, l_0 is given by

1. If $\alpha \in T_M(x)$ is final, then $l_0(\alpha) =_{\text{def}} \hat{g}(\alpha)$.
2. If $\alpha \in T_M(x)$ is not final, then $l_0(\alpha) =_{\text{def}} \perp$.

If $i > 0$, then we define l_i inductively by

1. If $\alpha \in T_M(x)$, $l_{i-1}(\alpha) \neq \perp$, then $l_i(\alpha) =_{\text{def}} l_{i-1}(\alpha)$.
2. If $\alpha \in T_M(x)$, $l_{i-1}(\alpha) = \perp$, $\hat{g}(\alpha) = \vee$, and $\beta_1, \ldots, \beta_s$ are the children of α in $T_M(x)$, then $l_i(\alpha) =_{\text{def}} 1$ if $\bigvee_{j=1}^{s} (l_{i-1}(\beta_j) = 1)$, and $l_i(\alpha) =_{\text{def}} \perp$ otherwise.
3. If $\alpha \in T_M(x)$, $l_{i-1}(\alpha) = \perp$, $\hat{g}(\alpha) = \wedge$, and $\beta_1, \ldots, \beta_s$ are the children of α in $T_M(x)$, then $l_i(\alpha) =_{\text{def}} 1$ if $\bigwedge_{j=1}^{s} (l_{i-1}(\beta_j) = 1)$, and $l_i(\alpha) =_{\text{def}} \perp$ otherwise.

Finally, l_M is defined to be l_ω, i.e., $l_M = l_k$ for the smallest k such that $l_k = l_{k+1}$. (That such a k always exists follows immediately from Tarski's Fixed Point Theorem (see [Dev93, 73]), but an elementary proof is also easy, see Exercise 2.9.)

An *accepting computation subtree* of M on input x is a tree $T \subseteq T_M(x)$ such that:

1. T contains the root of $T_M(x)$.
2. All nodes in T are labeled 1 by mapping l.
3. Every existential configuration in T has exactly one child, and every universal configuration has as many children in T as it has in $T_M(x)$.

A word $x \in \Sigma^*$ is *accepted* by M if the root of the computation tree of M on x is labeled 1. Say that $L(M) =_{\text{def}} \{ x \in \Sigma^* \mid M \text{ accepts } x \}$ is the language accepted by M. Observe that a word x is accepted by M if and only if there is an accepting subtree of M on x. We can use this idea to define complexity classes for alternating machines.

Let $f \colon \mathbb{N} \to \mathbb{N}$, and let M be an alternating machine as above. M is time-bounded by f if, for every $x \in L(M)$, there is an accepting computation subtree of M on x with depth at most $f(|x|)$. (As for deterministic machines in Appendix A4, we will assume without further statement that if the time bound is sub-linear, our machines work with random access to their input via an index tape.) M is space-bounded by f if, for every $x \in L(M)$, there is an accepting computation subtree T of M on x such that every configuration in T uses $O(f(|x|))$ work-tape cells. Besides time and space bounds, a third measure turns out to be of interest for alternating machines: it bounds the number of alternations between existential and universal states during a computation. An alternation here means that the machine changes during a transition from an existential state to a universal one or vice versa. M is alternation-bounded by f if, for every $x \in L(M)$, there is an accepting computation subtree of M on x in which there is no path with more than $f(|x|) - 1$ alternations between existential and universal states.

Let $a, s, t: \mathbb{N} \to \mathbb{N}$.

1. ATIME(t) is the class of sets B for which there is an alternating Turing machine M which is time-bounded by t such that $B = L(M)$.
2. ASPACE(s) is the class of sets B for which there is an alternating Turing machine M which is space-bounded by s such that $B = L(M)$.
3. ATIME-SPACE(t, s) is the class of sets B for which there is an alternating Turing machine M which is simultaneously time-bounded by t and space-bounded by s such that $B = L(M)$.
4. ATIME-ALT(t, a) is the class of sets B for which there is an alternating Turing machine M which is simultaneously time-bounded by t and alternation-bounded by a such that $B = L(M)$.
5. AALT-SPACE(a, s) is the class of sets B for which there is an alternating Turing machine M which is simultaneously alternation-bounded by a and space-bounded by s such that $B = L(M)$.

2.5.2 Relations to Deterministic Complexity

As it turns out there is a nice result comparing the power of alternating machines with that of deterministic ones: The time-space hierarchy $L \subseteq P \subseteq$ PSPACE $\subseteq$ EXPTIME $\subseteq \cdots$ is shifted by one level:

$$
\begin{array}{ccccc}
\vdots & & \vdots & & \vdots \\
\text{EXPTIME} & = & \text{DTIME}(2^{n^{O(1)}}) & = & \text{ASPACE}(n^{O(1)}) \\
\cup\! & & & & \\
\text{PSPACE} & = & \text{DSPACE}(n^{O(1)}) & = & \text{ATIME}(n^{O(1)}) \\
\cup\! & & & & \\
\text{P} & = & \text{DTIME}(n^{O(1)}) & = & \text{ASPACE}(\log n) \\
\cup\! & & & & \\
\text{L} & = & \text{DSPACE}(\log n) & \supseteq & \text{ATIME}(\log n)
\end{array}
$$

This follows from the following propositions. We give a proof for the first result since we will need it later. Complete proofs of all propositions can be found in the standard literature [BDG90, pp. 70ff.], see also Exercises 2.12–2.14.

Theorem 2.35. *Let $t(n) \geq \log n$. Then*

$$\text{DSPACE}(t(n)) \subseteq \text{ATIME}((t(n))^2).$$

Proof. If $A \in \mathrm{DSPACE}(t)$ via M then for every $x \in A$, $|x| = n$, there is an accepting computation of M on input x, which has length $2^{O(t(n))}$. (This is due to the fact that the number of possible configurations of M on input x is bounded by this term, see proof of Theorem 2.9.) Design a recursive divide-and-conquer procedure building on the fact that configuration β is reachable from α in at most 2^{t+1} steps if and only if there is a middle configuration γ such that (a) γ is reachable from α in 2^t steps, and (b) β is reachable from γ in 2^t steps. If $t(n)$ is space-constructible we can mark $t(n)$ cells on the work tape and then use existential branches to guess configuration γ in the marked space. Then we use a universal branch to verify in parallel (a) and (b). This leads to a recursion depth of $O(t(n))$, and the time needed in each level is also $O(t(n))$.

This procedure works if $t(n)$ is space-constructible. If this is not the case, then we let M guess existentially in unary a value $t(n) = 1, 2, 3, \ldots$. If M accepts x then $t(n)$ will be guessed in time $O(t(n))$. Hence the depth of an accepting subtree of the simulating alternating machine is again bounded by $O(t(n)^2)$. □

Proposition 2.36. *Let $t(n) \geq n$. Then*

$$\mathrm{ATIME}(t(n)) \subseteq \mathrm{DSPACE}(t(n)).$$

Proposition 2.37. *Let $s(n) \geq \log n$. Then*

$$\mathrm{ASPACE}(s(n)) \subseteq \mathrm{DTIME}(2^{O(s(n))}).$$

Proposition 2.38. *Let $s(n) \geq n$. Then*

$$\mathrm{DTIME}(s(n)) \subseteq \mathrm{ASPACE}(\log s(n)).$$

For later use we note the following consequence of the proof of Theorem 2.35:

Corollary 2.39. *Let $t(n) \geq \log n$. Then*

$$\mathrm{DSPACE}(t(n)) \subseteq \mathrm{ATIME\text{-}SPACE}((t(n))^2, t(n)).$$

The reader might wonder about the class ATIME($\log n$) and ask if it possibly collapses with L. We will come back to this point in the next section.

2.5.3 An Example: Addition

Let us consider as an example the problem ADD of adding two n-bit numbers. Of course it is not clear how to define a function to be computed by an alternating machine—above we only talked about acceptance of sets. Also, we have to specify how to give two numbers as inputs to such a machine. Therefore let us consider more precisely the problem to determine, given a string ab where a and b are n-bit numbers in binary, the high-order bit in the sum $a + b$. In other words, we want to check if $a + b \geq 2^n$.

Our machine M works as follows:

(1) First M determines the length of its input (see Exercise 2.16). If this is an odd number then reject. Otherwise compute n by integer division by 2.

From Sect. 1.1 it is clear that we have to determine the value c_n, the carry rippling into position n. We do this by a straightforward evaluation of the Boolean expression (1.1) defining c_n. Thus M continues:

(2) Branch existentially, writing down a number $j \in \{0, \dots, n-1\}$. To be more precise this has to be achieved in a sequence of existential moves, where each time we choose one of the values 0 or 1. In such a way the number j is constructed in binary.

(3) Check that g_j is on, i.e., that the jth bit in both a and b is on. For this we have to read the input bits at position j and $n+j$. If the test fails, halt rejecting.

(4) Branch universally on all numbers $k \in \{j+1, \dots, n-1\}$ bit by bit using the same technique as in step 2.

(5) Check that p_k is on by reading the kth bit in a and b, i.e., by reading input bits k and $n+k$.

What are the time requirements of this procedure? Step 1 can be carried out in logarithmic time. For step 2, observe that the length of j in binary is at most $\ell(n)$, thus step 2 requires as many existential transitions. In step 3 we have to read two input bits. Checking bit position j is easy: just copy value j to the index tape and check. For the other bit position we first have to compute $n+j$, but since the relevant numbers can be written down in binary with $\ell(n)$ bits, we can easily perform the required addition by the obvious sequential adding procedure in logarithmic time. Step 4 can be implemented like step 2. Step 5 can be implemented like step 3. All in all we see that alternating machines can solve the above problem in logarithmic time.

This example shows how Boolean formulas can be evaluated by alternating machines in a fairly straightforward way. It is our aim to generalize this example in the following sections.

2.5.4 Input Normal Form

In this subsection we want to show that alternating machines can be normalized with respect to the way they access their input.

Definition 2.40. An alternating Turing machine M with random access to its input (see Appendix A4) is in *input normal form*, if for every input x all accesses of M to x are of the following form: M with the number i written in binary on its index tape enters state $q_?^a$ (for some $a \in \{0,1\}$) and then immediately halts accepting if the i-th input bit is a, otherwise M halts rejecting.

Every random access input alternating machine can without loss of generality be assumed to be in the above normal form (as long as we do not count alternations), as the following lemma shows:

Lemma 2.41. *Let M be an alternating machine with random access to its input, accepting a set $A \subseteq \{0,1\}^*$ in time t using space s. Then there is an alternating machine M' in input normal form accepting A in time $O(t)$ using space s.*

Proof. When M accesses the i-th position of its input and continues the computation with this bit, M' simulates this in the following way (see Fig. 2.7): M' branches existentially, guessing a value $b \in \{0,1\}$ for the i-th input bit.

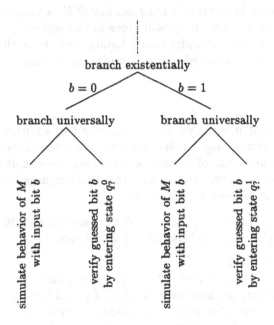

Fig. 2.7. Definition of machine M' in input normal form

For each outcome, M' branches universally as follows: On one branch M' continues with the simulation of M under the assumption that the guess was correct, i.e., that b is the actual value of the i-th input bit. On the other branch M' verifies that the guess was correct by actually reading the i-th input bit according to the normal form.

This leads to a constant factor slowdown without additional space requirements. $\square$

2.6 Simultaneous Resource Bounds

The constructions used to prove Corollaries 2.29 and 2.33 were designed to be optimal for the complexity measure under consideration, neglecting other measures. For example, the simulation of a machine running in time t leads to a circuit of size $O(t \log t)$, but the depth is immense as it is also $O(t \log t)$. When we simulated space-bounded machines by circuits of small depth, the circuit size became exponential in the depth. Conversely, simulation of a size-bounded circuit leads to a time-efficient machine but the space needed is proportional to the time, and the simulation of a depth-bounded circuit gives a space-efficient algorithm, but the time used is exponential in the depth. If we restrict both the size and depth of a circuit, can we do better? Is there a nice characterization in terms of Turing machines? If we consider alternating machines, the answer is yes, as we will prove in this section.

For simplicity we only consider circuit families over $\mathcal{B}_0$ in this section. By Proposition 1.16 this is as good as any bounded fan-in basis.

2.6.1 Uniformity

It will turn out that it is convenient to consider two additional uniformity conditions, when comparing circuits and alternating machines. Both will be defined via the complexity of a language encoding the circuit family under consideration. Above we used the direct connection language for this purpose, see Def. 2.24. Here we need an extension.

Definition 2.42. Let $C = (V, E, \alpha, \beta, \omega)$ be a circuit over $\mathcal{B}_0$. Let $v \in V$, and let $p \in \{0, 1\}^*$. Define $p(v)$ inductively as follows:

- If $p = \epsilon$ then $p(v) = v$.
- Let v be a non-input gate and $p \in \{0, 1\}$. If v is a fan-in 2 gate, then $p(v)$ is the first (second) predecessor gate to v, if $p = 0$ ($p = 1$, respectively). If v is a fan-in 1 gate then $p(v)$ is the predecessor gate to v. (First/second here refers to the order of predecessors induced by the ordering α on E.)
- If v is a non-input gate and $p = ap'$ for $a \in \{0, 1\}$, $p' \in \{0, 1\}^*$, then $p(v) = a(p'(v))$.

Note that these conditions leave $p(v)$ undefined, if v is an input gate of C and $p \neq \epsilon$.

Definition 2.43. Let $\mathcal{C} = (C_n)_{n \in \mathbb{N}}$ be a circuit family of size s over $\mathcal{B}_0$. Fix an admissible encoding scheme for $\mathcal{C}$. The *extended connection language* of $\mathcal{C}$ (with respect to the fixed encoding), $L_{EC}(\mathcal{C})$, is the set of all tuples

$$\langle y, g, p, b \rangle,$$

where for $n = |y|$

- g is the number of a gate v in C_n;
- $p \in \{0,1\}^*$, $|p| \leq \log s(n)$;
- if $p = \epsilon$ then b is the number of the function from $\mathcal{B}_0$ computed at v;
- if $p \neq \epsilon$ then b is the number of the gate $p(v)$.

For circuit families C over $\mathcal{B}_0$, certainly $L_{DC}(C) \subseteq L_{EC}(C)$.
We now define the following uniformity conditions:

Definition 2.44. Let $C = (C_n)_{n \in \mathbb{N}}$ be a bounded fan-in circuit family of size s and depth d.

1. C is U_E-uniform, if there is an admissible encoding scheme such that there is a deterministic Turing machine that accepts $L_{EC}(C)$ and runs in time $O(\log s(n))$ on inputs of the form $\langle y, \cdot, \cdot, \cdot \rangle$ where $|y| = n$.
2. C is U_E^*-uniform, if there is an admissible encoding scheme such that there is an alternating Turing machine that accepts $L_{EC}(C)$, uses space $O(\log s(n))$, and runs in time $O(d(n))$ on inputs of the form $\langle y, \cdot, \cdot, \cdot \rangle$ where $|y| = n$.

Thus U_E- and U_E^*-uniformity relate to the complexity of following paths up to length $\log s$ in a given circuit. The reader might wonder about the condition $|p| \leq \log s(n)$ in Def. 2.43. Certainly the definitions above are reasonable if we drop it. Unfortunately it is not clear whether then the results we give below remain valid.

An improvement of Lemma 2.25 can now be given as follows:

Lemma 2.45. *Let $C = (C_n)_{n \in \mathbb{N}}$ be a bounded fan-in circuit family of size s. Let $f: \mathbb{N} \to \mathbb{N}$, $f(n) \geq \log s(n)$. For all admissible encoding schemes the following holds: The map $1^n \mapsto \overline{C_n}$ is computable in space $O(f)$ if and only if $L_{EC}(C)$ is decidable by a deterministic Turing machine which needs space $O(f(n))$ on inputs $\langle y, \cdot, \cdot, \cdot \rangle$, where $|y| = n$.*

Proof. ($\Rightarrow$): Let $1^n \mapsto \overline{C_n} \in \text{FDSPACE}(f)$ via machine M. The machine described in Fig. 2.8 then decides $L_{EC}(C)$. The space requirements are given by the space needed to simulate M plus the space for the other variables, i.e., $O(\log s(n))$; hence the overall space needed is $O(f(n))$.

($\Leftarrow$): If $L_{EC}(C)$ is decidable by a deterministic Turing machine which needs space $O(f(n))$ on inputs $\langle y, \cdot, \cdot, \cdot \rangle$, where $|y| = n$, then $L_{DC}(C)$ can be decided in the same way. Hence by Lemma 2.25, we conclude that $1^n \mapsto \overline{C_n} \in \text{FDSPACE}(f)$. □

Remark 2.46. As in Remark 2.26 we observe that in the case that C is of polynomial size we have: C is logspace-uniform iff $L_{EC}(C) \in \text{DSPACE}(\log n)$.

Before we go on, let us clarify the implication structure between the different uniformity conditions, see also Fig. 2.9.

Theorem 2.47. *Let $C = (C_n)_{n \in \mathbb{N}}$ be a circuit family over $\mathcal{B}_0$ of size s and depth d. Then the following holds:*

> **input** $\langle y, g, p, b \rangle$:
>   **while** $p \neq \epsilon$ **do begin**
>   **let** $p = ap'$;
>   **if** $a = 0$ **then**
>   simulate M on input 1^n until a tuple $\langle g, \cdot, g', \cdot \rangle$ is printed
>   **else**
>   simulate M on input 1^n until a tuple $\langle g, \cdot, \cdot, g' \rangle$ is printed;
>   **if** no such tuple appears **then** reject;
> $p := p'$;
> $g := g'$
>   **end**;
>   simulate M on input 1^n until a tuple $\langle g, b, \cdot, \cdot \rangle$ is printed;
>   **if** no such tuple appears **then** reject **else** accept.

Fig. 2.8. Machine for $L_{EC}(\mathcal{C})$

1. If $\mathcal{C}$ is U_E-uniform, then $\mathcal{C}$ is logspace-uniform.
2. If $\mathcal{C}$ is U_E-uniform, then $\mathcal{C}$ is U_E^*-uniform.
3. If $\mathcal{C}$ is logspace-uniform, then $\mathcal{C}$ is ptime-uniform.
4. If $\mathcal{C}$ is logspace-uniform and $d(n) \geq \left(\log s(n)\right)^2$, then $\mathcal{C}$ is U_E^*-uniform.

Proof. 1. If $\mathcal{C}$ is U_E-uniform, then there is an admissible encoding scheme such that there is a deterministic Turing machine that accepts $L_{EC}(\mathcal{C})$ and runs in time $O(\log s(n))$ on inputs of the form $\langle y, \cdot, \cdot, \cdot \rangle$, where $|y| = n$. This machine certainly works within space $O(\log s(n))$, hence by Lemma 2.45 we conclude that $\mathcal{C}$ is U_L-uniform.

2. Suppose without loss of generality that $\mathcal{C}$ has no gates of fan-out 0 except the output gates. If $\mathcal{C}$ is U_E-uniform, then there is an admissible encoding scheme such that there is a deterministic Turing machine that accepts $L_{EC}(\mathcal{C})$ and runs in time $O(\log s(n))$ on inputs of the form $\langle y, \cdot, \cdot, \cdot \rangle$, where $|y| = n$. Hence there is an alternating machine M that accepts $L_{EC}(\mathcal{C})$ and runs in time $O(\log s(n))$ using space $O(\log s(n))$ on inputs of the form $\langle y, \cdot, \cdot, \cdot \rangle$, where $|y| = n$. (M is essentially the given deterministic machine.) However since $\mathcal{C}$ is of bounded fan-in, we must have $d(n) \geq \log s(n)$; thus M works in time $O(\log s(n))$ and space $O(d(n))$ and hence $\mathcal{C}$ is U_E^*-uniform.

3. If $\mathcal{C}$ is logspace-uniform, then there is an admissible encoding scheme such that $1^n \mapsto \overline{C_n} \in \text{FDSPACE}(\log s) \subseteq \text{FDTIME}(s^{O(1)})$; hence $\mathcal{C}$ is ptime-uniform.

4. If $\mathcal{C}$ is logspace-uniform, then there is an admissible encoding scheme such that $1^n \mapsto \overline{C_n} \in \text{FDSPACE}(\log s)$. Thus by Lemma 2.45, there is an admissible encoding scheme such that there is a deterministic Turing machine that accepts $L_{EC}(\mathcal{C})$ and runs in space $O(\log s(n))$ on inputs of the form $\langle y, \cdot, \cdot, \cdot \rangle$, where $|y| = n$. By Corollary 2.39 there is an alternating machine accepting $L_{EC}(\mathcal{C})$ using on inputs as above time $O\left((\log s(n))^2\right)$ and space $O(\log s(n))$, which shows that $\mathcal{C}$ is U_E^*-uniform in the case $d(n) \geq (\log s(n))^2$. □

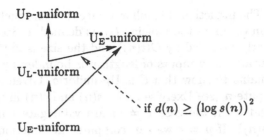

Fig. 2.9. Implication structure between uniformity conditions

2.6.2 Time and Space on Alternating Machines

Now we come to the relation between alternating Turing machines and Boolean circuits.

Theorem 2.48. *Let $s(n), t(n) \geq \log n$ such that the functions $x \mapsto s(|x|)$ and $x \mapsto t(|x|)$ are computable deterministically in time $O(s(|x|))$. Then*

$$\text{ATIME-SPACE}(t, s) \subseteq \text{U}_\text{E}\text{-SIZE-DEPTH}(2^{O(s)}, t).$$

Proof. Let $A \in \text{ATIME-SPACE}(t, s)$ via alternating machine M. We suppose without loss of generality that M is in input normal form and that M has for every configuration at most two possible successor configurations. We construct a circuit family $\mathcal{C} = (C_n)_{n \in \mathbb{N}}$ and an admissible encoding scheme for $\mathcal{C}$.

For technical simplicity, we will first consider the basis $\mathcal{B}_0' =_{\text{def}} \mathcal{B}_0 \cup \{0^0, 1^0, \text{id}^1\}$, i.e., we add to $\mathcal{B}_0$ the 0-ary constants 0 and 1, and the unary identity function. Start by numbering functions in $\mathcal{B}_0'$ arbitrarily. During the construction we will simultaneously determine a suitable gate numbering.

Fix an input length n. The circuit C_n will (besides the inputs) have gates numbered by $\langle k, \alpha \rangle$, where $0 \leq k \leq t(n)$, and α is (an encoding of) a configuration of M on inputs of length n using space of at most $s(n) + 1$. The function computed by the gate numbered $\langle k, \alpha \rangle$ will be existential (universal, constant 0, constant 1, respectively) if configuration α is existential (universal, rejecting, accepting, respectively). If α is a configuration with an input read state $q_?^a$ for $a \in \{0, 1\}$ and the value i on the index tape, then the function computed at $\langle k, \alpha \rangle$ will be the identity if $a = 1$ and the negation $\neg$, if $a = 0$. In both cases, $\langle k, \alpha \rangle$ has as input the i-th input gate x_i.

The gate numbered $\langle 0, \alpha \rangle$ where α is the initial configuration of M, is the output gate of C_n. The predecessor gates to $\langle k, \alpha \rangle$ for $k < t(n)$ will be all gates $\langle k + 1, \beta \rangle$ such that $\alpha \vdash \beta$. If for such a β the space required is more than $s(n)$ then its type will be set to the constant 0 function.

Now an easy induction shows that given an input x of length n, a configuration α in depth t is labeled by 1 under the mapping l_M, if and only if the gate numbered $\langle k, \alpha \rangle$ evaluates to 1. (For the case of input gates this is obvious. The induction step follows since the constructed circuit does nothing other than compute the labeling l_M as defined in Sect. 2.5.1.) The depth of C_n is clearly bounded by $O(t(n))$ and the size is $2^{O(s(n))}$, since the number of configurations on inputs of length n is bounded by this term.

It remains to show that C is U_E-uniform. Given an input $\langle y, g, p, b \rangle$, we first compute n (see Exercise 2.17), $s(n)$ and $t(n)$ in time $O(s(n))$. To check that g and b (in the case $p \neq \epsilon$) are valid gate numbers can be done in time $O(s(n))$. If $p = \epsilon$ we can compute the function computed by g from its number in time $O(s(n))$, and then compare this value with b. If $p \neq \epsilon$ we have to check that b is the number of the gate reached from g following path p. For this we simulate M starting with the configuration encoded in g for $|p|$ steps, where we always pick successor configurations deterministically according to p. In the end we check that the obtained configuration of M is consistent with gate number b. Since $|p| = O(s(n))$ this can be done in time $O(s(n))$.

These considerations show that the uniformity procedure needs time $O(s(n))$ which is logarithmic in the size of C_n. Using Proposition 1.16 we now transform C into a circuit family over $\mathcal{B}_0$. It is clear that this will not affect uniformity issues. $\square$

We now turn to a converse of the above theorem.

Theorem 2.49. *Let $d(n) \geq \log n$, $s(n) \geq n$. Let $x \mapsto \log s(|x|)$ be deterministically computable in time $O(\log s(|x|))$. Then*

$$U_E^*\text{-SIZE-DEPTH}(s, d) \subseteq \text{ATIME-SPACE}(d, \log s).$$

Proof. Let $A \in U_E^*\text{-SIZE-DEPTH}(s, d)$ via circuit family $C = (C_n)_{n \in \mathbb{N}}$. Fix an encoding such that $L_{EC}(C)$ is accepted by alternating machine M, using time $O(d(n))$ and space $O(\log s(n))$ on inputs of the form $\langle y, \cdot, \cdot, \cdot \rangle$, where $|y| = n$.

We use the recursive function value, as defined in Fig. 2.10, to compute the value of a gate in C. More precisely, value(n, g, p) will compute the value of the gate with number $p(g)$ in C_n. Given an input $x = a_1 \cdots a_n$ of length n, we first mark $c \log s(n)$ cells on a work tape using the machine that computes function $\log s$. Here $c \in \mathbb{N}$ is chosen in such a way that $c \log s(n)$ cells are sufficient to write down names of gates in C_n. Next we initialize a global variable m to 0, and then simply call value$(n, n + 1, \epsilon)$.

A few remarks on the correctness of the algorithm: The guessed t is intended to be the type of the current gate. Here we use the special symbol $t = ?$ in line 2 to indicate that the current gate is an input gate. Hence a corresponding h in line (3a) will simply be a number $h < n$, and the test for b is not necessary since we know that a gate with number $i < n$ will be the i-th input. For $t \neq ?$ a suitable b will be the number of a function from $\mathcal{B}_0$.

The idea in line (3b) is that we want to simulate $\wedge/\vee$ gates by universal/existential branches. $\neg$ gates will be handled by pushing the negations

function value(n, g, p): boolean;
{ returns true iff gate $p(g)$ in C_n evaluates to true }
begin
(1)   **if** $|p| < \log s(n)$ **then begin**
(2)   existentially guess $t \in B_0 \cup \{?\}$;
         universally branch on (3a) or (3b):
(3a)   verify t by guessing h and b and checking if
           $\langle x, g, p, h \rangle \in L_{EC}(\mathcal{C})$ and $\langle x, h, \epsilon, b \rangle \in L_{EC}(\mathcal{C})$;
(3b)   **if** $t = \neg$ **then** $m := 1 - m$
           else if $m = 0$ **then begin**
             if $t = ?$ **then begin**
               if $p = \epsilon$ **then return** a_g
               else begin
                   guess $i < n$;
                   universally branch on (A) or (B):
                   (A) verify $\langle x, g, p, i \rangle \in L_{EC}(\mathcal{C})$
                   (B) **return** a_i
                 end
               end
             else if $t = \wedge$ **then**
               return value($n, g, p0$) **and** value($n, g, p1$)
             else if $t = \vee$ **then**
               return value($n, g, p0$) **or** value($n, g, p1$)
           end
           else { $m = 1$ } **begin**
             if $t = ?$ **then begin**
               if $p = \epsilon$ **then return** $\neg a_g$
               else begin
                   guess $i < n$;
                   universally branch on (A) or (B):
                   (A) verify $\langle x, g, p, i \rangle \in L_{EC}(\mathcal{C})$
                   (B) **return** $\neg a_i$
                 end
               end
             else if $t = \wedge$ **then**
               return value($n, g, p0$) **or** value($n, g, p1$)
             else if $t = \vee$ **then**
               return value($n, g, p0$) **and** value($n, g, p1$)
           end
       end
       else { $|p| = \log s(n)$ } **begin**
(4)   existentially guess h;
         universally branch on (5a) or (5b):
(5a)   verify $\langle x, g, p, h \rangle \in L_{EC}(\mathcal{C})$;
(5b)   **return** value(n, h, ϵ)
       end
end;

Fig. 2.10. Function value(n, g, p)

down to the input gates, using de Morgan's laws. For this, we use variable m to keep track of the number of $\neg$ gates on the path from the current gate $p(g)$ to the output gate. If m is odd then this means (by de Morgan) that we have to simulate an $\wedge$ gate by an existential branch, an $\vee$ gate by a universal branch, and an access to an input by the negation of that input bit. In lines (4)–(5) we shorten the considered path p if we reach $|p| = \log s(n)$. We have to do this since the extended connection language only contains information about paths in C up to this length.

The space required by the simulation is clearly bounded by $O(\log s)$. For the time analysis, first observe that steps (3a) and (5a) do not lead to further recursive calls to procedure value, hence they contribute to the overall time only *additive* with $O(d)$. The recursion depth of the overall simulation is clearly $O(d)$. Each execution of a statement (3b) can be executed in constant time, except when an input has to be read—this needs time $O(\log n)$, but this occurs at most once on every computation path. Line (4) needs $O(\log s)$ time per execution, however it is only executed once in $O(\log s)$ consecutive recursion calls; thus the amortized cost per recursion call is $O(1)$. Line (5b) needs $O(1)$ time per call. Hence, the time requirement of the overall simulation is $O(d)$. $\qquad\qquad\qquad\qquad\qquad\qquad\qquad\qquad\qquad\qquad\qquad\qquad\quad\Box$

Corollary 2.50. *Let $s(n), t(n) \geq \log n$ such that the functions $x \mapsto s(|x|)$ and $x \mapsto t(|x|)$ are computable deterministically in time $O(s(|x|))$. Then*

$$
\begin{aligned}
\text{ATIME-SPACE}(t, s) &= \text{U}_{\text{E}}\text{-SIZE-DEPTH}(2^{O(s)}, t) \\
&= \text{U}_{\text{E}}^*\text{-SIZE-DEPTH}(2^{O(s)}, t).
\end{aligned}
$$

If moreover $t(n) \geq \left(s(n)\right)^2$, then $\text{U}_{\text{L}}\text{-SIZE-DEPTH}(2^{O(s)}, t)$ is equal to the above class.

Proof. By Theorem 2.48, $\text{ATIME-SPACE}(t, s) \subseteq \text{U}_{\text{E}}\text{-SIZE-DEPTH}(2^{O(s)}, t)$. By Theorem 2.47, $\text{U}_{\text{E}}\text{-SIZE-DEPTH}(2^{O(s)}, t) \subseteq \text{U}_{\text{E}}^*\text{-SIZE-DEPTH}(2^{O(s)}, t)$. By Theorem 2.49, $\text{U}_{\text{E}}^*\text{-SIZE-DEPTH}(2^{O(s)}, t) \subseteq \text{ATIME-SPACE}(t, s)$. By Theorem 2.47, $\text{U}_{\text{E}}\text{-SIZE-DEPTH}(2^{O(s)}, t) \subseteq \text{U}_{\text{L}}\text{-SIZE-DEPTH}(2^{O(s)}, t)$ and if $t(n) \geq \left(s(n)\right)^2$, we additionally get the inclusion $\text{U}_{\text{L}}\text{-SIZE-DEPTH}(2^{O(s)}, t) \subseteq \text{U}_{\text{E}}^*\text{-SIZE-DEPTH}(2^{O(s)}, t)$. $\qquad\qquad\qquad\qquad\qquad\Box$

Say that a circuit C is in *input normal form*, if all gates in C that compute the negation function are adjacent to an input gate.

Corollary 2.51. *Let $s(n) \geq n$ and $d(n) \geq \log n$ be such that the functions $x \mapsto \log s(|x|)$ and $x \mapsto d(|x|)$ are computable deterministically in time $O(\log s(|x|))$. Then every circuit family over $\mathcal{B}_0$ of size s and depth d can be simulated by a circuit family of size $s^{O(1)}$ and depth $O(d)$, which is in input normal form. This holds for U_{E}- and U_{E}^*-uniformity, and if $d(n) \geq (\log s(n))^2$ as well for U_{L}-uniformity.*

Proof. Simulate the given circuit family by an alternating machine as in Theorem 2.49. Then simulate this machine by another circuit family as in Theorem 2.48. The proof of this theorem shows that the constructed circuits have the desired property.   □

In Sect. 2.5.2 we presented a number of nice relations between classes defined by alternating machines and classes defined by deterministic machines. For the class ATIME($\log n$) we could only give as upper bound the class DSPACE($\log n$). From Theorems 2.32 and 2.47 we see that U_E-DEPTH($\log n$) is a subclass of DSPACE($\log n$). In fact we can now say more:

Corollary 2.52. U_E^*-DEPTH($\log n$) = U_E-DEPTH($\log n$) = ATIME($\log n$) $\subseteq$ L.

It is open whether this inclusion is strict.

2.7 Parallel Random Access Machines

In this section we want to consider a realistic model of a parallel computer: the parallel random access machine (PRAM).

A PRAM consists of an infinite sequence of processors $P_1, P_2, P_3, \ldots$. Each processor P_i has its local memory, realized as an infinite sequence $R_{i,0}, R_{i,1}, R_{i,2}, \ldots$ of registers. Additionally there is a common (or, global) memory, given by the infinite sequence $C_0, C_1, C_2, \ldots$ of registers. Each register can hold as value a natural number, stored in binary as a bit string.

A particular PRAM M is specified by a *program* and a *processor bound*. The program is a sequence of instructions $S_1, S_2, S_3, \ldots, S_s$. (We describe the particular types of instructions below.) The processor bound is a function $p \colon \mathbb{N} \to \mathbb{N}$.

M works as follows: Initially, the input is distributed over the lowest numbered global memory cells (we specify the precise way below). Then all processors $P_1, \ldots, P_{p(n)}$ start the execution of the program with the first instruction S_1. Each instruction S_m ($1 \leq m \leq s$) is of one of the following nine types. Let us consider a fixed processor $P_r, 1 \leq r \leq p(n)$. Let $c, i, j, k \in \mathbb{N}$ and $1 \leq l \leq s$. We describe the instructions and their effect in turn. If S_m is of one of the types (1)–(7), then after the execution of S_m processor P_r continues with instruction S_{m+1}; for instruction types (8)–(9) the number of the instruction to be executed next is determined as described below.

(1) $R_i \leftarrow c$: $R_{r,i}$ gets as value the constant c.
(2) $R_i \leftarrow \#$: $R_{r,i}$ gets as value the number of the processor, i.e., r.
(3) (Numerical operations) $R_i \leftarrow R_j + R_k$, $R_i \leftarrow R_j - R_k$: The result of adding the contents of $R_{r,j}$ and $R_{r,k}$ (subtracting the contents of $R_{r,j}$ from the contents of $R_{r,k}$, respectively) is stored in $R_{r,i}$. (Here, if the subtraction results in a negative number, then $R_{r,i}$ gets value 0.)

(4) (Bitwise operations) $R_i \leftarrow R_j \vee R_k$, $R_i \leftarrow R_j \wedge R_k$, $R_i \leftarrow R_j \oplus R_k$: The result of performing the bitwise OR (AND, PARITY) of the contents of $R_{r,j}$ and $R_{r,k}$ is stored in $R_{r,i}$. The bitwise OR of two numbers a and b is defined as follows: Let $l = \max\{\ell(a), \ell(b)\}$, $\mathrm{bin}_l(a) = a_1 \cdots a_l$, $\mathrm{bin}_l(b) = b_1 \cdots b_l$. For $1 \leq i \leq l$ define $c_i = a_i \vee b_i$. Then the bitwise OR of a and b is the number c for which $\mathrm{bin}_l(c) = c_1 \cdots c_l$. The other two operations are defined analogously.

(5) (Shift operations) $R_i \leftarrow \mathrm{shl}\ R_j$, $R_i \leftarrow \mathrm{shr}\ R_j$: If the contents of $R_{r,j}$ is a, $\mathrm{bin}(a) = a_1 \cdots a_l$, then register R_i will get as value that number b whose binary representation is $\mathrm{bin}(b) = a_1 \cdots a_l 0$ (for shl), and $\mathrm{bin}(b) = a_1 \cdots a_{l-1}$ (for shr).

(6) (Indirect addressing I) $R_i \leftarrow (R_j)$: The contents of that global memory register whose number can be found in $R_{r,j}$, is copied to $R_{r,i}$.

(7) (Indirect addressing II) $(R_i) \leftarrow R_j$: The contents of $R_{r,j}$ is copied to that global register whose number is given by the contents of $R_{r,i}$.

(8) (Conditional jump) IF $R_i < R_j$ GOTO l: If the contents of $R_{r,i}$ is smaller in value than the contents of $R_{r,j}$, then processor P_r continues with the execution of instruction S_l; otherwise it continues with the next instruction S_{m+1}.

(9) HALT: The computation of processor P_r stops.

The computation of M stops when all processors $R_1, \ldots, R_{p(n)}$ have halted.

Let $f \colon \mathbb{N}^* \to \mathbb{N}^*$. We say that a PRAM M computes f if the following holds: Given an input $(x_1, \ldots, x_n) \in \mathbb{N}^n$ (in this case, we say that n is the *length* of the input), the x_i are first distributed in global memory cells $C_1, \ldots, C_n$, i.e., C_i is initialized to x_i (for $1 \leq i \leq n$). Additionally, C_0 is initialized to n. Then the computation of M is started. Let m be the contents of register C_0 after the computation stops. Then $f(x_1, \ldots, x_n)$ is the vector from $\mathbb{N}^m$ whose components are the numbers in the global registers $C_1, \ldots, C_m$.

PRAMs computing number-theoretic functions as just defined are the most natural model from an algorithmic point of view, when one considers problems such as sorting, dictionary problems, graph-theoretic problems, and so on. However in this book we deal most of the time with functions defined on $\{0,1\}^*$, and hence we have to define how a PRAM M computes a function $f \colon \{0,1\}^* \to \{0,1\}^*$. Our input–output conventions for this case are as follows: Given an input $x \in \{0,1\}^n$, x is first distributed bit for bit over global memory cells $C_1, \ldots, C_n$, i.e., if $x = x_1 \cdots x_n$ then C_i is initialized to x_i (for $1 \leq i \leq n$). Additionally C_0 is initialized to n. Let C_0 contain m at the end of the computation. Then $f(x)$ is the word $f(x) = y = y_1 \cdots y_m$ such that y_i is 0 if the contents of C_i is 0, otherwise y_i is 1.

There is one point in the description of PRAMs which still deserves clarification. What happens if different processors want to write different values simultaneously into the same memory cell? (When different processors want to read the same value at the same time, no such problem occurs.) We stipulate that in such a situation the processor with the lowest number succeeds. (The values which other processors try to write are lost.) PRAMS of our type are commonly referred to as CRCW-PRAMS, where CRCW stands for concurrent read concurrent write. Of course, other variants are conceivable; see Exercises 2.19 and 2.20. From now on we use the term PRAM to refer to the model defined above. The model where we do not allow concurrent write operations is called CREW-PRAM (concurrent read, exclusive write). From a technical point of view, CREW-PRAMs are also not realizable. Thus, EREW-PRAMS (which additionally disallow concurrent read operations) have been studied. All these models are compared in Exercises 2.19–2.21.

We assume that all instructions above need one time-step to be executed. We say that M is time bounded by $t\colon \mathbb{N} \to \mathbb{N}$ if, for all inputs of length n, all processors halt after at most $t(n)$ steps.

Instruction types (1)–(9) present a typical collection of what is usually allowed in the literature. Often general multiplication and division are allowed as unit-time operations. (In this case, our instruction set above is referred to as the *restricted instruction set*.) However Theorem 2.56 below relating PRAMS to unbounded fan-in circuits does not hold for this extended model. This follows from results in Chap. 3 where we will show that no constant-depth circuit of polynomial size can compute multiplication or division. (We remark that allowing multiplication and division for numbers whose length is logarithmic in the input length does not affect Theorem 2.56, see Exercise 2.23.) Hence we keep the nine types of instructions we have described above.

We now give three examples for PRAM programs: The first one addresses a number-theoretic problem (a function from $\mathbb{N}^* \to \mathbb{N}^*$), while in the second and third example we consider word-theoretic functions (from $\{0,1\}^* \to \{0,1\}^*$).

Example 2.53. A PRAM, given n numbers $a_1, \ldots, a_n$, can compute the partial sums $s_i =_{\text{def}} \sum_{j=1}^{i} a_j$ (for all $1 \leq i \leq n$) in time $O(\log n)$. These sums are also called prefix sums, and the PRAM algorithm we give below is known as the *parallel prefix algorithm*. We describe the algorithm in Fig. 2.11 in a high-level programming language style, where we assume for simplicity that n is a power of 2.

The **for**-statements in lines (2) and (5) can be distributed over all processors, and hence can be executed in constant time. Division by 2 can be computed by right shift (shr). In line (3), we recursively compute the prefix sums $s_i' =_{\text{def}} \sum_{j=1}^{i} a_i'$ for $1 \leq i \leq \frac{n}{2}$. We conclude that the runtime of the algorithm is $O(\log n)$.

Observe that the above algorithm never uses any particular property of the operation of addition except associativity, hence it can be used to compute the values $a_1, a_1 * a_2, a_1 * a_2 * a_3, \ldots, a_1 * a_2 * \cdots * a_n$ for any associative operation $*$ that can be performed in constant time on a PRAM.

input: $a_1, \ldots, a_n$;
output: $s_1, \ldots, s_n$, where $s_i =_{\text{def}} \sum_{j=1}^{i} a_i$ for $1 \le i \le n$;
procedure pps $(a_1, \ldots, a_n)$;
(1)   **if** $n = 1$ **then** $s_1 := a_1$
       else begin
(2)   **for** $i = 1, \ldots, \frac{n}{2}$ **do in parallel** $a_i' := a_{2i-1} + a_{2i}$;
(3) $(s_1', \ldots, s_{\frac{n}{2}}') := $ pps$(a_1', \ldots, a_{\frac{n}{2}}')$;
(4) $s_0' := 0$;
(5)   **for** $i = 1, \ldots, \frac{n}{2}$ **do in parallel**
               begin
(6) $s_{2i} := s_i'$;
(7) $s_{2i-1} := s_{i-1}' + a_{2i-1}$
               end
       end;
       return $(s_1, \ldots, s_n)$.

Fig. 2.11. Parallel prefix algorithm

Example 2.54. A PRAM computing the logical OR of its inputs in constant time with n processors works as follows: Given $x = x_1 \cdots x_n$, processor P_i (for $1 \le i \le n$) first reads x_i and writes 0 to C_1. If now $x_i = 1$ then P_i writes 1 to C_1.

Example 2.55. A PRAM computing the logical AND of its inputs in constant time with n processors works as follows: Given $x = x_1 \cdots x_n$, processor P_i (for $1 \le i \le n$) first reads x_i and writes 1 to C_1. If now $x_i = 0$ then P_i writes 0 to C_1.

Note that in the preceding two examples it is essential that the different processors perform the assignments to cell C_1 *simultaneously*.

After these examples, we now want to prove a connection between the computational power of PRAMs and that of unbounded fan-in circuits. We start by showing how to simulate PRAMs by circuits.

Theorem 2.56. *Let* $f: \{0,1\}^* \to \{0,1\}^*$ *be a length-respecting function, computed by the CRCW-PRAM M in time $t(n)$. Let the processor bound of M be $p(n)$. Then there is a family C of circuits over B_1 that computes f, where the depth of C is $O(t(n))$ and the size is polynomial in $n + t(n) + p(n)$.*

Proof. Let the program of M consist of s instructions. Let k be the maximal index of a local register that appears in M's program. Fix an input length n.

First we want to determine the maximum length m of all contents of all registers during the computation of M on inputs of length n. Let c_m be the maximal constant occurring in M's instructions. Then all instructions $R_i \leftarrow c$ can produce a register content of length at most $\ell(c_m)$. All instructions of the form $R_i \leftarrow \#$ produce a register content of length $\ell(p(n))$. Addition and left shift can at most increase the maximal register length by 1 per computation step. All other instructions do not produce longer register contents. Hence $m \leq \max\{\ell(p(n)), \ell(c_m)\} + t(n)$.

Our circuit C_n will consists of $t(n) + 1$ subcircuits $L_0, L_1, \ldots, L_{t(n)}$ on top of each other, i.e., the wires out of L_i will lead into L_{i+1} for $0 \leq i < t(n)$. The idea behind the construction is that a subcircuit L_i computes the "configuration" of M after i computation steps. Such a configuration K_i must include

- for every processor P_i, $1 \leq i \leq p(n)$, an *instruction counter* $I_i \in \{1, \ldots, s\}$ indicating which instruction P_i has to execute in the next step;
- for every i, $1 \leq i \leq p(n)$, the contents of the registers $R_{i,0}, \ldots, R_{i,k}$;
- the contents of the global memory.

Every instruction counter can be encoded using $\ell(s)$ bits, hence we need $\ell(s)p(n)$ bits for all instruction counters. The contents of every local register has no more than m bits, hence we need $mkp(n)$ bits to encode all local memory cells. The global memory, however, cannot be handled in this easy way. Every single global register can be encoded using m bits, but we do not know in advance which global registers will be used by M. Note that the address (i.e., number) of every global register that will be used must appear somewhere in a local register, that is, the address has no more than m bits. This still leaves a set of 2^m possible global registers which might be used; this is an exponential number which is too large for the circuit size we are aiming at. But we know that, since M is time-bounded by $t(n)$, every processor cannot address more than $t(n)$ global memory cells during its runtime. At the beginning of the computation, $n + 1$ memory cells are already in use. Hence overall we know that we will use only $r =_{\text{def}} p(n) \cdot t(n) + n + 1$ global registers (taken from a possible set of 2^m).

Therefore we proceed as follows: We encode the global memory as a sequence of quadruples (a_i, v_i, t_i, b_i) (for $1 \leq i \leq r$), where a_i and v_i are m bit numbers, $0 \leq t_i \leq t(n)$, and $b_i \in \{0, 1\}$. The meaning of such a tuple is as follows: If $b_i = 1$ then global memory cell C_{a_i} has contents v_i at time step t_i, and we say that the quadruple is *valid*. If $b_i = 0$ then the quadruple carries no meaning; we say it is *invalid*.

Hence we encode the global memory with the help of r quadruples. Every quadruple can be encoded using $2m + \ell(t(n)) + 1$ bits. Hence a whole configuration of M can be encoded using $\ell(s)p(n) + mkp(n) + r(2m + \ell(t(n)) + 1)$ bits, a number which is polynomial in $n + t(n) + p(n)$. Every subcircuit L_i has as a sequence of gates whose values yield the encoding of the configuration K_i of M after i steps. It is clear that a constant-depth circuit can compute K_0, the initial configuration of M, given M's input. Furthermore it is clear how to compute the value computed by M: a constant-depth circuit can extract this from $K_{t(n)}$. Thus it only remains to design circuits that compute K_i from K_{i-1}.

This is achieved as follows: Depending on the value of the instruction counter in K_{i-1}, every processor P_j performs a particular instruction from

M's program. This instruction may change some of P_j's local registers as well as some global registers. Also the instruction counter has to be updated. We address the different types of instructions (see p. 67) and describe corresponding circuits.

(1)–(5): The contents of the local registers have to be updated according to the instruction. The necessary computations (addition, subtraction, bitwise operations) can be performed by constant-depth circuits (Theorem 1.15 and Exercise 1.5). The instruction counter has to be incremented by 1.

(6): The global memory in K_{i-1} is searched for a relevant valid quadruple. Its contents is copied to local memory. The instruction counter is incremented by 1.

(7): We use a new block of $p(n)$ quadruples in K_i to encode the global memory cells whose values change during time step i. Among all processors that try to write to the same cell the one with the lowest number is selected. All other quadruples are copied and the bit indicating whether the quadruple is still valid is updated if necessary. The instruction counter has to be incremented by 1.

(8): The comparison can be computed in constant depth (Exercise 1.4), and the instruction counter is updated according to the result.

(9): We do not make any changes. This means that in L_{i+1} (if $i \leq t(n)$) we again encounter the HALT instruction and nothing changes.

Hence we see that the computations necessary to determine K_i given K_{i-1} can be performed by constant-depth circuits. Observe that unbounded fan-in is needed for constant-depth circuits for addition, comparison, etc., but also for realizing instruction types (6) and (7), e.g., to determine the lowest numbered processor that tries to access a particular global memory cell. The size of the circuit family is clearly polynomial in $n + t(n) + p(n)$.   □

Now we want to turn to a converse relation. Of course we cannot simulate non-uniform circuits using a uniform machine model. Therefore we define:

Definition 2.57. A non-uniform PRAM is a PRAM M which has a different program for each input length n. We say that M is of *program size* $s(n)$ if the program for input length n can be written with $s(n)$ symbols from the alphabet $\{R, \leftarrow, \#, +, -, \vee, \wedge, \oplus, (,), <, 0, 1\}$ in the obvious way, using binary encoding for constants and register numbers. The other resources (processor bound, time) are defined as in the case of uniform PRAMs.

At first sight, the definition of a non-uniform PRAM might not be what the reader expected after studying Sect. 2.3, where we introduced non-uniform Turing machines by supplying regular Turing machines with a length-dependent advice. However, from a formal point of view, there is no difference between having a different program for each input length (as above), and getting an advice string for each input length as in Sect. 2.3, because it can be

$P_{(u,v)}$:
```
    if C_f = 1 then begin
        C_f := 0;
        R_0 := C(v);
        if type of v is ¬ then
            C(v) := ¬C(u)
        else if type of v is ∧ begin
            C(v) := 1;
            if C(u) = 0 then C(v) := 0
        end;
        else { type of v is ∨ } begin
            C(v) := 0;
            if C(u) = 1 then C(v) := 1
        end;
        if R_0 ≠ C(v) then C_f := 1
    end.
```

Fig. 2.12. PRAM simulation of a circuit

shown that one can use encodings to transform these different kinds of length-dependent information into one another. Since the existence of a universal PRAM can be established by standard techniques, this implies that the two models of non-uniform PRAMs are of the same power (at least when we ignore time restrictions). We do not develop this topic any further but continue by using the model introduced in Def. 2.57 to simulate (non-uniform) circuit families.

Theorem 2.58. *Let* $f: \{0,1\}^* \to \{0,1\}^*$ *be computed by a circuit family* $C = (C_n)_{n \in \mathbb{N}}$ *over* $\mathcal{B}_1$ *of depth* $d(n)$ *such that for all* $n \in \mathbb{N}$, C_n *has* $s(n)$ *wires. Then there is a non-uniform CRCW-PRAM* M *operating in time* $O(d(n))$ *that computes* f. M *'s processor bound and program size are both polynomial in* $s(n)$.

Proof. Fix an admissible encoding scheme for $\mathcal{C}$. Let $n \in \mathbb{N}$ and consider circuit C_n. Let (V, E) be the undirected acyclic graph underlying C_n. We define a PRAM program M_n as follows: M_n will use a global memory register $C(v)$ for every $v \in V$, more specifically: if $v \in V$ is the gate numbered i under the encoding scheme, then C_{i+1} is the register associated with v, i. e., $C(v) = C_{i+1}$. An additional global cell C_f is initialized to 1. (C_f will be used as a flag to determine when the computation has to stop.) M_n will have a processor $P_{(u,v)}$ for every edge $(u,v) \in E$. At the beginning, the input $x = x_1 \cdots x_n$ is distributed over $C_1, \ldots, C_n$ as usual. Then all processors $P_{(u,v)}$ perform the program in Fig. 2.12. Every processor $P_{(u,v)}$ performs at every time step the computation corresponding to wire (u, v). (Note that this information is length dependent, hence we obtain different programs M_n for different values of n.) If v is a $\neg$ gate the action is obvious. In the cases of an $\wedge$ and an $\vee$ gate we use the simulations from Examples 2.54 and 2.55. The global cell C_f is used for bookkeeping about value changes somewhere in the circuit. If no more changes occur, all gate values are computed correctly, i.e., cell C_{i+1} has as contents the value computed by the gate with number i. Thus, in the end it only remains to move the contents of register C_{n+1+i} to register C_{1+i} for $0 \leq i < |V|$. Recall that the output gates are numbered $n+1, n+2, \ldots$; thus this final step will move the output bits to registers $C_1, C_2, \ldots$ as required in our input–output conventions for PRAMs. $\quad\square$

There are other ways to bridge the gap between uniformity and non-uniformity, when comparing circuits and PRAMs, see Exercises 2.24–2.27.

Bibliographic Remarks

Examinations of circuit complexity compared with Turing machine complexity go back to Savage's textbook [Sav76]. The relationship between Turing machine time and circuit size presented in this chapter was given by [PF79], see also [Fis74, Weg87, Chap. 9.2]. The technique used to prove Lemma 2.5 goes back to a paper by Hennie and Stearns [HS66]. The relations between space and depth are from [Bor77], see also [BDG95, Chap. 5.4]. The paper [Bor77] is to our knowledge the first paper where uniformity issues are explicitly taken into account.

The formal framework for non-uniform complexity classes we use here is from [KL82]. Another approach to non-uniform Turing machines, based on so called *oracles*, has been suggested earlier in [Pip79], see also [Sch76].

Alternating Turing machines were introduced by A. K. Chandra, D. Kozen and L. J. Stockmeyer in [CKS81]. Propositions 2.35–2.38 were proved there. The relations between Boolean circuits and alternating Turing machines are from [Ruz81]. In this paper Ruzzo also discusses a non-uniform version of his correspondence of SIZE-DEPTH(s, d) to ATIME-SPACE$(d, \log s)$ (though these are not in the style of Karp/Lipton but use the oracle framework). With very similar techniques one can prove that the circuit class UnbSIZE-DEPTH(s, d) also has a corresponding class defined via Turing machines: it is AALT-SPACE$(d, \log s)$ [Coo85]. We come back to this relation in Sect. 4.2.

The precise definition of the PRAM model used in this chapter goes back to [SV81]. Parallelism in random access machines was considered for the first time in [FW78] and independently in [Gol82]. While the model from the latter paper is equivalent to CRCW-PRAMs from [SV81], the model from [FW78], later named CREW-PRAM [Vis83b], differs from the above by disallowing simultaneous write operations (but still allowing concurrent read operations), see Exercise 2.20. See [Fic93] for a detailed discussion of these and other models. The parallel prefix algorithm is from [LF80]. A number of survey articles presenting other fundamental PRAM programming techniques and algorithms are available, e.g., [KR90, May90]. The connection between Boolean circuits and parallel random access machines was given in [SV84], see also [KR90, Sect. 3.4] and [GHR95, Sect. 2.4]. Relations between other types of PRAMS and Boolean circuits are surveyed in [Fic93]

While we proved a circuit characterization of P/Poly in this chapter, we did not give an analogue of L/Poly in terms of circuits. Corollary 2.20 only gave a lower bound. In fact, it turns out that L/Poly corresponds to polynomial size *branching programs*, see Sect. 4.5.3.

Sect. 2.6 characterized simultaneous size- and depth-bounded circuit classes using alternating machines. The question whether Corollaries 2.29 and 2.33 hold simultaneously, that is, whether circuit size and depth correspond simultaneously to time and space on deterministic Turing machines, is still open. More about this will be said in the Bibliographic Remarks for Chap. 4 on p. 162.

Exercises

2.1. Show that for every length-respecting function $f: \{0,1\}^* \to \{0,1\}^*$, we have: $f \in \text{FSIZE-DEPTH}_{\mathcal{B}_1 \cup \{(\oplus)_{n \in \mathbb{N}}\}}(\frac{2^n}{n}, 1)$.

2.2. Let M be a k-tape Turing machine. Show how M can be simulated by a Turing machine with $2k$ pushdown stores.

2.3. Let $f: \mathbb{N} \to \mathbb{N}$ be such that

$$f(0) = O(1)$$
$$f(k) = 2 \cdot f(k-1) + O(2^k)$$

Prove: $f(k) = O(k \cdot 2^k)$.

2.4.* (1) Show that the circuit families designed in Chap. 1 for ITADD, MULT, MAJ, UCOUNT, BCOUNT, and SORT are logspace-uniform.
(2) Show that the circuit families designed in Theorem 1.40 and Exercise 1.19 are ptime-uniform. What is the obstacle which prevents us from claiming logspace-uniformity here?

2.5. Show that the circuit family constructed in Theorem 1.31 is logspace-uniform.

2.6. Prove Observation 2.13: The class $\text{DTIME}(n)/\text{F}(1)$ contains undecidable languages.

2.7. Let $f \in \mathbb{B}^n$. Prove: $f \in \text{FDTIME}(n)/\text{F}(2^n)$.

2.8. Prove:

(1) $\text{U}_\text{P}\text{-SIZE}(n^{O(1)}) = \text{DTIME}(n^{O(1)})$.
(2) $\text{U}_\text{E}\text{-SIZE}(n^{O(1)}) = \text{DTIME}(n^{O(1)})$.

2.9. Let M be an alternating Turing machine, and x be an input to M. Prove that the mapping l_M of the computation tree $T_M(x)$ always exists.

2.10.[*] Let M be an alternating Turing machine working in time $t(n) \geq n$ using space $s(n) \geq \log n$. Show that M can be simulated by a machine with random access to its input in time $O(t)$ using space $O(s)$.

2.11. Let M be an alternating Turing machine with random access to its input working in time $t(n) \geq n$ using space $s(n) \geq \log n$. Show that M can be simulated by a machine with a regular input tape in time $O(t)$ using space $O(s)$.

2.12.[*] Prove Proposition 2.36.
Hint: Design a deterministic procedure which performs a depth-first search in the computation tree of the given alternating machine and computes the configuration labeling.

2.13.[*] Prove Proposition 2.37.
Hint: Use a dynamic programming approach to compute the labeling of the alternating machine's computation tree.

2.14.[*] Prove Proposition 2.38.
Hint: Use divide-and-conquer similarly to Exercise 2.15.

2.15. Prove Corollary 2.39.

2.16. Show that a deterministic Turing machine with random access to its input can compute the length of its input in time $O(\log n)$.
Hint: Use one-sided binary search.

2.17. Let $t(n) \geq n$. Show that a deterministic Turing machine with random access to its input can compute n in time $O(t(n))$, given an input of the form $\langle y, u, v, w \rangle$, where $|y| = n$, and $|u|, |v|, |w| = O(t(n))$.

2.18.[*] Let $A \in$ AALT-SPACE$((\log n)^i, \log n)$ for some $i \geq 1$.

(1) Let M be an alternating machine accepting A. Show that M can be simulated by a machine M' which within one alternation makes at most $O(\log n)$ steps (in other words: at most $O(\log n)$ successive configurations are of the same type).
(2) Conclude that $A \in$ ATIME-SPACE$((\log n)^{i+1}, \log n)$

Hint: Observe that if configuration C' can be reached from configuration C, then this takes at most a polynomial number of steps. Simulate a path from C to C', whose configurations are all existential, as follows: When in configuration C, existentially guess configuration C'. Then branch universally: On one branch continue with the simulation of M; on the other branch verify that C' can be reached from C by a divide-and-conquer approach as in Theorem 2.35. Long paths consisting of universal configurations can be treated similarly.

2.19. In CRCW-PRAMs as introduced in Sect. 2.7 the processor with the lowest number succeeds when different processors try a concurrent write operation. This is called the PRIORITY version. Alternatively let us define the ARBITRARY version, where any one of the write operations succeeds and we do not know which one; in other words: our PRAM programs have to be designed in such a way that they work correctly no matter which processor succeeds during concurrent write operations. Finally, in the COMMON model, we require that, if different processors try a concurrent write, all of them try to write the same value (again, our programs have to be designed to ensure this).

- Show that an algorithm for a COMMON CRCW-PRAM will work unchanged on an ARBITRARY CRCW-PRAM. Show that an algorithm for an ARBITRARY CRCW-PRAM will work unchanged on a PRIORITY CRCW-PRAM.
- Show that any PRIORITY CRCW-PRAM working in time t using r processors can be simulated by a COMMON CRCW-PRAM using time $O(t)$ and $r^{O(1)}$ processors.

2.20.* A CREW-PRAM is a PRAM in which concurrent write operations are not allowed. Show that any CRCW-PRAM operating in time $t(n)$ using $p(n)$ processors can be simulated by a CREW-PRAM in time $t(n) \cdot \log p(n)$ using $p(n)$ processors. (Use the fact that a CREW-PRAM can sort r numbers in time $O(\log r)$ with r processors.)

2.21.** An EREW-PRAM is a PRAM in which neither concurrent write nor concurrent read operations are allowed. Show that any CRCW-PRAM operating in time $t(n)$ using $p(n)$ processors can be simulated by a EREW-PRAM in time $t(n) \cdot \log p(n)$ using $p(n)$ processors.

2.22.* Construct a CRCW-PRAM which computes the minimum of n numbers in constant time using $O(n^2)$ processors.

2.23. Show that Theorem 2.56 still holds when we expand the PRAM instruction set by allowing multiplication and division for numbers, whose length is logarithmic in the input length.

2.24. Show that Theorem 2.56 holds for non-uniform PRAMs as well; that is: Let $f: \{0,1\}^* \to \{0,1\}^*$ be computed by the non-uniform CRCW-PRAM M in time $t(n)$. Let the processor bound of M be $p(n)$. Let the program size of M be $s(n)$. Then there is a family $\mathcal{C}$ of circuits over $\mathcal{B}_1$ that computes f, where the depth of $\mathcal{C}$ is $O(t(n))$ and the size is polynomial in $n + t(n) + p(n) + s(n)$.

2.25. Show that the circuit family constructed in Theorem 2.56 is logspace-uniform.

2.26. Show that the transitive closure A^* of a given matrix A can be computed by PRAMS in logarithmic time.

2.27.* (1) Say that a PRAM M is logspace-uniform if, formally, M is non-uniform (i. e., there are different programs for different input lengths) but there is a deterministic Turing machine operating in logarithmic space that on input 1^n computes an encoding of M's program for inputs of length n.

Prove: Let $f: \{0,1\}^* \to \{0,1\}^*$ be computed by a logspace-uniform circuit family C of size $s(n)$ and depth $d(n)$. Then there is a logspace-uniform CRCW-PRAM M operating in time $O(d)$ that computes f. M's processor bound and program size are both polynomial in $s(n)$.

(2) Let $A \in \mathrm{NSPACE}(\log n)$. Show that there is a PRAM M that computes c_A in time $O(\log n)$, using a polynomial number of processors.

(3) Use the above to conclude that $A \in \mathrm{U_L}\text{-UnbSIZE-DEPTH}(n^{O(1)}, \log^k n)$ for $k \geq 1$, if and only if there is a PRAM (in the uniform sense, i. e., one program for all input lengths) that computes c_A in time $O(\log^k n)$ using a polynomial number of processors.

Hint: For (2), let $A \in \mathrm{NSPACE}(\log n)$ via nondeterministic Turing machine M. Let $x \in \{0,1\}^*$. Construct the transition matrix T_x and use Exercise 2.26.

Notes on the Exercises

2.12–2.14. Proofs were originally given in [CKS81, Theorems 3.2–3.4]. See also [BDG90, pp. 70ff.].

2.16. This is an observation credited to Martin Dowd in [Bus87]. Background on logtime machines can be found in [RV97].

2.19. This result is from [Kuč82], see also [KR90, 895].

2.20. This result is from [Vis83a], see also [KR90, 894–895]. The sorting algorithm mentioned is from [Col88].

2.21. This result is from [FRW88], see also [Fic93, 849–850].

2.22. This exercise is due to [SV81].

2.27. For (1), see [KR90, 900]. The result (2) goes back to [FW78], see [KR90, 906] or [Fic93, 863]. For (3), see also [GHR95, 34–35].

3. Lower Bounds

The computational model of Boolean circuits attracted much interest among complexity theorists since it is a model where non-trivial lower bounds for specific functions are known. We treat this topic in the present chapter, but we want to stress that we present only very few out of a large and still-growing body of results. The interested reader will find references at the end of the chapter.

Let us start by introducing some terminology and then note an easy lower bound that holds essentially for arbitrary Boolean functions. Recall the definition of $S(C)$ from p. 26.

Definition 3.1. Let B be a basis. For a Boolean function f, let $S_B(f) =_{\text{def}}$ $\min\{ S(C) \mid C$ is a circuit over B that computes $f \}$.

Definition 3.2. A Boolean function $f \in \mathbb{B}^n$ is *degenerated*, if there is an i, $1 \leq i \leq n$, such that for all $x_1, \ldots, x_{i-1}, x_{i+1}, \ldots, x_n$ we have $f(x_1, \ldots, x_{i-1}, 0, x_{i+1}, \ldots, x_n) = f(x_1, \ldots, x_{i-1}, 1, x_{i+1}, \ldots, x_n)$. In this case x_i is called an *inessential variable* of f.

When we want to give lower bounds for the complexity of a function $f \in \mathbb{B}^n$ measured in n, it suffices to consider only non-degenerated functions.

Theorem 3.3. Let $f \in \mathbb{B}^n$ be non-degenerated. Then $S_{\mathbb{B}^2}(f) \geq n - 1$.

Proof. Let C be a minimal-size circuit such that $f = f_C$, and let $S(C) = m$. The output gate in C will have fan-out zero; otherwise C would not be minimal. All other gates in C will have fan-out greater than zero; otherwise again, we get a contradiction to the minimality of C.

For every non-input gate $v \in C$ there are two input wires into v. Conversely every wire is input wire to some gate. Hence the number of wires in C is exactly $2m$.

On the other hand, every gate except the output gate will have at least one output wire. Therefore the number of wires in C is at least $n + m - 1$.

Hence, $2m \geq n + m - 1$ and $m = S_{\mathbb{B}^2}(C) \geq n - 1$. ∎

This result is optimal for the function $\oplus = (\oplus^n)_{n \in \mathbb{N}}$ for which there are obviously circuits of size $n - 1$ over $\mathbb{B}^2$.

Corollary 3.4. $S_{\mathbb{B}^2}(\oplus^n) = n - 1$.

3.1 The Elimination Method

In this section we want to consider a method, known as the *elimination method*, to obtain lower bounds. First we give an example, before we describe the general technique in a more abstract way.

3.1.1 A Lower Bound for the Threshold Function

In Sect. 1.4 we considered the threshold functions

$$T_m^n(a_1, \ldots, a_n) =_{\text{def}} \left[\!\!\left[\sum_{i=1}^n a_i \geq m \right]\!\!\right].$$

Theorem 3.5. *Let* $2 \leq k \leq n - 1$. *Then* $S_{\mathbb{B}^2}(T_k^n) \geq 2n - 4$.

Proof. By induction on n.

Let $n = 3$, then $k = 2$. T_2^3 is non-degenerated, hence by Theorem 3.3, $S_{\mathbb{B}^2}(T_2^3) \geq 2 = 2 \cdot 3 - 4$.

Now let $n \geq 4$, and we assume that for $3 \leq m < n$ and $2 \leq j \leq m - 1$ we have $S_{\mathbb{B}^2}(T_j^m) \geq 2m - 4$. Pick a k, $2 \leq k \leq n - 1$. Let C be a minimal-size circuit over $\mathbb{B}^2$ such that $T_k^n = f_C$. Certainly there is a gate α in C such that both predecessor gates of α are inputs to the circuit. Without loss of generality let x_1 and x_2 be those predecessors.

Now we replace the inputs x_1 and x_2 in C by constants; then the circuit thus obtained computes one of the following three different functions:

$$
\begin{array}{ll}
T_k^{n-2} & (\text{for } x_1 := 0, x_2 := 0) \\
T_{k-1}^{n-2} & (\text{for } x_1 := 0, x_2 := 1 \text{ or } x_1 := 1, x_2 := 0) \\
T_{k-2}^{n-2} & (\text{for } x_1 := 1, x_2 := 1)
\end{array}
$$

However the output of α is always one of the *two* values $\{0, 1\}$. Hence we conclude that there must be a gate $\beta \neq \alpha$ in C which is connected to either x_1 or x_2; let us assume without loss of generality that β is connected to x_1. Let γ be the other predecessor gate of β. γ can be any gate in C; in particular γ may be α or there may be a path from α to γ, but γ may also be an input gate. However if β is the output gate of C, then γ cannot be a circuit input since otherwise gate α is never needed, contradicting the minimality of C.

Now we replace only input x_1 by a constant $b \in \{0, 1\}$, see Fig. 3.1. (We fix b later.) Then gate α computes one of the unary functions $0, 1, x_2, \neg x_2$; and gate β computes one of the functions $0, 1, \gamma, \neg\gamma$. We now claim that in all these cases we can delete gates α and β from the circuit. If α computes x_2 and β computes γ this is clear. Suppose α computes $\neg x_2$, and suppose α is a left predecessor to a gate where the function $g \in \mathbb{B}^2$ is computed. Then we delete α and replace g by g' where $g'(x, y) = g(\neg x, y)$. Observe that $g' \in \mathbb{B}^2$. An analogous replacement can be made for all the other cases. Hence we

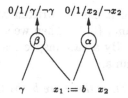

Fig. 3.1. Part of circuit C for threshold

may delete α and β, but we have to change the functions computed at those gates to which α and β are predecessor gates. If there is no gate to which β is predecessor (i.e., β is the output of C) then we argued above that γ is a non-input gate, and we can replace the function computed at γ in a suitable way. For example, if β computes $\overline{\gamma}$, we replace the function computed at γ by its negation.

This new circuit, which we denote by C^b, now computes one of the functions T_k^{n-1} (for $b = 0$) or T_{k-1}^{n-1} (for $b = 1$). If $k = 2$ then we pick $b := 0$, else $b := 1$. In any case, $2 \leq k - b \leq n - 2$ and C^b computes the function T_{k-b}^{n-1}. By induction hypothesis, C^b has size at least $2(n - 1) - 4$; hence the size of C cannot be less than $2n - 4$. $\qquad\Box$

The above proof is considered a typical example of an application of the "elimination method." The usual way to proceed is the following: Given a function $f = (f^n)_{n \in \mathbb{N}}$, start with a minimal-size circuit C for a fixed input length. Try to find out as much about the structure of the circuit as possible by considering different choices for constants instead of input gates. Then set some of the inputs of C to constants such that the resulting circuit computes the same function for a smaller arity. Eliminate unnecessary gates (usually a gate becomes unnecessary if it computes a constant, the identity or the negation). The number of gates that can be deleted determines the factor of n in the lower bound.

Let us consider another application in the next section.

3.1.2 A Lower Bound for the Parity Function

At the beginning of this chapter we proved that $S_{\mathbb{B}^2}(\oplus^n) = n - 1$. The upper bound was simply due to the fact that $\oplus^2 \in \mathbb{B}^2$. What if we disallow $\oplus^2$ as a basis function? Certainly we also have to disallow $\equiv^2$, where $\equiv(x, y) = \neg\oplus(x, y)$. Let us choose

$$\mathcal{U}_2 = \left\{ (x^a \wedge y^b)^c \mid a, b, c \in \{0, 1\} \right\},$$

where we set $x^0 = \overline{x}$ and $x^1 = x$ for $x \in \{0, 1\}$. (This definition x^a is different from the one used in Sect. 1.5.2.) Observe that $\{\oplus^2, \equiv^2\} \cap \mathcal{U}_2 = \emptyset$, more specifically, $\mathbb{B}^2 = \mathcal{U}_2 \cup \{\oplus^2, \equiv^2\} \cup \{x^a, y^a \mid a \in \{0, 1\}\} \cup \{0, 1\}$. Say that a function $f \in \mathbb{B}^n$ is *affine*, if it can be written as a $\oplus$-sum of the variables $\{x_1, \ldots, x_n\}$ and the constants $\{0, 1\}$. Then we see that $\mathcal{U}_2$ is the basis of all binary non-affine functions. By a case inspection (see Exercise 3.6) one can now prove the following lemma:

Lemma 3.6. *Let* $f \in \mathcal{U}_2$. *Then there are* $b_1, b_2 \in \{0, 1\}$ *such that the functions* $f_1(x) =_{\text{def}} f(x, b_1)$ *and* $f_2(x) =_{\text{def}} f(b_2, x)$ *are constant.*

Theorem 3.7. $S_{\mathcal{U}_2}(\oplus^n) = 3(n - 1)$.

Proof. We first prove the upper bound. Certainly $x \oplus y = (x \wedge \overline{y}) \vee (\overline{x} \wedge y)$, hence $S_{\mathcal{U}_2}(\oplus^2) \leq 3$. Since $\oplus^n(x_1, \ldots, x_n)$ can be computed with $n - 1$ gates of type $\oplus^2$, we immediately get $S_{\mathcal{U}_2}(\oplus^n) \leq 3(n - 1)$.

It remains to show that $S_{\mathcal{U}_2}(\oplus^n) \geq 3(n-1)$. We give a proof by induction.

Let $n = 2$. Since $\oplus^2 \notin \mathcal{U}_2$ we have $S_{\mathcal{U}_2}(\oplus^2) \geq 2$. Suppose $S_{\mathcal{U}_2}(\oplus^2) = 2$. Then $\oplus^2$ is computed by a circuit of the form given in Fig. 3.2. Since $\alpha \in \mathcal{U}_2$,

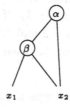

Fig. 3.2. Hypothetical circuit for $\oplus^2$

by Lemma 3.6 there is a value $b \in \{0, 1\}$ such that the circuit we get when we set $x_2 := b$ computes a (formally unary) constant. However, this property is not shared by the function $\oplus^2$. Hence the above circuit cannot compute $\oplus^2$, and therefore $S_{\mathcal{U}_2}(\oplus^2) \geq 3$. Since we will need the result shortly, we note that our argument also shows that the circuit as given in Fig. 3.2 cannot compute the function $\equiv^2$.

Now let $n > 2$. Let C be a minimal-size circuit for $\oplus^n$. Let α be a gate in C whose predecessor gates are both inputs to the circuit; without loss of generality let these be x_1 and x_2. By Lemma 3.6 there is a b such that if we set $x_2 := b$ then α will compute a constant. However, $\oplus(x_1, b, x_3, \ldots, x_n)$ still depends on x_1. Hence x_1 is predecessor to another gate β. Let γ be the other predecessor gate of β. Until now we have identified a subpart of C which is identical to the one we had in the proof of the previous theorem, cf. Fig. 3.1. As above we note that if β is the output of C then γ must not be an input gate.

But we want to find out more about the structure of C. First, clearly, α cannot be the output gate of C. Suppose now that α has only one output wire leading into β, i.e., $\alpha = \gamma$. Then β computes the value

$$\beta(\alpha(x_1, x_2), x_2)).$$

(Here and in what follows we take the liberty to identify gates with the functions from $\mathcal{U}_2$ that they compute.) By the argument given above for $n = 2$ we see then that β computes a function from $\mathbb{B}^2 \setminus \{\oplus^2, \equiv^2\}$; thus this must be either a function from $\mathcal{U}_2$, or a constant, the identity, or the negation. If β computes a function $g \in \mathcal{U}_2$ then α and β can be replaced by one gate directly computing g. If β computes a constant, the identity, or the negation, then we can delete α and β and modify the function computed at the gate to which β is predecessor. (Such a gate must exist as β cannot be the circuit output, since $\oplus^n$ is not degenerated.) Therefore, in both cases we get a contradiction to the minimality of C; hence our assumption that α has only one output wire leading into β must be wrong.

From the above we conclude that there is a third gate δ to which α is predecessor. Let η be the second predecessor of δ. We now have the situation given in Fig. 3.3. As above we observe that if δ is the output of C then η

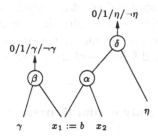

Fig. 3.3. Part of hypothetical circuit C for parity

cannot be an input of C (otherwise β is never needed).

Since $\alpha \in \mathcal{U}_2$, by Lemma 3.6 there is a $b \in \{0, 1\}$ such that α computes a constant if we set $x_1 := b$. Then β computes 0, 1, γ, or $\neg\gamma$, and δ computes 0, 1, η, or $\neg\eta$. Thus we can delete all three gates α, β, and γ (where we treat negations and constants as before, see Exercise 3.5). The resulting circuit will compute $\oplus^{n-1}$ or $\equiv^{n-1}$, hence it must have at least $3(n-2)$ gates. Thus $S(C) \geq 3(n-1)$.   □

As we see, the general problem with the elimination method is that it is only possible to examine the circuit very "near" to the input gates. The effect that setting some of the inputs to a constant has on a gate somewhere deep in the circuit, is hard to understand. Up to now the elimination method has only led to linear lower bounds.

On the other hand, no non-linear lower bound for circuit size of a set or function in P (or even NP) is known in the unrestricted circuit model (cf. the discussion at the beginning of Sect. 6.2). The best lower bound obtained so far over a complete basis is of order $3n$ and was given in [Blu84]. Thus the elimination method is still among the best tools we have.

3.2 The Polynomial Method

As we pointed out at the end of the previous section, the lower bounds which one could prove so far for general circuits are very weak. Hence researchers considered circuits of more restricted forms.

Both out of theoretical and practical considerations, circuits of constant depth are very important. Observe that the circuits constructed in Lupanov's Theorem (Theorem 1.51) have small, constant depth. Constant-depth circuits are a theoretical model of programmable logic arrays (PLA's) which are used in practice inside microprocessors to represent many functions, see [MC81]. With today's technology only circuits of very small depth can be constructed, hence any result improving our knowledge about the power of this type of circuit will be very useful. On the other hand it seems that restricting our gate types to the standard basis $(\wedge, \vee, \neg)$ is not justified from practical considerations. On the contrary, gates like parity are easy to implement. Therefore below we will consider circuits of constant depth, where as basis gates we allow gates for parity (and some generalizations) in addition to the standard gates. It is clear that if we restrict our attention to constant-depth circuits, we have to allow unbounded fan-in—otherwise we would only consider finite circuit families.

3.2.1 Perceptrons and Polynomial Representations

Before we examine lower bounds for constant-depth circuits over the standard unbounded fan-in basis in Sect. 3.3 below, let us first consider circuits of a very particular type, the so-called perceptrons. These circuits gain their importance from the fact that they are the basic components of neural networks. Upper and lower bounds for perceptrons translate to results on the computational power of neural networks, see [MP88].

Definition 3.8. A *perceptron* with n inputs is a circuit C of depth 2 over basis $\{(\wedge^n)_{n \in \mathbb{N}}, (T_t^n)_{n \in \mathbb{N}}\}$ (for some $t \in \mathbb{N}$) of the following form:

- On level 0 we have the inputs $x_1, \ldots, x_n$.
- On level 1 we have $\wedge$ gates of arbitrary fan-in. Predecessors to these gates are only level-0 gates.
- On level 2 we have a threshold gate. Predecessors to this gate are only $\wedge$ gates from level 1.

The input wires to the threshold gate on top are *weighted* with some integer. If gate v is connected via a wire with weight w to the threshold gate, then this gate v counts as w gates. The perceptron outputs 1 iff the weighted sum of the predecessors of the threshold gate is at least t. More formally, let us fix an input x, and let $v_1, \ldots, v_k$ be those $\wedge$ gates on level 1 that evaluate to 1 on input x. For $1 \leq i \leq k$ let w_i be the weight of the edge connecting v_i with the output threshold gate. Then the circuit outputs 1 iff $\sum_{i=1}^{k} w_i \geq t$.

   C is of *size s*, if the number of $\land$ gates in C is at most s. C is of *order r*, if the $\land$ gates are of fan-in at most r. C is of *weight W*, if the absolute values of all weights in C are bounded by W.

   Strictly speaking a perceptron is not a circuit, since in circuits we do not allow weights on the edges. (Note that in particular a weight can be negative.) However the following holds (see Exercise 3.6):

Lemma 3.9. *Let C be a perceptron with n inputs of size $s \geq n$ and weight W. Then C can be simulated by a constant-depth circuit of size $O(W \cdot s)$ over basis $\mathcal{B}_1 \cup \{\text{MAJ}\}$.*

   The core technique of the *polynomial method* is to represent Boolean functions and circuits by polynomials over some fixed ring $\mathcal{R} = (R, 0, 1, +, \cdot)$. (See Appendix A8.) Given a function f, one then uses algebraic methods to prove that polynomials needed to compute or approximate f have a certain complexity (e. g., large degree or size of coefficients). This is exploited to deduce lower bounds for circuits that compute f.

   In order to represent Boolean functions by polynomials we first have to define how we represent Boolean values by ring elements. This is done via a one-one mapping $c \colon \{0, 1\} \to R$. In particular we will consider two possible representation schemes. In the *standard representation* we represent the Boolean values 0 and 1 by the ring elements 0 and 1 ($c^{\text{standard}}(0) = 0$, $c^{\text{standard}}(1) = 1$). In the *Fourier representation* we represent the Boolean values 0 and 1 by 1 and -1 ($c^{\text{Fourier}}(0) = 1$, $c^{\text{Fourier}}(1) = -1$).

   Second, we have to define how we interpret the result of a polynomial evaluation as a Boolean value. If we work over an ordered ring the most common method is *sign representation*, where we use the function sign defined as $\text{sign}(m) = [\![m > 0]\!]$ for all ring elements m.

   For the rest of this section we fix the ring of the integers and use standard representation for the inputs and sign representation for the outputs. Given a Boolean function $f \in \mathbb{B}^n$, we say that polynomial p *represents* f, if for all $x_1, \ldots, x_n \in \{0, 1\}$ we have:

$$f(x_1, \ldots, x_n) = \text{sign}\left(p\left(c^{\text{standard}}(x_1), \ldots, c^{\text{standard}}(x_n)\right)\right)$$

Call this the *standard/sign representation scheme*.

   Under this scheme, the following connection between polynomials and perceptrons holds:

Theorem 3.10. *Let $f \in \mathbb{B}^n$, and use standard/sign representation. The following statements are equivalent:*

1. *There is a perceptron of size s, order d, and weight W, that computes f.*
2. *There exists a polynomial p of degree d with s monomials and coefficients whose absolute values are bounded by W, that represents f.*

Proof. To prove the direction (1) $\Rightarrow$ (2), let v be a gate $\wedge(x_1, \ldots, x_k)$. Represent v by the monomial $x_1 \cdots x_k$. The coefficient of this monomial in p is the weight of the edge from v to the output. The polynomial p now is simply the sum of these monomials plus the absolute term $1 - t$, where t is the threshold of the output gate of C. For the other direction, monomials of positive degree can be directly "computed" by weighted $\wedge$ gates, and the sign test can be performed by a threshold gate, where the actual threshold is determined by the value of the one monomial in p of degree 0. $\square$

Thus in particular, a lower bound for the degree of a polynomial representing f translates to a lower bound for the order of a perceptron computing f. Given $f \in \mathbb{B}^n$, let $d^{\text{sign}}(f)$ denote the minimal degree of a polynomial representing f in standard/sign representation.

3.2.2 Degree Lower Bounds

In this section we give two examples for lower bounds for the degree of a polynomial representing a Boolean function. The general idea will be that we want to show that the representing polynomial has a large number of zeroes, from which we then conclude that the degree must also be large. However the well-known relation that the number of zeroes (counting multiplicities) is equal to the degree only holds for polynomials in one variable. Fortunately there is a result that says that for symmetric polynomials in n variables we find a related polynomial in only one variable:

Lemma 3.11. *Let p be a polynomial in n variables of degree d which, restricted to the domain $\{0,1\}^n$, is symmetric. Then there is a polynomial q in one variable of degree d, such that for all $x_1, \ldots, x_n \in \{0,1\}$,*

$$p(x_1, \ldots, x_n) = q(x_1 + \cdots + x_n).$$

Proof. For $1 \leq i \leq n$, if $x_i \in \{0,1\}$ we have $x_i^2 = x_i$. Thus we may replace in p all powers x_i^k for $k > 1$ by x_i, and we only have to deal with a multi-linear polynomial. Therefore p can be written as $p(x_1, \ldots, x_n) = \sum_{j=1}^{m} c_j \prod_{i \in S_j} x_i$ for suitable $m \geq 0$, $S_1, \ldots, S_m \subseteq \{1, \ldots, n\}$. Since p is symmetric, we must have $c_j = c_{j'}$ if $|S_j| = |S_{j'}|$. (A formal proof proceeds by induction on $|S_j| = |S_{j'}|$.) Certainly $|S_j| \leq d$ for all j. Hence there are numbers $c'_0, \ldots, c'_d$ such that

$$p(x_1, \ldots, x_n) = \sum_{k=0}^{d} c'_k \binom{x_1 + \cdots + x_n}{k}.$$

$\square$

As a first example, we turn again to the parity function.

Theorem 3.12. $d^{\text{sign}}(\oplus^n) = n.$

Proof. For the upper bound, observe that for all functions $f \in \mathbb{B}^n$, we have $d^{\text{sign}}(f) \leq n$, see Exercise 3.7.

For the lower bound, let p be a polynomial of degree d that represents $\oplus^n$. We have to show that $d \geq n$. We would like to use Lemma 3.11 to go to a polynomial in one variable, but p does not have to be symmetric. We only know that $p(a_1, \ldots, a_n) > 0 \iff p(b_1, \ldots, b_n) > 0$ for all $a_1, \ldots, a_n, b_1, \ldots, b_n \in \{0,1\}$ such that $\sum_{i=1}^n a_i = \sum_{i=1}^n b_i$. This is weaker than symmetry which requires that under the above we always have $p(a_1, \ldots, a_n) = p(b_1, \ldots, b_n)$. However, we can construct from p a symmetric polynomial that represents $\oplus^n$: Let $\sigma \in S_n$ be a permutation on n elements. Then certainly also $p^\sigma(x_1, \ldots, x_n) =_{\text{def}} p(x_{\sigma(1)}, \ldots, x_{\sigma(n)})$ represents $\oplus^n$, and p^σ is again a polynomial of degree d. Define

$$P(x_1, \ldots, x_n) =_{\text{def}} \sum_{\sigma \in S_n} p^\sigma(x_1, \ldots, x_n).$$

Then P has degree d, P represents $\oplus^n$, and P is symmetric on $\{0,1\}^n$. By Lemma 3.11 there is a polynomial Q in one variable of degree d such that $P(x_1, \ldots, x_n) = Q(x_1 + \cdots + x_n)$, hence $\oplus(x_1, \ldots, x_n) = 1 \iff Q(x_1 + \cdots + x_n) > 0$. Now since $Q(0) \leq 0$, $Q(1) > 0$, $Q(2) \leq 0$, $Q(3) > 0, \ldots$, if we look at Q as a polynomial over the rationals it has $n - 1$ local minima or maxima. Hence $d \geq n$. $\quad\square$

We note that the proof also shows that $d^{\text{sign}}(\equiv^n) = n$.

Corollary 3.13. *Perceptrons that compute $\oplus^n$ or $\equiv^n$ must have order n.*

As a second example we turn to a simple "AND-of-OR" predicate. A perceptron for this necessarily has high order.

Theorem 3.14. *Let $n = 4m^3$, $A_i = \{(i-1)4m^2 + 1, \ldots, i \cdot 4m^2\}$ for $i = 1, \ldots, m$. Define $f \in \mathbb{B}^n$ by*

$$f(x_1, \ldots, x_n) = \bigwedge_{i=1}^m \bigvee_{j \in A_i} x_j.$$

Then $d^{\text{sign}}(f) \geq m$.

Proof. Let p be a polynomial of degree d representing f. Again we want to make use of Lemma 3.11. Therefore, let S_n' be the set of those permutations $\sigma \in S_n$ for which $\sigma(i) \in A_j$ for all $1 \leq j \leq m$ and $i \in A_j$. Let $p^\sigma(x_1, \ldots, x_n) =_{\text{def}} p(x_{\sigma(1)}, \ldots, x_{\sigma(n)})$ and define

$$P(x_1, \ldots, x_n) =_{\text{def}} \sum_{\sigma \in S_n'} p^\sigma(x_1, \ldots, x_n).$$

Then P has degree d and represents f. Furthermore, if $(x_1, \ldots, x_n)$ and $(x_1', \ldots, x_n')$ are such that for every j, $1 \leq j \leq m$, $\sum_{i \in A_j} x_i = \sum_{i \in A_j} x_i'$, then

$f(x_1, \ldots, x_n) = f(x'_1, \ldots, x'_n)$. (Intuitively we might say that P is symmetric in the variables with indices from A_j.) Generalizing the ideas of Lemma 3.11 (see Exercise 3.8) now shows that there is a polynomial Q of degree d in m variables such that

$$P(x_1, \ldots, x_n) = Q\left(\sum_{j \in A_1} x_j, \ldots, \sum_{j \in A_m} x_j\right),$$

hence $Q(x_1, \ldots, x_m) > 0 \iff \bigwedge_{i=1}^m (x_i > 0)$ for $0 \le x_1, \ldots, x_m \le 4m^2$.

We will now use Q to design a low-degree polynomial for the negation of parity which, as we know by Theorem 3.12, cannot exist. In a sense we *reduce* the negation of parity to Q.

Define $Q'(x_1, \ldots, x_m) =_{\text{def}} Q(x_1^2, \ldots, x_m^2)$. The degree of Q' is $2d$, and for $-2m \le x_1, \ldots, x_m \le 2m$ we have $Q'(x_1, \ldots, x_m) > 0 \iff \bigwedge_{i=1}^m (x_i \neq 0)$. Given now $x_1, \ldots, x_{2m} \in \{0, 1\}$, let $z_k = x_1 + \cdots + x_{2m} - k$ for $k = 1, 3, \ldots, 2m - 1$. Then $-2m < z_k < 2m$ and $z_k \neq 0 \iff \sum_{i=1}^{2m} x_i \neq k$. Hence $Q'(z_1, z_3, \ldots, z_{2m-1}) > 0 \iff \sum_{i=1}^{2m} x_i \equiv 0 \pmod 2$. This shows that $Q'(z_1, \ldots, z_{2m-1})$ represents the negation of the parity of $x_1, \ldots, x_{2m}$, hence by the above result must have degree at least $2m$, and we conclude $d \ge m$. $\qquad\square$

Corollary 3.15. *Let $f \in \mathbb{B}^n$ be as in Theorem 3.14. Perceptrons that compute f must have order $\sqrt[3]{\frac{n}{4}}$.*

Minsky and Papert found this example remarkable, since an OR can easily be computed by perceptrons with order 1, but an AND-of-OR needs unbounded order. In their book [MP88] they also presented an example of a disjunction of particular predicates, where each of these predicates can be computed by perceptrons of constant order but the disjunction needs unbounded order.

So far we have seen that the polynomial method is well suited for lower bounds for the order of perceptrons. One might argue that the perceptron is maybe not a very important circuit model, and the order is not a very natural computational resource. A possible reply to this is that the results presented above are important to judge the power of neural networks (which was the main motivation for Minsky and Papert). Also, lower bounds on order of perceptrons translate immediately to lower bounds on the so-called *bottom fan-in* of depth 2 circuits. (The bottom fan-in of a circuit is the fan-in of the gates connected directly to the inputs.) These lower bounds can in turn be used to make statements about the power of alternating logarithmic time-bounded Turing machines, see [CC95, CCH97]. Moreover we will see in the next section that the polynomial method is also useful for proving lower bounds for general constant-depth circuits.

3.3 Lower Bounds for Constant-Depth Circuits

In this section we will consider constant-depth circuits of unbounded fan-in over the standard unbounded fan-in basis. Thus central to our examination will be the class UnbSIZE-DEPTH($n^{O(1)}, 1$), i.e., constant-depth circuits of polynomial size over $\mathcal{B}_1$. However we will also study circuits of the same size where we allow additional gates. As we pointed out at the beginning of Sect. 3.2 this is also justified by practical considerations.

Definition 3.16. For $p \in \mathbb{N}$, define $\text{MOD}_p = \left(\text{mod}_p^n\right)_{n \in \mathbb{N}}$ by

$$\text{mod}_p^n(x_1, \ldots, x_n) = 1 \Longleftrightarrow_{\text{def}} \sum_{i=1}^{n} x_i \equiv 0 \pmod{p}.$$

Definition 3.17. $\mathcal{B}_1(p) =_{\text{def}} \mathcal{B}_1 \cup \{\text{MOD}_p\}$.

We will consider classes of the form SIZE-DEPTH$_{\mathcal{B}_1(p)}(n^{O(1)}, 1)$ in this section. Observe that $\text{MOD}_p \leq_{\text{cd}} \text{MAJ}$ easily for all $p \geq 2$ (see Exercise 3.12), hence all problems from the above classes reduce to majority, and are included in DEPTH($\log n$). The main result we will present below is that the function MOD_q is not in SIZE-DEPTH$_{\mathcal{B}_1(p)}(n^{O(1)}, 1)$ if p is a prime number and q is relatively prime to p. For this we first prove a remarkable upper bound.

3.3.1 Depth Reduction for Constant-Depth Circuits

We will show that all classes SIZE-DEPTH$_{\mathcal{B}_1(p)}(n^{O(1)}, 1)$ can be simulated by *probabilistic circuits* of very small depth. For technical simplicity we only consider circuits with one output, but we remark that (under a reasonable definition of probabilistic computation of functions) all results can be generalized to the case of computation of non-characteristic functions.

Definition 3.18. A *probabilistic circuit* C over basis B with n inputs and r random bits is a circuit over B with $n+r$ inputs. These inputs will usually be denoted by $x_1, \ldots, x_n, u_1, \ldots, u_r$, and we say that $x_1, \ldots, x_n$ are the regular inputs while $u_1, \ldots, u_r$ are the random inputs. (If it is clear from the context what we mean, we will often simply use *input* instead of regular input.) Given a (regular) input $x = x_1 \cdots x_n \in \{0,1\}^n$ we randomly pick a string $u = u_1 \cdots u_r \in \{0,1\}^r$ and then determine the outcome of C on input xu. We use a uniform probability distribution on $\{0,1\}^r$. For each $x \in \{0,1\}^n$ the output of C thus becomes a random variable with range $\{0,1\}$. For $a \in \{0,1\}$,

$$\text{prob}_{u \in \{0,1\}^r}\left[C(xu) = a\right]$$

denotes the probability that C with regular input x produces outcome a. Given a language $A \subseteq \{0,1\}^n$, the *error probability* of C on x with respect to A is given by $\text{prob}_{u \in \{0,1\}^r}\left[C(xu) \neq c_A(x)\right]$.

Given a family $C = (C_n)_{n \in \mathbb{N}}$ of probabilistic circuits where C_n uses $r(n)$ random bits, and a language $A \subseteq \{0,1\}^*$, A is said to be *accepted* by C if there are $k, n_0 \in \mathbb{N}$ such that for all $n \geq n_0$ and all $x \in \{0,1\}^n$, the error probability of C_n on x with respect to $A^n =_{\text{def}} A \cap \{0,1\}^n$ is bounded above by n^{-k}.

For explicitness we will sometimes use the term *deterministic circuit* to refer to standard, i.e., non-probabilistic, circuits.

Constant-depth circuits can be simulated by depth-3 probabilistic circuits of a particular form, as we prove in the following theorem. The term "depth reduction" was coined in [AH94] to refer to this phenomenon.

Definition 3.19. A MOD_p-$\wedge$-*circuit* with n inputs is a depth-3 circuit C over $\mathcal{B}_1(p)$ of the following form:

- The output gate of C is a MOD_p gate.
- The predecessors to the MOD_p gate are $\wedge$ gates.
- The predecessors to the $\wedge$ gates are either circuit inputs or negations of circuit inputs.

C is of *order* r if all the $\wedge$ gates are of fan-in at most r.

A probabilistic MOD_p-$\wedge$-circuit is of the same form as above, but now we distinguish between regular and random inputs. Acceptance is defined as for general probabilistic circuits.

Say that a circuit C is *layered* if the following holds: The set of gates of C can be partitioned into a number of so-called *levels* $L_0, \ldots, L_d$. In level L_0 we have only input gates and negation gates whose predecessors are input gates. For $i = 1, 2, \ldots, d$, the following holds:

- All gates in L_i are of the same type.
- The predecessors of all gates in L_i are in L_{i-1}.

If $v \in L_i$ then we also say that v is *of level* i.

All MOD_p-$\wedge$-circuits are layered by definition, but circuits of a more general form also have layered equivalents, see Exercise 3.10.

Theorem 3.20. *Let p be prime, and suppose $A \in \{0,1\}^*$ is accepted by a circuit family over $\mathcal{B}_1(p)$ of polynomial size and constant depth. For every k there is a family $C = (C_n)_{n \in \mathbb{N}}$ of probabilistic MOD_p-$\wedge$-circuits of size $2^{(\log n)^{O(1)}}$, order $(\log n)^{O(1)}$, using $n^{O(1)}$ random bits, that accepts A with error probability at most n^{-k} for inputs of length n.*

Proof. Let $A \in \{0,1\}^*$ be accepted by circuit family $\mathcal{D} = (D_n)_{n \in \mathbb{N}}$. We construct $C = (C_n)_{n \in \mathbb{N}}$ with the claimed properties in a sequence of four transformations.

Fix an input length n, and consider D_n.

Step 1: Replace all $\wedge$ gates by $\vee$ and $\neg$ gates using de Morgan's laws.

Step 2: Now our circuit has only $\vee$, $\neg$, and MOD_p gates. Replace all $\neg$ gates by mod_p^1 gates. The resulting circuit is equivalent to D_n since for $x \in \{0,1\}$ certainly $\neg x \iff x \equiv 0 \pmod{p}$.

Step 3: Now our circuit has only $\vee$ and MOD_p gates. The fan-in of all gates may be polynomial in n. Going from the input gates to the output gate, replace all $\vee$ gates as follows:

Let $g = z_1 \vee \cdots \vee z_m$ be a gate in the circuit such that all gates on paths from circuit inputs to this gate have already been treated. For $1 \leq i \leq m$ define $\hat{z}_i = z_i \wedge u_i$ where u_i is a new random bit. Let $d(z_1, \ldots, z_m) = \text{mod}_p^m(\hat{z}_1, \ldots, \hat{z}_m)$.

Then the following holds:

- If $g(z_1, \ldots, z_m) = 0$ then $\hat{z}_1 = \cdots = \hat{z}_m = 0$, hence $d(z_1, \ldots, z_m) = 1$, independent of the random inputs.
- If $g(z_1, \ldots, z_m) = 1$, then let, without loss of generality, $z_1 = \cdots = z_e = 1$, $z_{e+1} = \cdots = z_m = 0$. Certainly $\hat{z}_{e+1} = \cdots = \hat{z}_m = 0$, and $d(z_1, \ldots, z_m) = 1$ if and only if the random bits $u_1, \ldots, u_e$ are chosen in such a way that among the $\hat{z}_1, \ldots, \hat{z}_e$ the number of 1's is divisible by p. For the case $p = 2$ this happens with probability

$$\frac{\left|\left\{ u_1, \ldots, u_e \in \{0,1\} \mid \sum_{i=1}^{e} u_i \equiv 0 \pmod{2} \right\}\right|}{2^e} = \frac{1}{2},$$

and for $p > 2$ the probability cannot be higher.

These considerations show that d computes $\neg g$ with error probability at most $\frac{1}{2}$.

Now define $h(z_1, \ldots, z_m)$ to be a mod_p^1 gate, applied to the conjunction of $l \cdot \log n$ many different copies of the above gate d, where for each copy we pick different random inputs (l will be determined below.) Then the following holds:

$$g(z_1, \ldots, z_m) = 0 \implies h(z_1, \ldots, z_m) = 0$$
$$g(z_1, \ldots, z_m) = 1 \implies h(z_1, \ldots, z_m) \neq 1$$

$$\text{with probability} \leq \left(\tfrac{1}{2}\right)^{l \cdot \log n} = \tfrac{1}{n^l}.$$

If we have to treat altogether n^{k_0} gates in step 3, then we choose $l = k + k_0$. Since we use new random bits for each replacement of an $\vee$ gate, the error probability of the whole construction is then bounded by $n^{k_0} \cdot n^{-l} = n^{-k}$.

Step 4: Now our circuit has only $\wedge$ and MOD_p gates. Observe that the $\wedge$ gates introduced in Step 3 have their fan-in bounded by $O(\log n)$. By repeatedly applying the operations

▷ swap an "AND-of-MODs" to a "MOD-of-ANDs", and

▷ merge connected $\wedge$ and MOD_p gates,

the circuit is now transformed into the required shape. These two operations are described in detail as follows. The fan-in of $\wedge$ gates will increase up to $(\log n)^{O(1)}$ and the size of the circuit will become $2^{(\log n)^{O(1)}}$.

First we turn our attention to the *swap* operation. Let $g = \wedge(g_1, \ldots, g_q)$ where $q = (\log n)^{O(1)}$, and $g_i = \operatorname{mod}_p^{n_i}(z_{i,1}, \ldots, z_{i,n_i})$, $n_i = n^{O(1)}$. (In the beginning of course we must have $q = O(\log n)$, but as we will see, the possible fan-in of $\wedge$ gates will increase during the construction. Therefore we prove the result directly for values of q bounded by a function from $(\log n)^{O(1)}$.) First we want to bring the MOD$_p$ operation "on top." This is achieved as follows:

By Fermat's Theorem (see Appendix A7), $1 - a^{p-1} \equiv 1 \pmod{p}$ if $a \equiv 0 \pmod{p}$, and $1 - a^{p-1} \equiv 0 \pmod{p}$ if $a \not\equiv 0 \pmod{p}$. Hence,

$$g = 1 \iff \bigwedge_{i=1}^{q} \left[\sum_{j=1}^{n_i} z_{i,j} \equiv 0 \pmod{p} \right]$$

$$\iff \bigwedge_{i=1}^{q} \left[1 - \left(\sum_{j=1}^{n_i} z_{i,j} \right)^{p-1} \equiv 1 \pmod{p} \right]$$

$$\iff \underbrace{1 - \prod_{i=1}^{q} \left[1 - \left(\sum_{j=1}^{n_i} z_{i,j} \right)^{p-1} \right]}_{(*)} \equiv 0 \pmod{p}.$$

The question now is how to implement the power of the sum below the MOD$_p$ gate. Fix an i, $1 \leq i \leq q$, and let $\bar{z}_1, \bar{z}_2, \bar{z}_3, \ldots, \bar{z}_{n_i^{p-1}}$ be an enumeration of all $(p-1)$-tuples of the variables $z_{i,1}, \ldots, z_{i,n_i}$. For such a tuple $\bar{z}$ let $\wedge\bar{z}$ be the conjunction of all gates in $\bar{z}$, i.e., if $\bar{z} = (z_{i,k_1}, \ldots, z_{i,k_{p-1}})$ then $\wedge\bar{z} = z_{i,k_1} \wedge \cdots \wedge z_{i,k_{p-1}}$. If now $\sum_{j=1}^{n_i} z_{i,j} = e$ then there are e^{p-1} tuples $\bar{z}$ out of the n_i^{p-1} tuples above for which we have $\wedge\bar{z} = 1$. Hence,

$$\left(\sum_{j=1}^{n_i} z_{i,j} \right)^{p-1} = \sum_{j=1}^{n_i^{p-1}} \wedge\bar{z}_j.$$

Using this identity we can transform the term $(*)$ above as follows, where for clarity we only give the principal structure of the expressions:

$$1 - \prod_{(\log n)^{O(1)}} \left[1 - \left(\sum_{n^{O(1)}} (\cdots) \right)^{p-1} \right] = 1 - \prod_{(\log n)^{O(1)}} \left(1 - \sum_{n^{O(1)}} \prod_{O(1)} (\cdots) \right).$$

Using the distributive laws we now multiply out and bring the product of sums into the form sum of products; hence we obtain a term of the form:

$$1 - \left(\sum_{n^{(\log n)^{O(1)}}} c_l \prod_{(\log n)^{O(1)}} (\cdots) \right).$$

Observe that since we calculate modulo p, all the coefficients c_l are in $\{0, \ldots, p-1\}$. We can now replace each c_l by its additive inverse modulo p, and then we get the following expression equivalent to $(*)$:

$$\sum_{2^{(\log n)^{O(1)}}} c_l' \prod_{(\log n)^{O(1)}} (\cdots).$$

Gate g evaluates to 1 if and only if this latter expression gives a value which is divisible by p. Hence we can replace g by a MOD_p gate of fan-in $2^{(\log n)^{O(1)}}$, whose predecessors are $\wedge$ gates of fan-in $(\log n)^{O(1)}$. This concludes our discussion of the swap transformation.

Our second task is to *merge* connected gates of the same type. For $\wedge$ gates this is obvious: A conjunction of $(\log n)^{O(1)}$ many $\wedge$ gates where all gates have fan-in $(\log n)^{O(1)}$ can immediately be replaced by one $\wedge$ gate of fan-in $(\log n)^{O(1)}$. Suppose now that $g = \mathrm{mod}_p^q(g_1, \ldots, g_q)$, where $g_i = \mathrm{mod}_p^{n_i}(z_{i,1}, \ldots, z_{i,n_i})$ for $1 \leq i \leq q$ $(q, n_1, \ldots, n_q \in n^{O(1)})$. Again by Fermat's Theorem, we obtain:

$$g = 1 \iff \sum_{i=1}^{q} \left[1 - \left(\sum_{j=1}^{n_i} z_{i,j} \right)^{p-1} \right] \equiv 0 \pmod{p}.$$

We compute the power of the sum as a sum of $\wedge$ gates of constant fan-in as for the swap operation above. Also, we can eliminate the subtraction as for the swap operation above. Hence we can replace g by one MOD_p gate whose predecessors are $\wedge$ gates of constant fan-in.

Steps 1 to 4 transform D_n into a circuit with one MOD_p gate on top, whose fan-in is $2^{(\log n)^{O(1)}}$ and whose predecessors are $\wedge$ gates. Observe that before we start to swap and merge, the $\wedge$ gates have logarithmic fan-in. Both operations keep this fan-in polylogarithmic. Hence the theorem follows. $\square$

The theorem above shows how we can simulate any constant-depth circuit family $\mathcal{D}$ by a *probabilistic* family $\mathcal{C}$ of depth 3. Next we want to develop a deterministic circuit family $\mathcal{D}'$ of depth 3 which *approximates* the original family. Approximation here means that $\mathcal{D}'$ behaves as $\mathcal{D}$ for "most" inputs.

First we prove how a probabilistic circuit family generally yields a deterministic approximation.

Theorem 3.21. *Let $\mathcal{C} = (C_n)_{n \in \mathbb{N}}$ be a probabilistic circuit family over basis B with r random bits, of size s and depth d, which accepts a language $A \subseteq \{0,1\}^*$ with error probability at most $e(n)$ for inputs of length n. Then there* *is a deterministic circuit family $\mathcal{D} = (D_n)_{n \in \mathbb{N}}$ over $B \cup \{0^0, 1^0\}$ (the 0-ary constants) of size $s+r$ and depth d, which for every n computes $c_A(x)$ correctly for at least $2^n(1 - e(n))$ inputs x of length n.*

Proof. Fix an input length n, an enumeration $x_1, x_2, \ldots, x_{2^n}$ of all inputs of length n, and an enumeration $u_1, u_2, \ldots, u_{2^{r(n)}}$ of all random inputs of length $r(n)$. Define the matrix $M = (m_{i,j})_{\substack{1 \le i \le 2^{r(n)} \\ 1 \le j \le 2^n}}$ as follows:

$$m_{i,j} = 1 \text{ if and only if } C_n \text{ on input } x_j u_i \text{ computes } c_A(x_j).$$

The following property holds: For every j the probability that, for a randomly chosen $u \in \{0,1\}^{r(n)}$, we have $f_C(x_j u) \ne c_A(x)$ is at most $e(n)$, i.e.,

$$\frac{\left|\{ i \mid 1 \le i \le 2^{r(n)} \wedge m_{i,j} = 0 \}\right|}{2^{r(n)}} \le e(n),$$

in other words, the maximal number of zeroes in a column is bounded by $m_2 =_{\text{def}} 2^{r(n)} e(n)$. Let m_0 be the minimal number of zeroes in a row of M and m_1 be the number of all zeroes in M. Then $2^{r(n)} \cdot m_0 \le m_1 \le 2^n \cdot m_2$, hence $m_0 \le 2^n e(n)$. Pick an i_0, $1 \le i_0 \le 2^{r(n)}$ such that $\left|\{ j \mid 1 \le j \le 2^n \wedge m_{i_0, j} = 0 \}\right| = m_0$. Then u_{i_0} is an "optimal" random string, since it errs for only m_0 inputs. Obtain the circuit D_n from C_n by hardwiring u_{i_0} into the circuit instead of the random inputs, i.e., $f_{D_n}(x) = f_{C_n}(x u_{i_0})$ for all $x \in \{0,1\}^n$. D_n will work correctly for at least $2^n - m_0 \ge 2^n(1 - e(n))$ inputs of length n. $\Box$

Corollary 3.22. *Let p be prime, and let A be a language from the class* $\text{SIZE-DEPTH}_{\mathcal{B}_1(p)}(n^{O(1)}, 1)$. *For every k there is a family $C = (C_n)_{n \in \mathbb{N}}$ of* $\text{MOD}_p\text{-}\wedge\text{-circuits of size } 2^{(\log n)^{O(1)}}$ *and order $(\log n)^{O(1)}$ such that for every n, C_n computes c_A correctly for at least $2^n(1 - n^{-k})$ inputs of length n.*

Proof. Immediate from Theorems 3.20 and 3.21. Observe that in the particular case of a $\text{MOD}_p\text{-}\wedge$-circuit C, in order to hardwire the optimal random string u_{i_0} determined in the proof of Theorem 3.21 into C, we do not need gates for the constants. If a constant 1 has to be input into an $\wedge$ gate, we can safely ignore this. If a constant 0 has to be input into an $\wedge$ gate, we may delete the whole gate. Thus here we get circuits over basis $\mathcal{B}_1(p)$. $\Box$

Remarkably we can even state that every probabilistic circuit family can be simulated by a deterministic family without error. We only have to accept a very small increase of the depth.

Theorem 3.23. *Let $A \subseteq \{0,1\}^*$ such that for every k there is a probabilistic circuit family over basis B of size s and depth d, that accepts A with error probability n^{-k} for inputs of length n. Then there is a family of deterministic circuits over basis $B \cup \mathcal{B}_1 \cup \{0^0, 1^0\}$ of size $s^{O(1)}$ and depth $d + 2$ that accepts A.*

Proof. Pick k such that $\frac{1}{n^k} < \frac{1}{2n^2}$ for large n. Let $\mathcal{D} = (D_n)_{n \in \mathbb{N}}$ be a probabilistic circuit family that accepts A with error probability n^{-k} for inputs of length n. For every n, define D'_n as follows: The output gate of D'_n is an $\wedge$ gate of fan-in n^2. As predecessors of this gate we take n^2 different copies of D_n, where for every copy we use new random input bits. Now,

- if $x \notin A$ then $f_{D'_n}(x) = 1$ iff all the copies of D_n output 1 on input x, hence the error probability is less than $\left(\frac{1}{2n^2}\right)^{n^2} \le \left(\frac{1}{2}\right)^{n^2} = \frac{1}{2^{n^2}}$;
- if $x \in A$ then $f_{D'_n}(x) = 0$ iff at least one copy of D_n outputs 0 on input x, hence the error probability in this case is less than $\frac{n^2}{2n^2} = \frac{1}{2}$.

Let C'_n be a disjunction of n copies of D'_n, each with different random bits. Now,

- if $x \notin A$ then $f_{C'_n}(x) = 1$ iff at least one copy of D'_n outputs 1, hence the error probability is less than $\frac{n}{2^{n^2}}$; hence there is some $n_0 \in \mathbb{N}$ such that the error probability is less than $\frac{1}{2^n}$ for all $n \ge n_0$;
- if $x \in A$ then $f_{C'_n}(x) = 0$ iff all copies of D'_n output 0, hence the error probability is less than $\frac{1}{2^n}$.

Hence $\mathcal{C}'$ accepts A with error probability (strictly) less than $\frac{1}{2^n}$.

Theorem 3.21 now tells us that there is a deterministic circuit family $\mathcal{C} = (C_n)_{n \in \mathbb{N}}$ which for every n computes c_A correctly for more than $2^n \left(1 - \frac{1}{2^n}\right)$ inputs of length n. Hence $\mathcal{C}$ accepts A. The size of C_n is clearly $s^{O(1)}$, and the depth is $d + 2$. $\square$

Corollary 3.24. *Let p be prime, and let A be a language from the class* SIZE-DEPTH$_{\mathcal{B}_1(p)}(n^{O(1)}, 1)$. *Then there is a family $\mathcal{C} = (C_n)_{n \in \mathbb{N}}$ of circuits over $\mathcal{B}_1(p)$ of size $2^{(\log n)^{O(1)}}$ and depth 5 that accepts A.*

Proof. Immediate from Theorems 3.20 and 3.23. Constants are implemented as in the proof of Corollary 3.22. $\square$

It should be remarked that the deterministic family produced in Theorem 3.23 from a *uniform* probabilistic circuit family is not necessarily uniform under the same condition. The reason is that to find the optimal random string u_{i_0} in the proof of Theorem 3.21 can be very complex (though it is certainly always computable). If we perform the depth reduction of Corollary 3.24 we will therefore also arrive at a weaker uniformity condition.

3.3.2 Approximating Circuits by Polynomials

In Theorem 3.10 we proved a close connection between perceptrons and polynomials over the ring of the integers. Above we considered circuits with a MOD$_p$ gate instead of a threshold gate on top. As one might expect, these circuits also have a correspondence in terms of polynomials, this time over the finite field GF(p).

Theorem 3.25. *Let $A \subseteq \{0,1\}^n$, and let p be prime. If there is a MOD_p-$\wedge$-circuit of order d that accepts A, then there exists a multi-linear polynomial q over $\text{GF}(p)$ in n variables of degree $(p-1)d$ such that for all $x = x_1 \cdots x_n \in \{0,1\}^n$ we have*

$$c_A(x_1 \cdots x_n) = q(x_1, \ldots, x_n).$$

Proof. The proof is very similar to the proof of Theorem 3.10. Let C be the circuit that accepts A. If g is an $\wedge$ gate in C, and $1 \le i \le n$, define

- $t_{g,i} = 1$, if neither x_i nor $\neg x_i$ are predecessors to g;
- $t_{g,i} = 0$, if both x_i and $\neg x_i$ are predecessors to g;
- $t_{g,i} = x_i$, if x_i but not $\neg x_i$ is a predecessor to g;
- $t_{g,i} = (1 - x_i)$, if $\neg x_i$ but not x_i is a predecessor to g.

Let $t_g(x_1, \ldots, x_n) =_{\text{def}} \prod_{i=1}^n t_{g,i}$, $t =_{\text{def}} \sum_{\substack{g \in C \\ g \text{ is } \wedge \text{ gate}}} t_g$. Finally let

$$q(x_1, \ldots, x_n) =_{\text{def}} 1 - t_g(x_1, \ldots, x_n)^{p-1}.$$

For every input $x = x_1 \cdots x_n$, $t_g(x_1, \ldots, x_n)$ will clearly be the value computed by gate g. Hence C computes value 1 if and only if $t(x_1, \ldots, x_n) \equiv 0 \pmod{p}$. Thus we conclude: $f_C(x_1, \ldots, x_n) = q(x_1, \ldots, x_n)$. The degree of t is bounded by the fan-in of $\wedge$ gates in C, and the degree of q is $p-1$ times the degree of t. □

Corollary 3.26. *Let $A \in \text{SIZE-DEPTH}_{B_1(p)}(n^{O(1)}, 1)$. For every n and k there is a multi-linear polynomial q in n variables over $\text{GF}(p)$ of degree $(\log n)^{O(1)}$, such that for at least $2^n(1 - n^{-k})$ inputs $x_1 x_2 \cdots x_n$ of length n we have:*

$$c_A(x_1 x_2 \cdots x_n) = q(x_1, x_2, \ldots, x_n).$$

Proof. Immediate from Corollary 3.22 and Theorem 3.25. □

Thus every language $A \in \text{SIZE-DEPTH}_{B_1(p)}(n^{O(1)}, 1)$ for prime number p can be approximated by low-degree polynomials over $\text{GF}(p)$. This very strong upper bound for $\text{SIZE-DEPTH}_{B_1(p)}(n^{O(1)}, 1)$ will lead below to the result that MOD_q cannot be computed by such circuits if p and q are relatively prime.

3.3.3 Smolensky's Theorem

One of the most successful applications of the polynomial method is Smolensky's Theorem, proving that $\text{MOD}_r \notin \text{SIZE-DEPTH}_{B_1(p)}(n^{O(1)}, 1)$ if p and r are relatively prime. By Corollary 3.26 it is sufficient to show that MOD_r cannot be well approximated by low degree polynomials over $\text{GF}(p)$. We prove this below, but first we need two lemmas.

We start with a combinatorial result.

Lemma 3.27. *There is an $n_0 \geq 0$ such that for all $n > n_0$, the number of multi-linear monomials in n variables of degree at most $\frac{n}{2} + \frac{\sqrt{n}}{2}$ is bounded by $\frac{9}{10} \cdot 2^n$.*

Proof. A multi-linear monomial of degree exactly l consists of l variables from the set $\{x_1, \ldots, x_n\}$. Thus there are $\binom{n}{l}$ such monomials. Hence to prove the lemma, we have to show:

$$\sum_{l=0}^{\lfloor \frac{n}{2} + \frac{\sqrt{n}}{2} \rfloor} \binom{n}{l} \leq \frac{9}{10} \cdot 2^n.$$

We first want to give an upper bound for the quantity $\binom{n}{\lceil \frac{n}{2} \rceil}$.

Let $S(n) =_{\text{def}} \left(\frac{n}{e}\right)^n \sqrt{2\pi n}$. By Stirling's formula, $n! \approx S(n)$, more precisely, $n! = S(n) \cdot \left(1 + \frac{1}{12n} + \frac{1}{288n^2} + \cdots\right)$. If we let $\epsilon(n) =_{\text{def}} \frac{n!}{S(n)}$, i.e., $n! = S(n) \cdot \epsilon(n)$, then we have that $\epsilon(i) < \epsilon(j)$ whenever $i > j$.

First, let n be even. Then we have in particular $\epsilon\left(\frac{n}{2}\right) > \epsilon(n) > 1$. Hence,

$$\binom{n}{\lceil \frac{n}{2} \rceil} = \binom{n}{\frac{n}{2}} = \frac{n!}{\frac{n}{2}! \cdot \frac{n}{2}!}$$

$$= \frac{S(n)\epsilon(n)}{S\left(\frac{n}{2}\right)\epsilon\left(\frac{n}{2}\right) \cdot S\left(\frac{n}{2}\right)\epsilon\left(\frac{n}{2}\right)} < \frac{S(n)}{S\left(\frac{n}{2}\right) \cdot S\left(\frac{n}{2}\right)}$$

$$= \frac{\frac{n^n}{e^n}\sqrt{2\pi n}}{\frac{\left(\frac{n}{2}\right)^{\frac{n}{2}}}{e^{\frac{n}{2}}}\sqrt{2\pi\frac{n}{2}} \cdot \frac{\left(\frac{n}{2}\right)^{\frac{n}{2}}}{e^{\frac{n}{2}}}\sqrt{2\pi\frac{n}{2}}} = 2^n \cdot \sqrt{\frac{2}{\pi n}}$$

$$= \frac{2^n}{\sqrt{\pi\frac{n}{2}}}.$$

If n is odd, then

$$\binom{n}{\lceil \frac{n}{2} \rceil} = \frac{n!}{\lceil \frac{n}{2} \rceil! \cdot \lfloor \frac{n}{2} \rfloor!} = \frac{n!}{\frac{n+1}{2}! \cdot \frac{n-1}{2}!} = \frac{1}{2} \cdot \frac{(n+1)!}{\frac{n+1}{2}! \cdot \frac{n+1}{2}!}$$

Observe that now $n + 1$ is even. Hence by the above we continue:

$$\binom{n}{\lceil \frac{n}{2} \rceil} < \frac{1}{2} \cdot 2^{n+1} \sqrt{\frac{2}{\pi(n+1)}} = \frac{2^n}{\sqrt{\pi\frac{n+1}{2}}}$$

In both cases we thus have

$$\binom{n}{\lceil \frac{n}{2} \rceil} \leq \frac{2^n}{\sqrt{\pi\lceil \frac{n}{2} \rceil}} \qquad\qquad (i)$$

Furthermore, since $\sum_{l=0}^{n} \binom{n}{l} = 2^n$, we have

$$\sum_{l=0}^{\lceil \frac{n}{2} \rceil - 1} \binom{n}{l} \leq 2^{n-1} \qquad (ii)$$

and if $\lfloor \frac{n}{2} \rfloor < l_0 < l_1$, then

$$\binom{n}{l_0} > \binom{n}{l_1}. \qquad (iii)$$

From (i)–(iii) we now conclude:

$$\sum_{l=0}^{\lfloor \frac{n}{2} + \frac{\sqrt{n}}{2} \rfloor} \binom{n}{l} \leq 2^{n-1} + \left(\frac{\sqrt{n}}{2} + 1 \right) \cdot \frac{2^n}{\sqrt{\pi \lceil \frac{n}{2} \rceil}}$$

$$\leq 2^n \left(\frac{1}{2} + \frac{1}{\sqrt{2\pi}} + \frac{1}{\sqrt{\pi \frac{n}{2}}} \right)$$

Clearly $\frac{1}{\sqrt{\pi \frac{n}{2}}} \to 0$ as $n \to \infty$, hence there is an $n_0 \geq 0$ such that for all $n \geq n_0$,

$$\sum_{l=0}^{\lfloor \frac{n}{2} + \frac{\sqrt{n}}{2} \rfloor} \binom{n}{l} \leq 2^n \cdot \frac{9}{10}.$$

□

The following result is well known from field theory, see, e.g., [Jac85, pp. 229ff.]:

Lemma 3.28. *Let p be prime, $r \neq 1$ be relatively prime to p. Then there is a $k > 0$ such that in $\mathrm{GF}(p^k)$ there is an element $\omega \neq 1$ for which $\omega^r = 1$.*

Proof. Let q be a polynomial in one variable over a field $\mathcal{K}$ such that q has a multiple root a. Then there is a polynomial $\hat{q}$ over some field extending $\mathcal{K}$ such that $q(x) = (x - a)^2 \hat{q}(x)$. The derivative of q is then $q'(x) = 2(x - a)\hat{q}(x) + (x - a)^2 \hat{q}'(x)$, and a is a root of q' as well. Hence we have shown: For any polynomial q over a field $\mathcal{K}$ with a multiple root, the derivative of q has the same root.

Now consider the polynomial $q(x) = x^r - 1$ over $\mathrm{GF}(p)$. Then $q'(x) = rx^{r-1}$. Since p and r are relatively prime, $q'(x) = 0$ only if $x^{r-1} = 0$. Hence the only root of q' is $x = 0$ (a field possesses no zero-divisors). Since this is not a root of q we conclude that q has no multiple zeroes. Thus there is a $k \geq 1$ such that in $\mathrm{GF}(p^k)$, q factors into r (pairwise different) linear factors. Therefore we conclude q has r different roots. Let ω be one of them which is not the identity. □

We now define the following generalizations of the function MOD_r:

Definition 3.29. For $r \geq 1$ and $0 \leq i < r$, let $\mathrm{MOD}_{r,i} = \left(\mathrm{mod}_{r,i}^n\right)_{n \in \mathbb{N}}$ be defined for $x = x_1 \cdots x_n \in \{0,1\}^n$ by

$$\mathrm{mod}_{r,i}^n(x) =_{\mathrm{def}} \begin{cases} 1, & \text{if } x_1 + \cdots + x_n \equiv i \pmod{r}, \\ 0, & \text{otherwise.} \end{cases}$$

The main step to prove the lower bound for MOD_r now is the following non-approximability result.

Main Lemma 3.30. *Let p be prime and $r \neq 1$ be relatively prime to p. Let $k \in \mathbb{N}$ be large enough that $\mathrm{GF}(p^k)$ has an element $\omega \neq 1$ for which $\omega^r = 1$. Then there is an $n_0 \in \mathbb{N}$ such that for all $n > n_0$ the following holds: Let $F_0, F_1, \ldots, F_{r-1}$ be multi-linear polynomials in n variables over $\mathrm{GF}(p^k)$ of degree at most $\sqrt{n}$. Then there are at least $\frac{2^n}{10}$ words $x \in \{0,1\}^n$, for which there is an i, $0 \leq i < r$ such that $F_i(x) \neq \mathrm{MOD}_{r,i}(x)$.*

Proof. Let n_0 be as in Lemma 3.27. Fix an $n > n_0$. In the sequel of the proof, $x = x_1 \cdots x_n$ will always be a vector of Boolean values, and $y = y_1 \cdots y_n$ will always be a vector over $\{1, \omega\}$. Define

$$A =_{\mathrm{def}} \left\{ x \in \{0,1\}^n \mid F_i(x) = \mathrm{MOD}_{r,i}(x) \text{ for all } i, 0 \leq i < r \right\}.$$

A is the set of those inputs for which all the $F_0, F_1, \ldots, F_{r-1}$ produce correct results. We want to show that $|A| \leq \frac{9}{10} 2^n$.

For all $x \in \{0,1\}^n$, let

$$F(x) =_{\mathrm{def}} \sum_{i=0}^{r-1} \omega^i \cdot F_i(x).$$

Given any $x \in A$, choose l, $0 \leq l < r$, and b such that $\sum_{j=1}^n x_j \equiv l \pmod{r}$, more specifically: $\sum_{j=1}^n x_j = l + b \cdot r$. Then $\omega^{\sum_{j=1}^n x_j} = \omega^l$. By definition of the $\mathrm{MOD}_{r,i}$, we have $\mathrm{MOD}_{r,l}(x) = 1$ and $\mathrm{MOD}_{r,i}(x) = 0$ for $i \neq l$. From the definition of A we conclude $\mathrm{MOD}_{r,i}(x) = F_i(x)$ for all i, $0 \leq i < r$. Hence by definition of F, we get

$$F(x) = \omega^l = \omega^{\sum_{j=0}^n x_j} \text{ for all } x \in A. \tag{3.1}$$

Now define

$$\tilde{A} =_{\mathrm{def}} \left\{ y \in \{1, \omega\}^n \mid \left(\tfrac{y_1 - 1}{\omega - 1}, \cdots, \tfrac{y_n - 1}{\omega - 1}\right) \in A \right\}.$$

If we take the representation of Boolean values $c: \{0,1\} \to \mathrm{GF}(p^k)$, defined by $c(0) = 1$, $c(1) = \omega$, then $c(A) = \tilde{A}$. Observe that $c^{-1}(\zeta) = \frac{\zeta - 1}{\omega - 1}$ for $\zeta \in \{1, \omega\}$. Define

$$\tilde{F}(y) =_{\text{def}} F\left(c^{-1}(y_1), \ldots, c^{-1}(y_n)\right).$$

For $y \in \tilde{A}$, we have $\tilde{F}(y) = \omega^{\sum_{j=1}^{n} c^{-1}(y_j)} = \omega^{|\{j \mid y_j = \omega\}|}$ because of (3.1), hence

$$\tilde{F}(y) = \prod_{j=1}^{n} y_j \quad \text{for all } y \in \tilde{A}. \tag{3.2}$$

Each F_i by assumption is of degree at most $\sqrt{n}$, hence also F and $\tilde{F}$ are of degree at most $\sqrt{n}$.

Let $\mathrm{GF}(p^k)^{\tilde{A}}$ be the set of all functions from $\tilde{A}$ to $\mathrm{GF}(p^k)$, and let $\mathbf{G}$ be the set of all multi-linear polynomials over $\mathrm{GF}(p^k)$ of degree at most $\frac{n}{2} + \frac{\sqrt{n}}{2}$.

Let us suppose for the moment that $\left|\mathrm{GF}(p^k)^{\tilde{A}}\right| \leq |\mathbf{G}|$. Then we can argue as follows: Clearly $\left|\mathrm{GF}(p^k)^{\tilde{A}}\right| = |\mathrm{GF}(p^k)|^{|\tilde{A}|}$. Any polynomial from $\mathbf{G}$ is determined by giving one coefficient from $\mathrm{GF}(p^k)$ for each possible monomial. The number of these monomials is bounded by $\frac{9}{10}2^n$ by Lemma 3.27. Thus $|\mathbf{G}| \leq |\mathrm{GF}(p^k)|^{\frac{9}{10}2^n}$. Combining this with the above yields $|\tilde{A}| \leq \frac{9}{10}2^n$, and the lemma is proven.

It remains to show $\left|\mathrm{GF}(p^k)^{\tilde{A}}\right| \leq |\mathbf{G}|$. We will construct a one-to-one mapping from $\mathrm{GF}(p^k)^{\tilde{A}}$ to $\mathbf{G}$. Let $g \in \mathrm{GF}(p^k)^{\tilde{A}}$, $g\colon \tilde{A} \to \mathrm{GF}(p^k)$. Since g is a finite function we can write g as a multi-linear polynomial using Lagrange interpolation as follows: The condition $[\xi = y]$ for fixed $\xi = \xi_1 \cdots \xi_n$ can be expressed by the polynomial

$$[\xi = y] = \prod_{j=1}^{n} \begin{cases} c^{-1}(y_j), & \text{if } \xi_j = \omega, \\ 1 - c^{-1}(y_j), & \text{otherwise.} \end{cases}$$

If we now define

$$\hat{g}(y) = \sum_{\xi \in \tilde{A}} g(\xi) \cdot [\xi = y],$$

then $\hat{g}(y) = g(y)$ for all $y \in \tilde{A}$. The set $\tilde{A}$ is finite, hence $\hat{g}$ is a multi-linear polynomial. However, the degree of $\hat{g}$ can be large (any integer value $\leq n$).

If $\hat{g} \notin \mathbf{G}$, let $m(y) = a \cdot \prod_{i \in I} y_i$ for $a \in \mathrm{GF}(p^k)$, $I \subseteq \{1, \ldots, n\}$ be a monomial in $\hat{g}$ with degree too high, i.e., $|I| > \frac{n}{2} + \frac{\sqrt{n}}{2}$. Define

$$m'(y) =_{\text{def}} a \cdot \prod_{j=1}^{n} y_j \prod_{i \in \bar{I}} \frac{\omega + 1 - y_i}{\omega},$$

where $\bar{I} = \{1, \ldots, n\} \setminus I$. Observe that for $\zeta \in \{1, \omega\}$, $\frac{\omega + 1 - \zeta}{\omega}$ is the multiplicative inverse of ζ, hence $\zeta \cdot \frac{\omega + 1 - \zeta}{\omega} = 1$, and thus $m(y) = m'(y)$ for all $y \in \tilde{A}$. If we now define

$$\tilde{m}(y) = a \cdot \tilde{F}(y) \cdot \prod_{i \in \bar{I}} \frac{\omega + 1 - y_i}{\omega},$$

then it follows from (3.2) that $\tilde{m}(y) = m'(y)$ for $y \in \tilde{A}$. $\tilde{F}$ is a polynomial, hence $\tilde{m}$ is obtained from $\tilde{F}$ by multiplying each monomial in $\tilde{F}$ by $\prod_{i \in \bar{I}} \frac{\omega + 1 - y_i}{\omega}$. If it then happens that a variable y_i occurs both in the original monomial and in the product on the right, then we may delete both y_i and $\frac{\omega + 1 - y_i}{\omega}$, since their product is 1. This shows that $\tilde{m}$ is multi-linear. However the degree of $\tilde{m}$ is at most $\sqrt{n} + n - |I| \leq \sqrt{n} + n - \frac{n}{2} - \frac{\sqrt{n}}{2} = \frac{n}{2} + \frac{\sqrt{n}}{2}$.

Now obtain $\tilde{g}$ from $\hat{g}$ by replacing each monomial m with degree too high by the corresponding $\tilde{m}$. Then $\tilde{g} \in \mathbf{G}$. Since $g(y) = \tilde{g}(y)$ for all $y \in \tilde{A}$ it follows that the correspondence between $g \in GF(p^k)^{\tilde{A}}$ and $\tilde{g} \in \mathbf{G}$ is one-to-one. This proves $\left| GF(p^k)^{\tilde{A}} \right| \leq |\mathbf{G}|$. $\qquad\square$

Now we prove the main result of this section:

Theorem 3.31 (Smolensky). *Let p be prime and $r > 1$ be relatively prime to p. Then $\mathrm{MOD}_r \notin \mathrm{SIZE\text{-}DEPTH}_{\mathcal{B}_1(p)}(n^{O(1)}, 1)$.*

Proof. Let k be as in Lemma 3.28. Suppose for contradiction that $\mathrm{MOD}_r \in \mathrm{SIZE\text{-}DEPTH}_{\mathcal{B}_1(p)}(n^{O(1)}, 1)$. Then by Corollary 3.26, for every n there is a multi-linear polynomial q_n in n variables over $GF(p)$ such that for $2^n(1-n^{-k})$ words $x \in \{0, 1\}^n$ we have $q_n(x) = 1 \iff \mathrm{mod}_r(x) = 0$. The degree of q_n is $(\log n)^c$ for some $c \in \mathbb{N}$. Since for every k, $GF(p)$ is isomorphic to a subfield of $GF(p^k)$ we can look at q_n as a polynomial over $GF(p^k)$.

Let now n_0 and k be as in Lemma 3.30. Pick $n_1 > n_0$ such that $2^{n_1}(1 - n_1^{-1}) \geq \frac{9}{10} 2^{n_1}$ and $(\log n_1)^c \leq \sqrt{n_1}$. Let $n > n_1$. Define $F_0, \ldots, F_{r-1}$ as follows:

$$F_0(x_1, \ldots, x_n) = q_n(x_1, \ldots, x_n)$$
$$F_i(x_1, \ldots, x_n) = q_{n+r-i}(x_1, \ldots, x_n, \underbrace{1, \ldots, 1}_{r-i}) \text{ for } 1 \leq i < r.$$

For all $n > n_1$ we now conclude that $F_i(x) = \mathrm{MOD}_{r,i}(x)$ for $2^n(1 - n^{-k})$ words $x \in \{0, 1\}^n$. This is a contradiction to Lemma 3.30. Hence $\mathrm{MOD}_r \notin \mathrm{SIZE\text{-}DEPTH}_{\mathcal{B}_1(p)}(n^{O(1)}, 1)$. $\qquad\square$

Smolensky's Theorem even allows us to conclude that MOD_r cannot be in $\mathrm{SIZE\text{-}DEPTH}_{\mathcal{B}_1(p)}(n^{O(1)}, 1)$ if r has at least one prime divisor different from p (see Exercise 3.14).

We note two immediate consequences. First Theorem 3.31 shows as a special case that all the MOD_p-functions cannot be computed over the standard unbounded fan-in basis.

Corollary 3.32. *The function $\mathrm{MOD}_p = \left(\mathrm{mod}_p^n\right)_{n \in \mathbb{N}}$ is not contained in the class $\mathrm{UnbSIZE\text{-}DEPTH}(n^{O(1)}, 1)$.*

Second, we can conclude that all the problems proved in Chap. 1 to be equivalent to the MAJ-problem cannot be contained in one of the classes SIZE-DEPTH$_{\mathcal{B}_1(p)}(n^{O(1)}, 1)$.

Corollary 3.33. *If* MAJ $\leq_{cd} A$ *for* $A \subseteq \{0,1\}^*$, *then for all prime numbers* p, $A \notin$ SIZE-DEPTH$_{\mathcal{B}_1(p)}(n^{O(1)}, 1)$.

Proof. Let $r > 1$ be relatively prime to p. Certainly MOD$_r \leq_{cd}$ MAJ (Exercise 3.12), and therefore MOD$_r \leq_{cd} A$. If $A \in$ SIZE-DEPTH$_{\mathcal{B}_1(p)}(n^{O(1)}, 1)$ then also MOD$_r \in$ SIZE-DEPTH$_{\mathcal{B}_1(p)}(n^{O(1)}, 1)$, which contradicts Smolensky's Theorem.   □

Corollary 3.34. *For all prime* p, SIZE-DEPTH$_{\mathcal{B}_1(p)}(n^{O(1)}, 1)$ *contains none of the problems* MAJ, UCOUNT, BCOUNT, ITADD, MULT, ITMULT, SORT.

Bibliographic Remarks

An excellent resource for the development of lower bounds in circuit complexity up to 1987 is Wegener's monograph [Weg87]. The elimination method is discussed there in detail with more examples, see [Weg87, Sects. 5.1–5.3]. Our examples above were originally given by [Sch74].

The result that $(\oplus^n)_{n \in \mathbf{N}}$ is not in UnbSIZE-DEPTH$(n^{O(1)}, 1)$ (Corollary 3.32) was first proved by Furst, Saxe, and Sipser in [FSS84]. Their proof relies on an approach completely different from the one presented here, the so-called *probabilistic restriction method*, which also found applications for other classes of constant-depth circuits. This method can be seen as a probabilistic version of the elimination method. The actual size bounds achieved by Furst et al. were later substantially improved [Yao85, Hås86]. All this is also discussed in Wegener's book [Weg87, Chap. 11], as well as in the monograph [Hås88]. Since Furst, Saxe, and Sipser's result is subsumed by Smolensky's Theorem (Theorem 3.31) and since a number of textbooks available today [Dun88, SP98, Imm99, CK99] contain a presentation of this result and its original proof, we decided not to cover the probabilistic restriction method. A different proof for the PARITY lower bound using methods from finite model theory was given in [Ajt83].

The special case of Corollary 3.33 that MAJ cannot be computed by constant-depth polynomial size circuits over $\mathcal{B}_1 \cup \{\oplus\}$ was given in [Raz87]. Barrington [Bar86] noticed that this can be extended to MOD$_p$ for any prime p instead of $\oplus$. The breakthrough that even MOD$_r$ cannot be computed by SIZE-DEPTH$_{\mathcal{B}_1(p)}(n^{O(1)}, 1)$ circuits if p and r are relatively prime, was given in [Smo87]. It was noticed subsequently by different authors that a famous theorem of Toda about the power of the classes PP and #P [Tod91] can be related to Smolensky's Theorem. This led to the re-formulation of the proof, which we chose to follow in this chapter. Our presentation was much inspired by [All95, Lects. 6–9]. (A presentation of the proof closer to Smolensky's original outline can be found in [CK99]; see also [Str94, Sect. VIII.3] and [SP98, Chap. 12].)

The connection between Smolensky's Theorem and Toda's paper chiefly concerns the way to achieve depth reduction, see Corollaries 3.22 and 3.24. These results were given by [AH94]. Improved upper bounds along this avenue for the class SIZE-DEPTH$_{B_1(p)}(n^{O(1)}, 1)$ were later given in [Yao90, BT94] and elsewhere; refer to the discussion in [All96].

It was shown by Jin-Yi Cai [Cai89] that the lower bounds for $\oplus$ and MOD$_r$ can be strengthened to show that these functions cannot even be approximated by constant-depth circuits over the standard unbounded fan-in basis. More precisely, Cai showed that these circuits err on about half of their inputs when computing the MOD$_r$ function.

The de-randomization of probabilistic circuits (Theorem 3.23) goes back to [Adl78] and [Gil77].

A very good survey of the polynomial method with more applications and many pointers to the literature is [Bei93]; some newer results are surveyed in [Reg97]. The results presented in Sect. 3.2 were given in [MP88, Sects. 3–4]. Our presentation follows [Bei93]. More degree lower bounds can be found in [MP88]. Deeper connections between circuits and neural networks are presented in [Par94]. A number of recent results related to polynomial representations of Boolean functions can be found in the different contributions of [MT94].

Another field of lower bounds which we do not consider in this textbook is the examination of monotone circuits, i.e., circuits over a monotone basis (a basis consisting only of monotone functions). As is still the case for unrestricted circuits, the best-known lower bounds for monotone circuit size of problems in NP had been linear for a long time until Razborov achieved a breakthrough, proving that the clique problem requires monotone circuits of superpolynomial size [Raz85]. It is known that lower bounds for this model are strongly related to lower bounds in communication complexity. For results and an extensive bibliography we refer the interested reader to the monograph [Hro97]; see also [Weg87, Dun88].

Many more results on lower bounds in circuit complexity can be found in [BS90]. However, it should be noted that the rapid development of techniques for proving lower bounds in circuit complexity in the 1980s has slowed down a little recently. We still do not know today if there is any language in NP that cannot be accepted by polynomial size circuits of depth 3 consisting only of MOD$_6$ gates. A survey of the state of the art can be found in [All96]; see also the discussion on p. 223 of this book.

We return to the topic of lower bounds in Sects. 6.1.2–6.1.4 and 6.2.

Exercises

3.1. Let $f \in \mathbb{B}^n$ be non-degenerated, and let C be a minimal-depth circuit over $\mathbb{B}^2$ such that $f = f_C$. Prove that the depth of C cannot be less than $\lceil \log n \rceil$.

3.2. Determine $S_{\mathbb{B}^2}(T_1^n)$ and $S_{\mathbb{B}^2}(T_n^n)$.

3.3. Let $n \geq 3$. A function $f \in \mathbb{B}^n$ belongs to Q^n, if the following two conditions hold:

– For all $i, j \in \{1, \ldots, n\}$, $i \neq j$, if we set the inputs x_i and x_j to constants, then we get at least three sub-functions of f.
– If $n \geq 4$ then there is an $i \in \{1, \ldots, n\}$ such that, if we set the input x_i to an appropriate constant, then we get as sub-function an element of Q^{n-1}.

Prove that for $f \in Q^n$ we have $S_{\mathbb{B}^2}(f) \geq 2n - 4$.

3.4. Prove Lemma 3.6.

3.5. Determine in detail the appearance of the circuit in Theorem 3.7 obtained from deleting the gates α, β, γ from C. Use ideas similar to those developed in Theorem 3.5 to deal with the case that one of the gates computes a constant or the negation.

3.6. Prove Lemma 3.9.
Hint: First make all weights non-negative by adding a suitable natural number w_0 to all weights. The threshold gate then has to be replaced by a comparison (see Exercise 1.18). Second, obtain C' from C'' by replacing every $\wedge$ gate v where the edge to the output is weighted by $w \geq 0$, with $w \wedge$ gates $v_1, \ldots, v_k$, whose predecessors are exactly the predecessors of v.

3.7. Prove: For all functions $f \in \mathbb{B}^n$, we have $d^{\mathrm{sign}}(f) \leq n$.
Hint: Start with the disjunctive normal form for f to produce an appropriate polynomial.

3.8. Show that the polynomial Q with the properties claimed in the proof of Theorem 3.14 exists.
Hint: Generalize the proof of Lemma 3.11. Use induction on the degree of monomials in P.

3.9.** A *probabilistic polynomial* is a polynomial p over the integers that has a number n of (regular) variables and a number r of probabilistic variables. Given a regular input $x = x_1 \cdots x_n$, a probabilistic input $u_1 \cdots u_r$ is chosen uniformly at random from $\{0,1\}^r$. Thus $p(x_1, \ldots, x_n)$ becomes a random variable with range $\{0,1\}$. Say that p *computes* $f \colon \{0,1\}^n \to \{0,1\}$ *with error* ϵ if for every $x \in \{0,1\}^n$, the probability that $p(x) \neq f(x)$ is at most ϵ.
    Prove:

(1) Let $n \in \mathbb{N}$. For every $\epsilon > 0$ there exists a probabilistic polynomial of degree $\log \frac{1}{\epsilon} (\log n)^{O(1)}$ that computes $\vee^n$ with error ϵ.

(2) For every circuit C with n inputs over $\mathcal{B}_1$ of size s and depth d and for every $\epsilon > 0$ there is a probabilistic polynomial of degree $\left(\log \frac{1}{\epsilon} \log n \right)^{O(1)}$ that computes f_C with error ϵ.

(3) For every language $A \in \text{UnbSIZE-DEPTH}(n^{O(1)}, 1)$ there is a family of probabilistic polynomials of degree $(\log n)^{O(1)}$ that computes c_A with error at most $\frac{1}{n}$ for inputs of length n.

Hint: For (1) use the following theorem of [VV86]:

Let $S \subseteq \{0,1\}^k$, $S \neq \emptyset$. Let $w_1, \dots, w_k \in \{0,1\}^k$ be chosen uniformly at random. Define

$$S_0 = S$$
$$S_i = \{ w \in S \mid w \cdot w_j \equiv 0 \pmod{2} \text{ for all } 0 \leq j \leq k \},$$

where $(a_1, \dots, a_k) \cdot (b_1, \dots, b_k) =_{\text{def}} \sum_{i=1}^k a_i \cdot b_i$ for all $a_1, \dots, a_k, b_1, \dots, b_k \in \{0,1\}$. Then the probability that there is an i, $0 \leq i \leq k$, such that $|S_i| = 1$ is at least $\frac{1}{4}$.

Start with S as the set of all predecessors to the $\vee$ gate that compute 1. Express the cardinalities of the S_i as polynomials.

For (2) replace $\wedge$ gates using de Morgan's laws and $\vee$ gates as above. The resulting polynomial will be the composition of the polynomials for the single gates.

3.10. Let $\mathcal{C} = (C_n)_{n \in \mathbb{N}}$ be a circuit family of depth d and size s over basis $\mathcal{B}_0$ ($\mathcal{B}_1$, $\mathcal{B}_1(p)$ for prime p, respectively). Show that there is an equivalent layered circuit family $\mathcal{C}'$ over the same basis of size $s^{O(1)}$ and depth $O(d)$.

3.11. Let $e > 0$. Prove:

$$\frac{\left| \left\{ u_1, \dots, u_e \in \{0,1\} \mid \sum_{i=1}^e u_i \equiv 0 \pmod{2} \right\} \right|}{2^e} = \frac{1}{2}.$$

3.12. Show that for $p > 1$, $\text{MOD}_p \leq_{\text{cd}} \text{MAJ}$.

3.13. Let p be prime, r be relatively prime to p, and $i > 1$. Show that $\text{MOD}_r \leq_{\text{cd}} \text{MOD}_{rp^i}$.

3.14. Prove that the statement of Smolensky's Theorem (Theorem 3.31) holds, if p is prime and r is not a power of p.

3.15. Show that a CRCW-PRAM with a polynomially bounded number of processors, that operates in constant time, cannot compute parity, multiply integers, divide integers, or sort a sequence of integers.

Notes on the Exercises

3.3. This result is from [Sch74], see also [Weg87, 122–123].

3.8. This exercise (as well as Lemma 3.11) are special cases of the *Group-Invariance Theorem,* stated and proved as Theorem 2.3 in [MP88].

3.9. These results were given in [BRS91] and independently in [Tar93]. See also [Bei93].

4. The NC Hierarchy

In this chapter we want to examine a class which has turned out to be of immense importance in the theory of efficient parallel algorithms. What should be considered to be a "good" parallel algorithm? First the number of processors should certainly not be too high, i.e., still polynomial in the input length. Second, the algorithm should achieve a considerable speed-up compared to efficient sequential algorithms, which typically run in polynomial time for a polynomial of small degree. These requirements led researchers to consider the class NC of all problems for which there are circuit families that are simultaneously of polynomial size and polylogarithmic (i.e., $(\log n)^{O(1)}$) depth. By the results presented in Sect. 2.7 this corresponds to polynomial processor number and polylogarithmic time on parallel random access machines. NC is the class of problems solvable with very fast parallel algorithms using a reasonable amount of hardware.

Certainly when we talk about parallel *algorithms* we have to consider *uniform* circuit families. However which uniformity condition to choose is not so clear. This will be clarified below.

4.1 Bounded Fan-in Circuits

Let us start by formally introducing the NC hierarchy. NC is Cook's mnemonic for "Nick's class" in recognition of [Pip79]. It was this paper where the class SIZE-DEPTH $\left(n^{O(1)}, (\log n)^{O(1)}\right)$ was first identified as a class of interest.

NC is defined using bounded fan-in circuit families. We consider the corresponding class for unbounded fan-in circuits in Sect. 4.2.

Definition 4.1. For $i \geq 0$, define

$$\mathrm{NC}^i =_{\mathrm{def}} \mathrm{SIZE\text{-}DEPTH}\left(n^{O(1)}, (\log n)^i\right),$$

and let

$$\mathrm{NC} =_{\mathrm{def}} \bigcup_{i \geq 0} \mathrm{NC}^i.$$

The following connections to time- and space-bounded classes can be given:

Corollary 4.2. $U_L\text{-}NC^1 \subseteq L \subseteq NL \subseteq U_L\text{-}NC^2 \subseteq U_L\text{-}NC^3 \subseteq \cdots \subseteq U_L\text{-}NC \subseteq P \cap DSPACE\left((\log n)^{O(1)}\right).$

Proof. Follows from Corollary 2.29 and Theorems 2.31 and 2.32.   ☐

Recalling the results given in Sect. 2.6 it is no surprise that the NC^i classes have a nice characterization in terms of alternating Turing machines.

Corollary 4.3.   *1.* $U_E\text{-}NC^1 = U_E^*\text{-}NC^1 = ATIME(\log n) \subseteq U_L\text{-}NC^1.$
2. $U_E\text{-}NC^i = U_E^*\text{-}NC^i = U_L\text{-}NC^i = ATIME\text{-}SPACE\left((\log n)^i, \log n\right)$ *for all*
   $i \geq 2.$
3. $U_E\text{-}NC = U_E^*\text{-}NC = U_L\text{-}NC = ATIME\text{-}SPACE\left((\log n)^{O(1)}, \log n\right).$

Proof. Follows from Corollary 2.50.   ☐

For later use we mention the following result:

Corollary 4.4. *If* $A \in U_E\text{-}NC^i$ *for* $i \geq 1$, *then there is an* $U_E\text{-}NC^i$ *circuit family* C *accepting* A *which is in input normal form.*

Proof. Follows from Corollary 2.51.   ☐

4.2 Unbounded Fan-in Circuits

What if we consider circuits of polynomial size and polylogarithmic depth, but over an unbounded fan-in basis? Formally we look at the following classes:

Definition 4.5. For $i \geq 0$, define

$$AC^i =_{def} UnbSIZE\text{-}DEPTH\left(n^{O(1)}, (\log n)^i\right),$$

and let

$$AC =_{def} \bigcup_{i \geq 0} AC^i.$$

A connection to the classes of the NC hierarchy can be stated as follows:

Lemma 4.6. *For all* $i \geq 0$, *we have* $NC^i \subseteq AC^i \subseteq NC^{i+1}$; *hence* $NC = AC.$

Proof. The first inclusion follows from Proposition 1.17; the second inclusion is immediate from the definition.   ☐

It can be seen easily that this lemma holds for all uniformity conditions we have considered so far (see Exercise 4.1).

The question now is whether there is a nice characterization of the classes AC^i on alternating Turing machines, similar to Corollary 4.3. We will show that the answer is yes, and this result is at least partially responsible for the (not quite consistent) naming of the classes. The "A" in AC stands for "alternation."

Theorem 4.7. $U_L\text{-}AC^i = AALT\text{-}SPACE\left((\log n)^i, \log n\right)$ *for all* $i \geq 1$.

Proof. First we observe that an AC^i circuit family is U_L-uniform if and only if its direct connection language can be recognized in space $O(\log n)$ (see Exercise 4.2).

Now the proof follows using the constructions given in Theorems 2.48 and 2.49. For the inclusion $U_L\text{-}AC^i \subseteq AALT\text{-}SPACE\left((\log n)^i, \log n\right)$ we use the machine given in Fig. 2.10 (p. 65). If the given circuit has depth d then on every path from an input to the output there will be at most d alternations between $\vee$ and $\wedge$ gates. Thus the simulating machine will be alternation-bounded by d. The machine witnessing uniformity now runs in space $O(\log n)$ deterministically, i.e., this does not add to the alternation bound.

For the inclusion $AALT\text{-}SPACE\left((\log n)^i, \log n\right) \subseteq U_L\text{-}AC^i$, we first assume that the given alternating machine M makes an at most logarithmic number of steps within one alternation (see Exercise 2.18). We now modify M as follows: When it starts a new alternation, it keeps the first configuration on a particular tape. Call this configuration the *head* of the alternation. On another special purpose tape M stores the path leading from the head to the current configuration. This needs only logarithmic space since the time within an alternation by the above is logarithmic. Now use the construction of the proof of Theorem 2.48 with the following modification: When M makes a sequence of branches of the same type (universal or existential), the circuit will simulate this by just one $\wedge$ or $\vee$ gate of appropriate fan-in. U_L-uniformity of this construction follows since to check whether gates are connected in the circuit we just simulate the alternating machine for at most one alternation. This can be done deterministically since by the modification above any configuration of M carries with it the path on which it can be reached from the head of its alternation. Hence types of gates and connections among gates in the circuit can be computed in space $O(\log n)$. $\qquad\square$

This theorem yields the following corollary, similar to Corollary 4.4 in the case of bounded fan-in circuits:

Corollary 4.8. *If* $A \in U_L\text{-}AC^i$ *for* $i \geq 1$, *then there is an* $U_L\text{-}AC^i$ *circuit family* C *accepting* A *which is in input normal form.*

Using the results in Sect. 2.7 we see that $U_L\text{-}AC^i$ corresponds to time $(\log n)^i$ and a polynomial number of processors on CRCW-PRAMs. Hence a language is in NC if and only if it is accepted by a CRCW-PRAM with a polynomial number of processors running in polylogarithmic time.

Looking back at Theorem 4.7, the question arises if AC^0 has a similar characterization. For this we first have to identify a suitable uniformity condition for this class. Logspace-uniformity is not strict enough as we will argue in Sect. 4.5.1 below.

4.3 Semi-Unbounded Fan-in Circuits

In this section we will consider classes over the basis $B_s =_{\mathrm{def}} \{\wedge^2, (\vee^n)_{n\in\mathsf{N}}\}$. Since in circuits over B_s, $\wedge$ gates have bounded fan-in while $\vee$ gates have unbounded fan-in, these circuits are called *semi-unbounded fan-in circuits*. Basis B_s consists only of monotone functions, therefore circuits over B_s can obviously compute only monotone functions. This is not what we want. But if we allow $\neg$ gates arbitrarily in our circuits then B_s will be as good as the basis B_1 (since we can simulate unbounded fan-in $\wedge$ gates by unbounded fan-in $\vee$ gates and $\neg$ gates, using the laws of de Morgan). Therefore we only consider circuits in input normal form and define:

Definition 4.9. For $i \geq 0$, define SAC^i to be the class of all languages A for which there is a circuit family $C = (C_n)_{n\in\mathsf{N}}$ that accepts A, with the following properties:

– C is of polynomial size and depth $O((\log n)^i)$.
– Every C_n is a circuit over $B_s \cup \{\neg\}$ in input normal form, that is, for every $\neg$ gate in C_n, the predecessor of that gate is an input gate.

Let $SAC =_{\mathrm{def}} \bigcup_{i\geq 0} SAC^i$.

A connection to the classes of the NC hierarchy examined above can be stated as follows:

Lemma 4.10. *For all $i \geq 0$, we have $NC^i \subseteq SAC^i \subseteq AC^i \subseteq NC^{i+1}$; hence $NC = AC = SAC$.*

Proof. The first inclusion follows easily from the laws of de Morgan, the second inclusion is obvious from the definition, and the last inclusion was given in Lemma 4.6. $\qquad\qquad\square$

It can be seen easily that this lemma holds for all uniformity conditions we have considered so far (see Exercise 4.1).

There is a characterization of (uniform) semi-unbounded fan-in circuit classes on alternating Turing machines along the lines of Corollary 4.3 and Theorem 4.7 (see the Bibliographic Remarks for this chapter and Exercise 4.3). We will turn to another characterization of these circuit classes in the next subsection, using the so-called *tree-size* of alternating Turing machines.

4.3.1 Polynomial Tree-size

In the previous chapters we considered the resources time, space, and number of alternations for alternating Turing machines. We will now introduce a fourth complexity measure.

Remember the notion of an accepting computation subtree for alternating machines. Such a machine M accepts its input x exactly if there is an accepting computation subtree in the computation tree produced during the work of M on x. We will now consider the size of this subtree.

Definition 4.11. Let M be an alternating Turing machine, and let $t\colon \mathbb{N} \to \mathbb{N}$. We say that M is *tree-size bounded* by t if, for every $x \in L(M)$, there is an accepting computation subtree of M on input x which has at most $t(|x|)$ nodes.

Let $s, t\colon \mathbb{N} \to \mathbb{N}$. Then we define ASPACE-TREESIZE(s, t) to be the class of all sets B for which there is an alternating Turing machine M which is space bounded by s and tree-size bounded by t, such that $B = L(M)$.

As always, particular attention is paid to the case of polynomial resource bounds. In the case of tree-size, we get a remarkable connection with the power of semi-unbounded fan-in circuits.

Theorem 4.12. U_L-SAC^1 = ASPACE-TREESIZE($\log n, n^{O(1)}$).

Proof. ($\subseteq$): Let $\mathcal{C} = (C_n)_{n \in \mathbb{N}}$ be an SAC^1 circuit family. $\mathcal{C}$ is logspace-uniform, hence we conclude that $L_{DC}(\mathcal{C})$ can be decided in logarithmic space (see Exercise 4.2).

Now we design an alternating Turing machine M that simulates family $\mathcal{C}$ as in the proof of Theorem 2.49, see Fig. 2.10 on p. 65. Without loss of generality we may assume that every configuration of M has at most two successor configurations. The space requirements of M are clearly logarithmic. How do accepting computation subtrees of M appear? For an input x that is accepted by C_n, an accepting computation subtree of M on input x is a tree of logarithmic depth where universal configurations have at most two successor configurations while existential configurations have (by definition) exactly one successor configuration. The simulation of the machine witnessing uniformity of $\mathcal{C}$ leads to polynomially long, deterministic paths. We may arbitrarily assign type universal or existential to these deterministic configurations. In any case they will have at most one successor. Altogether we see that the tree-size is polynomial in $|x|$.

($\supseteq$): Let $A \in$ ASPACE-TREESIZE($\log n, n^{O(1)}$), $A = L(M)$ for the alternating Turing machine M. Again, we assume that every configuration of M has at most two successor configurations. First we design an alternating machine M' simulating M with the property that computation trees produced by M will be 'shallow.'

We start by introducing some notation. Fix an input x, $|x| = n$. Let $T_M(x)$ be the computation tree of M on input x. A *fragment* of the computation of

M on input x is a pair (r, Λ), where r is a configuration of M on x using space $O(\log n)$ and Λ is a set of such configurations. A fragment (r, Λ) is said to be *realizable*, if there is a subtree T' of $T_M(x)$ with the following properties:

1. The root of T' is r.
2. Each leaf of T' is either accepting or an element from Λ.
3. No element in Λ is accepting, and all elements in Λ appear as leaves in T'.

The *size* of a fragment (r, Λ) is the minimal number of nodes in a subtree $T' \subseteq T_M(x)$ witnessing that (r, Λ) is realizable.

Observe that $x \in L(M)$ if and only if $(K_0, \emptyset)$ is realizable, where K_0 is the initial configuration of M on input x.

We now claim that a fragment (r, Λ) is realizable if and only if one of the following two statements hold:

(a) There is a tree T' with at most 3 nodes witnessing that (r, Λ) is realizable.
(b) (r, Λ) can be divided into two fragments $(r, \Lambda' \cup \{s\})$ and (s, Λ''), that are both realizable and strictly of smaller size than (r, Λ), where $\Lambda = \Lambda' \cup \Lambda''$.

We first argue the correctness of the claim:
($\Leftarrow$): In case (a), nothing has to be proved. In case (b), let T' and T'' be the trees witnessing that $(r, \Lambda' \cup \{s\})$ and (s, Λ'') are realizable. If we now replace every leaf s in T' by the tree T'', then we obtain a tree witnessing that (r, Λ) is realizable.
($\Rightarrow$): Let T' be a tree witnessing that (r, Λ) is realizable. If T' has at most 3 nodes, then statement (a) holds. If T' has more than 3 nodes then there is a node $s \in T'$ which is neither the root nor a leaf of T'. If we now pick Λ' to be those leaves in Λ which are not descendants of s and Λ'' to be those leaves in Λ which are descendants of s then statement (b) holds. This finishes the proof of the claim.

Now machine M' works as follows: M' tries to prove that $(K_0, \emptyset)$ is realizable. For this, M' will have a procedure realize(r, Λ) which returns true if and only if (r, Λ) is realizable. This procedure works as follows: M' existentially picks one of the two following possibilities:

1. M' tries to show that (r, Λ) is realizable by constructing a witnessing tree of size at most 3.
2. M' existentially guesses s, Λ', Λ'', verifies that $\Lambda = \Lambda' \cup \Lambda''$ and universally starts the two recursive calls realize$(r, \Lambda' \cup \{s\})$ and realize(s, Λ'').

The correctness of this simulation is obvious from the claim above. We now analyze the resources needed by M'. For this we make two observations.

(A) We note that it is sufficient to restrict our attention to computation subtrees with at most 3 non-accepting leaves. If the realizability of some (r, Λ) is witnessed by T' with more than 3 leaves then there is a node s in T' such that in the subtree below s there will be exactly two of the leaves from T'. Thus we can divide (r, Λ) into two fragments $(r, \Lambda' \cup \{s\})$ and (s, Λ'')

which are both realizable if and only if (r, Λ) is realizable, and moreover, $|\Lambda'|, |\Lambda''| \leq 2$. This shows that in each accepting computation tree of M' there will be an accepting subtree where all occurring fragments (r, Λ) have $|\Lambda| \leq 3$. Hence the space requirements of M' are given by $O(\log n)$.

(B) We can always divide a fragment (r, Λ) of size $m > 3$ into two fragments $(r, \Lambda' \cup \{s\})$ and (s, Λ'') which are both realizable if and only if (r, Λ) is realizable, and moreover the size of both of these fragments is at most a constant fraction of m, more precisely at most $\frac{m}{12}$. (See Exercise 4.4.)

From both observations it follows that for every accepted input x of length n, there is an accepting subtree of $T_{M'}(x)$ in which the recursion depth (with respect to procedure realize) is bounded by $O(\log n)$. This can be achieved as follows: As long as the sets Λ under consideration have fewer than three elements, we choose a split of the current fragment according to observation (B). This will reduce the size of the fragment by a constant fraction. If Λ gets too large, we choose a split of the fragment according to observation (A). This will not increase the fragment size. Since in this way at least every other split will reduce the fragment size to a constant fraction, we see that a logarithmic recursion depth will suffice.

Since the recursion depth is logarithmic and the time needed per recursion step is also logarithmic, we see that the time needed by M' can be bounded by $O((\log n)^2)$. However we are interested in another property: The existential branches during the execution of M' are needed to guess s, Λ', Λ''. The only universal branches that M' makes are the two calls realize$(r, \Lambda' \cup \{s\})$ and realize(s, Λ''). That is, on any path in an accepting computation subtree, the number of universal nodes between two existential nodes is bounded by a constant. Now we simulate M' as in the proof of the inclusion AALT-SPACE$(\log n, \log n) \subseteq AC^1$ in Theorem 4.7. Then the resulting circuit family will be of depth $O(\log n)$, and will have unbounded fan-in $\vee$ and bounded fan-in $\wedge$ gates. Hence we have proved that $L(M') \in U_L\text{-SAC}^1$. $\square$

4.3.2 Closure under Complementation

A complexity class $\mathcal{K}$ is said to be closed under complementation, if whenever $A \in \mathcal{K}$ then also $\overline{A} =_{\text{def}} \{0, 1\}^* \backslash A \in \mathcal{K}$. The classes NC^k and AC^k are trivially closed under complementation. A circuit family for $\overline{A}$ can be obtained from a family for A by adding a negation gate on top of every circuit. This is not possible for SAC^k circuits, because the negation gates are only allowed to be used near the inputs. If we try to complement an SAC^k circuit using the laws of de Morgan, then this results in a circuit with bounded fan-in $\vee$ and unbounded fan-in $\wedge$ gates, which is not an SAC circuit. We find only that the complement of any language from SAC^k is in AC^k. One might hope that by recalling Theorem 4.12 it is possible to show that SAC^1 is closed under complementation using the model of alternating Turing machines. However a transformation according to de Morgan's laws transforms a small tree-size size into a large one. So the question whether the classes SAC^k are closed under complement is not obvious. However, if $k \geq 1$ then SAC^k *is* closed under complementation as we show next.

First we prove a lemma. Recall the definition of the functions T_k^n from p. 20.

Lemma 4.13. *For every* $n \geq 0$, $0 \leq k \leq n$, *the threshold function* T_k^n *can be computed by circuits over* $\mathcal{B}_s$ *of size* $n^{O(1)}$ *and depth* $O(\log n)$.

Proof. The circuits are constructed by divide-and-conquer according to the following property of the threshold functions:

$$T_k^n(x_1, \ldots, x_n) = \bigvee_{j=0}^{k} \left(T_j^{\lfloor \frac{n}{2} \rfloor}(x_1, \ldots, x_{\lfloor \frac{n}{2} \rfloor}) \wedge T_{k-j}^{\lceil \frac{n}{2} \rceil}(x_{\lfloor \frac{n}{2} \rfloor + 1}, \ldots, x_n) \right).$$

□

Theorem 4.14. *The classes* SAC^k *for* $k \geq 1$ *are closed under complementation.*

Proof. The proof relies on the method of "inductive counting." We will show how we can compute how many gates at depth d in a given SAC^k circuit evaluate to 1, if we know how many gates at depth $d - 1$ evaluate to 1. (The *depth of a gate* v is the length of a longest path from an input gate to v.)

Let $k \geq 1$ and $A \in \mathrm{SAC}^k$ be accepted by circuit family $\mathcal{C} = (C_n)_{n \in \mathbb{N}}$. Convert $\mathcal{C}$ into an equivalent circuit family $\mathcal{D} = (D_n)_{n \in \mathbb{N}}$ with the following properties: For every n, let $d = d(n)$ be the depth of C_n; then there is an $s = s(n) \in \mathbb{N}$ such that:

(1) D_n consists of $2d + 1$ levels.
(2) On level 0 we have the inputs and their negations, on level $2d$ we have fan-out 0 gates, one of which is designated to be the output gate of D_n.
(3) All predecessors of nodes from level i are from level $i - 1$.
(4) For every level $i > 0$, the number of gates in D_n at level i is exactly s.
(5) For $i > 0$, all gates at level $2i - 1$ are $\vee$ gates and all gates at level $2i$ are $\wedge$ gates.

This normal form can be established as follows: In each level $i > 0$ of D_n we have one copy of each gate v in C_n. The type of each copy of v will be set according to condition (5). If v is an input gate in C_n then all predecessors of the copy of v on level i are the copy of v on level $i - 1$. If v is an $\wedge$ gate in C_n but the copy is required to be an $\vee$ gate because of condition (5) then all predecessors of the copy of v on level i will be the copy of v on level $i - 1$. (If $i = 1$ then the predecessors of the copy of v may be arbitrary gates from level 0.) Analogously we proceed in the case of a copy of an $\vee$ gate on an $\wedge$ level. If v is of type $\vee$ and i is odd then the predecessors of the copy of v are the copies of the predecessors of v in C_n; and analogously for an $\wedge$ gate on an even level. If $i = 1$ and the predecessors of v in C_n are not all inputs or negations of inputs then the predecessors of the copy of v may be chosen arbitrarily from level 0.

The following property can be proved easily (by induction on i): Let $0 \le i \le d$. Let $v \in C_n$, and let i be the depth of v. If $v' \in D_n$ is a copy of v on level $j \ge 2i$ then v' will compute the same function as v.

We now construct a circuit family $\mathcal{D}' = (D'_n)_{n \in \mathbb{N}}$ that accepts $\overline{A}$. As a first step, the circuits D'_n will be circuits with unbounded fan-in $\lor$ gates, bounded fan-in $\land$ gates, negations (but only at the inputs), and additionally unbounded fan-in threshold gates. (In the end we will show how to eliminate these threshold gates.)

Fix an input length n. For $0 \le i \le d$ we define

$$E_i(x) =_{\mathrm{def}} \{ v \mid v \text{ is a gate of } D_n \text{ on level } 2i$$
$$\text{that evaluates to 1 on input } x\}.$$

Let $N_i(x)$ denote the set of all gates of D_n on level $2i$ that do not belong to $E_i(x)$.

Circuit D'_n will be constructed in several stages. We start by setting $D'_n \leftarrow D_n$ and then add consecutive gates to D'_n, until finally we see how the complement of A can be accepted. D'_n will result from three such enhancement steps. During the construction we will refer to the gates that stem from D_n as "original gates."

Enhancement step 1. Let $1 \le i \le d$, and let g be an original gate on level $2i$ or $2i - 1$. Let $0 \le c \le s$. Add a gate g^c to D'_n as follows:

- If g is an $\land$ gate (that is, g is on level $2i$) with predecessors g_0 and g_1, then g^c is an $\lor$ gate with predecessors g_0^c and g_1^c.
- If g is an $\lor$ gate (that is, g is on level $2i - 1$), then g^c is a T_c-threshold gate whose inputs are all original gates on level $2i - 2$ that are not inputs to g.

We call the gates g^c the *c-contingent complement* gates for g.

The idea of these gates is that for an original gate g on level $2i$ or $2i - 1$, the gate g^c for $c = |E_{i-1}(x)|$ computes exactly the negation of g. This holds because of the following considerations: If g is an $\lor$ gate then g evaluates to 1 if and only if at least one of its predecessors evaluates to 1. Hence g evaluates to 0 if and only if all its predecessors belong to $N_{i-1}(x)$ if and only if among the gates on level $2i - 2$ that are not inputs of g there are c gates that evaluate to 1. The threshold gate g^c tests exactly this latter condition; hence g^c computes the negation of g. If g is an $\land$ gate with predecessors g_1 and g_2, then by the above (since g_1 and g_2 are $\lor$ gates) g_1^c and g_2^c compute the negations of g_1 and g_2. Hence by the laws of de Morgan g^c computes the negation of g.

Enhancement step 2. Let $0 \le i < d$, and let $0 \le c \le s$. Add a gate h_i^c to D'_n as follows: For $i = 0$, h_0^n is the constant 1, and all h_0^c for $c \ne n$ are constant 0 gates. For $i \ge 1$ we first define a few auxiliary gates: Let γ_i^c be a

T_c^s-threshold gate whose inputs are all original gates on level $2i$. Moreover, for all $0 \le j \le s$, let $\eta_i^{c,j}$ be a T_c^s-threshold gate whose inputs are all the j-contingent complement gates for original gates on level $2i$, i.e., all gates g^j for original gates g on level $2i$. Finally h_i^c is defined to be

$$\bigvee_{j=0}^{k} \left(h_{i-1}^j \wedge \gamma_i^c \wedge \eta_i^{s-c,j} \right). \tag{4.1}$$

We call the gates h_i^c *counting gates*.

We next prove by induction on i that h_i^c evaluates to 1 if and only if $|E_i(x)| = c$. The case $i = 0$ is obvious, since on level 0 we have exactly $2n$ gates, half of which evaluate to 1. Let now $i > 0$. By induction hypothesis h_{i-1}^j evaluates to true if and only if $j = |E_{i-1}(x)|$. Hence the only term in the disjunction (4.1) that possibly evaluates to true is the term for $j = |E_{i-1}(x)|$. All gates g^j for an original gate g on level $2i$ then compute the negation of g. Thus γ_i^c evaluates to true if and only if on level $2i$ at least c gates evaluate to true, and $\eta_i^{s-c,j}$ evaluates to true if and only if on level $2i$ at least $s - c$ gates evaluate to false. Hence their conjunction $\gamma_i^c \wedge \eta_i^{s-c,j}$ expresses that on level $2i$ exactly c gates evaluate to true, i.e., that $|E_i(x)| = c$. This finishes the induction.

Enhancement step 3. We add an output gate to D_n', defined as follows: Let g be the output of D_n. Then g is on level $2d$. The output gate of D_n' will be a gate computing

$$\bigvee_{c=0}^{k} \left(h_{d-1}^c \wedge g^c \right).$$

Then certainly D_n' computes the complement of the function computed by D_n. D_n' has gates from the basis $\mathcal{B}_s$, negation gates (but only at the inputs) and threshold gates. The size of D_n' is certainly polynomial in n, and the depth is $O(\log n)$.

Observe that on every path from a contingent complement gate to the circuit inputs we clearly have at most one threshold gate, since the threshold gates used in the definition of a contingent complement gate have as predecessors only original gates. Similarly on every path from a gate γ_i^c to the inputs we will only find at most one threshold gate. However, on a path from a gate $\eta_i^{s-c,j}$ to the inputs we will find up to two threshold gates, hence the same is true for the counting gates h_i^d. Altogether we conclude that on no path from an input gate to the output gate we have more than two threshold gates. Hence if we replace the threshold gates by the monotone SAC^1 circuits we constructed in Lemma 4.13, we get an SAC^i circuit family that accepts $\overline{A}$. □

The proof above does not work for constant depth, since the monotone circuits for the threshold function constructed in Lemma 4.13 require loga-

rithmic depth. So the question whether the class SAC^0 is closed under complement is left open in Theorem 4.14. In fact SAC^0 does not share this property
as we will prove below. First we define the notion of accepting subtrees of
Boolean circuits, analogous to accepting computation subtrees for alternating
Turing machines.

Definition 4.15. Let C be a circuit with n inputs in input normal form.
Unwind C into a tree $T(C)$ (see p. 44). Fix an input $x \in \{0,1\}^n$. An *accepting
subtree* H of C with input x is a subtree of $T(C)$ with the following properties:

- H contains the output gate.
- For every $\wedge$ gate v in H, all the predecessors of v are in H.
- For every $\vee$ gate v in H, exactly one predecessor of v is in H.
- All gates in H evaluate to 1 on input x.

Obviously, an input x is accepted by C if and only if there is an accepting
subtree of C with input x.

Theorem 4.16. SAC^0 *is not closed under complementation.*

Proof. Let $A \subseteq \{0,1\}^*$ be defined as $A = \{ a_1 a_2 \cdots a_m b_1 b_2 \cdots b_m \mid m \in$
$\mathbb{N}$ and there is an i, $1 \leq i \leq m$, such that $a_i \neq b_i \}$. Obviously, $A \in \text{SAC}^0$.

Let $C = (C_n)_{n \in \mathbb{N}}$ be an arbitrary, semi-unbounded fan-in circuit family of
constant depth. Fix an input length n, and suppose that there is an input x
of length n that is accepted by C_n. Let T be an accepting subtree of C_n with
input x. Let v be a gate in T at depth d. Then the subtree of T rooted at v
has at most 2^d leaves (remember that $\wedge$ gates in T have fan-in ≤ 2).

Hence there is a constant $c \in \mathbb{N}$ such that for every n, the output gate
of C_n depends on at most c of the input bits. Therefore, if C_n accepts one
input, then in fact it accepts at least 2^{n-c} many inputs of length n.

Now consider the complement $\overline{A}$ of A, $\overline{A} = \{0,1\}^* \setminus A$. If n is odd, then
$\overline{A}$ contains all strings of length n, but if $n = 2m$ is even then among the
2^n strings of length n, exactly $2^m = 2^{\frac{n}{2}}$ belong to $\overline{A}$. Hence by the above
property of semi-unbounded fan-in constant-depth circuits, we conclude that
$\overline{A} \notin \text{SAC}^0$. ∎

We conclude this section by noting that the two results above also hold
in the case of logspace-uniformity.

Corollary 4.17. *The classes* $U_L\text{-SAC}^k$ *for* $k \geq 1$ *are closed under complementation.*

Proof. Let $A \in U_L\text{-SAC}^k$ via circuit family C. Look at the proof of Theorem 4.14. The circuit family $\mathcal{D}$ will certainly be logspace-uniform if C is. The
circuit family $\mathcal{D}'$ consists of $\mathcal{D}$ plus a few, very regular enhancements which
are easily seen to be logspace-uniform (see Exercise 4.11). ∎

Corollary 4.18. U_L-SAC^0 *is not closed under complementation.*

Proof. The set A considered in the proof of Theorem 4.16 is certainly in U_L-SAC^0. However the complement is not even in non-uniform SAC^0, as proved there. $\qquad\Box$

4.4 The Decomposability of the NC Hierarchy

In this section we will consider the following structural question concerning the NC hierarchy: Is it possible to "decompose" any level NC^i of the hierarchy into components taken from lower levels?

The idea that a circuit for a problem consists of subcircuits solving certain subproblems was developed in Sect. 1.4. There we restricted our attention to constant-depth circuits with gates powerful enough to solve certain subproblems. Here we want to consider NC^i as well as AC^i circuits with gates that can solve membership subproblems for any fixed language $B \subseteq \{0,1\}^*$. Formally we define:

Definition 4.19. Let $i \geq 1$, and let $B \subseteq \{0,1\}^*$. Then $NC^i(B)$ is the class of all languages A for which there is a circuit family $C = (C_n)_{n\in\mathbb{N}}$ that accepts A, such that the following holds for every $n \in \mathbb{N}$:

- C_n is a circuit over the basis $\mathcal{B}_0 \cup \{c_B\}$.
- The size of C_n is bounded by a polynomial in n. Here, a gate in C_n computing c_B for words of length k contributes to the size of C_n with the term k.
- The depth of C_n is $O(\log^i n)$. In the calculation of the depth of C_n, a gate computing c_B for words of length k by definition contributes with the term $\lceil \log k \rceil$.

If $\mathcal{K}$ is a class of languages, then $NC^i(\mathcal{K}) =_{\text{def}} \bigcup_{B\in\mathcal{K}} NC^i(B)$.

Definition 4.20. Let $i \geq 0$, and let $B \subseteq \{0,1\}^*$. Then $AC^i(B)$ is the class of all languages A for which there is a circuit family $C = (C_n)_{n\in\mathbb{N}}$ that accepts A, such that the following holds for every $n \in \mathbb{N}$:

- C_n is a circuit over the basis $\mathcal{B}_1 \cup \{c_B\}$.
- The size of C_n is bounded by a polynomial in n. Here, a gate in C_n computing c_B for words of length k contributes to the size of C_n with the term k.
- The depth of C_n is $O(\log^i n)$, where gates computing c_B count as usual with depth 1.

If $\mathcal{K}$ is a class of languages, then $AC^i(\mathcal{K}) =_{\text{def}} \bigcup_{B\in\mathcal{K}} AC^i(B)$.

In the two definitions above, the gates that compute the function c_B are also called *oracle gates*. The language B is called an *oracle*. A circuit with oracle gates c_B can obtain information about B almost for free.

The definition of constant-depth reducibility is a special case of Def. 4.20, in the sense that $A \leq_{cd} B$ if and only if $A \in AC^0(B)$. For this reason, we also refer to constant-depth reductions as AC^0 *reductions*.

The stipulation that fan-in k oracle gates add $\log k$ to the depth for bounded fan-in circuits and only 1 for unbounded fan-in circuits has been made in this way to parallel the fact that, e.g., a disjunction or conjunction of k bits can be computed in depth 1 by unbounded fan-in circuits, while bounded fan-in circuits need depth $\log k$.

In the following the class $\mathcal{K}$ from the definitions above will be a class from the NC-hierarchy. This will lead to a decomposition of the hierarchy, as given by the following two theorems.

Theorem 4.21. *Let $i, j \geq 1$. Then $NC^i(NC^j) = NC^{i+j-1}$.*

Proof. ($\subseteq$): Let $B \in NC^j$ and $A \in NC^i(B)$ via circuit family $\mathcal{C} = (C_n)_{n \in \mathbb{N}}$ of depth at most $c(\log n)^i$ for $c \in \mathbb{N}$. Let B be accepted by the circuit family $\mathcal{D} = (D_n)_{n \in \mathbb{N}}$, where the depth of D_n is bounded by $d(\log n)^j$ for $d \in \mathbb{N}$. Fix an input length n and consider a path in C_n from the inputs to the output. Let $g_1, \ldots, g_k$ be the oracle gates on this path. Let n_m be the fan-in of g_m ($1 \leq m \leq k$). The length of the path is at most $c \log^i n$ for some $c \in \mathbb{N}$, therefore by Def. 4.19 we must have

$$\sum_{m=1}^{k} \log n_m \leq c(\log n)^i.$$

Since the size of C_n is polynomial, every n_m is bounded by a polynomial in n, and hence there is a constant $d' \in \mathbb{N}$ such that $\log n_m \leq d' \log n$ for $1 \leq m \leq k$. We now replace every g_m in C_n by the circuit D_{n_m}. The sum of the depths of these circuits is bounded by

$$\sum_{i=m}^{k} d(\log n_m)^j = d \sum_{m=1}^{k} (\log n_m)^{j-1} \cdot \log n_m$$

$$\leq d \sum_{m=1}^{k} (d' \log n)^{j-1} \cdot \log n_m$$

$$= d(d' \log n)^{j-1} \cdot \sum_{m=1}^{k} \log n_m$$

$$\leq d(d' \log n)^{j-1} \cdot c(\log n)^i$$

$$\leq c'(\log n)^{i+j-1}$$

for a suitable $c' \in \mathbb{N}$. Hence replacing in C_n all gates for c_B with corresponding circuits from $\mathcal{D}$ adds to the depth with the term $O((\log n)^{i+j-1})$. Hence the overall depth is also bounded by this, and the resulting family shows that $A \in NC^{i+j-1}$.

($\supseteq$): Let $A \in NC^{i+j-1}$ be accepted by circuit family $C = (C_n)_{n\in N}$. Let the depth of C_n be bounded by $c(\log n)^{i+j-1}$. The idea is that we will split every C_n into "slices" of depth $(\log n)^j$. There will be $(\log n)^{i-1}$ many such slices on top of each other. The computation within each slice defines a language B which (due to the depth restriction for each slice) is from NC^j. Thus there is a circuit family C' for A which uses gates for c_B to simulate the computation of family C. Since in C'_n we need $(\log n)^{i-1}$ many layers of c_B gates, this leads to an NC^i family with oracle B that accepts A. This is essentially a proof which works for the non-uniform case. However we will describe this next in more detail, using a slightly more complicated way of dividing up every C_n. The reason is that this proof will also work for the uniform case that we examine below.

Let the depth of C_n be bounded by $c(\log n)^{i+j-1}$. Fix an input length n. Let s be the size of C_n. First we transform C_n into a circuit D_n exactly as in the first step of the proof of Theorem 4.14. Thus D_n consists of $d \leq 2c(\log n)^{i+j-1} + 1$ levels, and in every level l in D_n (except level 0) we find copies of all s gates from C_n.

Now we split D_n into subcircuits $D_n^0, \ldots, D_n^{2c(\log n)^{i-1}}$, where D_n^0 is level 0 and D_n^l consists of levels $(l-1)(\log n)^j + 1, \ldots, l(\log n)^j$. Each circuit D_n^l computes a polynomially length-bounded function f_l^n, that is, every f_l^n takes an s bit input word and produces an s bit output word, except f_1^n which takes as input all gates from level 0, i.e., a $2n$ bit input. Since the depth of each D_n^l is restricted by $(\log n)^j$, the language $B_l =_{def} \{ \langle 1^n, y, k \rangle \mid$ the kth bit of $f_l^n(y)$ is $1 \}$ is in NC^j. All these languages can be combined into one language

$$B =_{def} \{ \langle 1^n, y, k, l \rangle \mid \text{the } k\text{th bit of } f_l^n(y) \text{ is } 1 \},$$

which is then also in NC^j.

Finally construct a circuit family $C' = (C'_n)_{n\in N}$ as follows: Every C'_n consists of $2c(\log n)^{i-1} + 1$ layers. In layer 0 we compute the negations of all input bits. In layer 1 we compute f_1 applied to C'_n's inputs and their negations, using gates for language B. The result of this computation is then used to form an input y_1 for the function f_2. Again using gates for B, $f_2(y_1)$ is computed. The result is used to produce y_2, and $f_3(y_2)$ is computed, and so on. In this way we compute all the bits y_i that are computed in level i and are fed into the gates on level $i + 1$ ($1 \leq i < d$). The length of all the intermediate results y_i is polynomial in n. Hence the gates for B contribute to the depth of C'_n with $O(\log n)$. Thus we see that A can be accepted in depth $O((\log n)^i)$ using oracle gates for a language in NC^j, which shows that $A \in NC^i(NC^j)$.   □

Theorem 4.22. *Let $i, j \geq 1$. Then $AC^i(AC^j) = AC^{i+j}$.*

Proof. ($\subseteq$): As in the proof of Theorem 4.21 replace gates for AC^j languages with their corresponding circuits. This results in the required AC^{i+j} circuit family.

($\supseteq$): Again we proceed as in the proof of Theorem 4.21. First transform the given family C into a regular, leveled family $\mathcal{D}$, still of depth $O((\log n)^{i+j})$. Then this family is split into $O((\log n)^i)$ subcircuits of depth $(\log n)^j$. Let f_l^n be the function computed by the lth of these subcircuits. The language

$$B =_{\text{def}} \left\{ \langle 1^n, y, k, l \rangle \mid \text{the } k\text{th bit of } f_l^n(y) \text{ is } 1 \right\}$$

is now in AC^j. Hence we get an $AC^i(AC^j)$ circuit family. $\square$

The two theorems above have shown how every class NC^i and AC^i can be considered as being built up from components from lower levels of the NC/AC hierarchy. By simple modifications of the proofs above, we can present results about "mixed" classes, i.e., NC^i circuits with AC^j oracles or vice versa.

Corollary 4.23. *Let $i, j \geq 1$. Then $AC^i(NC^j) = NC^{i+j}$.*

Proof. The left to right inclusion follows from $AC^i(NC^j) \subseteq NC^{i+1}(NC^j) \subseteq NC^{i+j}$ by Theorem 4.21. Observe that $AC^k(\mathcal{K}) \subseteq NC^{k+1}(\mathcal{K})$ for all $k \geq 1$ and all language classes $\mathcal{K}$, as can be proved in the same way as $AC^k \subseteq NC^{k+1}$ (see Lemma 4.6).

The right to left inclusion follows using the same technique as in Theorem 4.21, i.e., first leveling the circuit and then subdividing the result. The circuit family C' from the above is now regarded as an unbounded fan-in family. Hence the oracle gates contribute with depth 1, and since there are $O((\log n)^i)$ many of those, the depth of C' is also $O((\log n)^i)$. $\square$

Corollary 4.24. *Let $i, j \geq 1$. Then $NC^i(AC^j) = AC^{i+j-1}$.*

Proof. For the left to right inclusion we now have to replace gates for an AC^j language with AC^j circuits. This results in an unbounded fan-in circuit family. The calculation of the depth is as in Theorem 4.21.

The right to left inclusion follows from $AC^{i+j-1} \subseteq AC^{i-1}(AC^j) \subseteq NC^i(AC^j)$ by Theorem 4.22 (and the obvious inclusion $AC^j \subseteq AC^0(AC^j)$ for the case $i = 0$). Again recall that $AC^k(\mathcal{K}) \subseteq NC^{k+1}(\mathcal{K})$ for all $k \geq 1$ and all language classes $\mathcal{K}$. $\square$

Let us now turn to uniformity considerations. We will use logspace-uniformity, but the results hold for U_E-uniformity as well (see Exercise 4.12).

Corollary 4.25. *Let $i, j \geq 1$. Then the following holds.*

1. $U_L\text{-}NC^i(U_L\text{-}NC^j) = U_L\text{-}NC^{i+j-1}$.
2. $U_L\text{-}AC^i(U_L\text{-}AC^j) = U_L\text{-}AC^{i+j}$.
3. $U_L\text{-}AC^i(U_L\text{-}NC^j) = U_L\text{-}NC^{i+j}$.
4. $U_L\text{-}NC^i(U_L\text{-}AC^j) = U_L\text{-}AC^{i+j-1}$.

Proof. We prove statement 1. The proofs of the other three statements are completely analogous.

The combination of an NC^j and an NC^i circuit family, needed to prove the left to right inclusion in Theorem 4.21, is clearly logspace-uniform if the starting circuit families are.

For the right to left inclusion, let us first consider the transformation of family C into a leveled family D. For every D_n there is a number $s \in N$ such that all levels in D_n (except the input level) consist of s gates. If $n_1, \ldots, n_s$ are the numbers of the original gates in the encoding of C_n, we now use the numbers $\langle l, n_1 \rangle, \ldots, \langle l, n_s \rangle$ to encode the gates in D_n on level l. Now it is easy to see that if C is logspace-uniform, then so is D. Moreover since gates carry their depth in D_n in their name, the language B from the proof of Theorem 4.21 is easily seen to be in $U_L\text{-}NC^j$. The NC^i family computing A with the help of oracle gates for B is obviously logspace-uniform. $\square$

It is not known if the NC hierarchy is proper, i.e., consists of infinitely many different levels. However, from the above characterizations we can now obtain the following structural result:

Corollary 4.26. *For all $i \geq 1$, if $NC^i = NC^{i+1}$ then $NC = NC^i$. This holds non-uniformly as well as uniformly.*

Proof. We prove by induction on j that, under the assumptions of the corollary, $NC^{i+j} = NC^i$. The case $j = 1$ is exactly the assumption. Let now $j > 1$, and assume $NC^{i+j-1} = NC^i$. By Theorem 4.21, $NC^{i+j} = NC^2(NC^{i+j-1}) = NC^2(NC^i) = NC^{i+1} = NC^i$. $\square$

With virtually the same proof one can show:

Corollary 4.27. *For all $i \geq 1$, if $AC^i = AC^{i+1}$ then $NC = AC^i$. This holds non-uniformly as well as uniformly.*

4.5 Classes within NC^1

In this section we want to look in more detail at the fine structure of NC^1. After discussing uniformity we will examine several subclasses and give a number of characterizations for them.

4.5.1 Uniformity within NC^1

If we look at a circuit class, what is the "right" uniformity condition to use? First, the computational power of the uniformity machine should not be too high to obscure the power of the circuit itself. Thus logspace-uniformity for NC^1 is bad since under $U_E\text{-}$, $U_E^*\text{-}$, U_L-uniformity and also non-uniformly, NC^1 is a (possibly strict) subclass of L. On the other hand, $U_E^*\text{-}NC^1$ can in a sense be rephrased as "NC^1-uniform NC^1," since U$_{\text{E}}^*$-uniformity requires that the extended connection language is in ATIME($\log n$) which is equal to U$_{\text{E}}^*$-NC^1. Thus in the case of U$_{\text{E}}^*$-NC^1 the power of the uniformity machine is not higher than the power of the circuit family. For SAC^1 and above, logspace-uniformity is acceptable since DSPACE($\log n$) $\subseteq$ ASPACE-TREESIZE($\log n, n^{O(1)}$) = SAC^1.

Second, the resulting circuit class should be *robust* in the sense that slight variations in the model do not affect the power of the class, and that the class has many "natural" characterizations on different computation models. From this point, U$_{\text{E}}^*$-uniformity is again a good choice for NC^1 since it is the weakest uniformity condition we know of for which Corollary 4.3 still holds. (We will add more evidence to this in Sect. 4.5.3.)

The subclasses of NC^1 that we examine below will always be of the form SIZE-DEPTH$_B(n^{O(1)}, 1)$ for some unbounded fan-in basis B. If we want to proceed with U$_{\text{E}}$- or U$_{\text{E}}^*$-uniformity we first have to adapt Def. 2.43 to the case of unbounded fan-in bases. This can be achieved as follows: First, if $p = \langle \text{bin}(i_1), \ldots, \text{bin}(i_k) \rangle$ and g is a gate in a circuit C over B, define $p(g)$ to be the gate that we reach when we follow path p from g to the inputs, i. e., $\langle \text{bin}(i) \rangle (g)$ is the ith predecessor of g, $\langle \text{bin}(i), \text{bin}(j) \rangle (g)$ is the ith predecessor of the jth predecessor of g, and so on. Now given a circuit family $C = (C_n)_{n \in \mathbb{N}}$ and fixing an admissible encoding scheme, $L_{\text{EC}}(C)$ consists of all tuples $\langle y, g, p, b \rangle$, where b is the type of gate g in $C_{|y|}$ if $p = \epsilon$, or the number of gate $p(g)$ in $C_{|y|}$ if $p \neq \epsilon$, and $|p|$ is at most logarithmic in the size of $C_{|y|}$. Observe that for SIZE-DEPTH$_B(n^{O(1)}, 1)$ circuit families C we do not have to consider restrictions on the length of p, since all paths of any circuit in C are of constant length.

As in the case of bounded fan-in, we say that C is U$_{\text{E}}$-uniform if $L_{\text{EC}}(C) \in$ DTIME($\log s(n)$). (We do not define U$_{\text{E}}^*$-uniformity. A formal translation of Def. 2.44 into our context here, where we study only classes of the form SIZE-DEPTH$_B(n^{O(1)}, 1)$, would require alternating machines to run in constant time which is unreasonable.) Fortunately it turns out that for constant-depth circuits, U$_{\text{E}}$-uniformity is equivalent to a more tractable condition.

Definition 4.28. Let $C = (C_n)_{n \in \mathbb{N}}$ be a circuit family of size s and depth d over some basis B. We say that C is U$_{\text{D}}$-uniform (or, logtime-uniform), if there is an admissible encoding scheme such that there is a deterministic Turing machine that accepts $L_{\text{DC}}(C)$ and runs in time $O(\log s(n))$ on inputs of the form $\langle y, \cdot, \cdot, \cdot \rangle$ where $|y| = n$.

The following observation follows immediately from the definition:

Observation 4.29. Let $C = (C_n)_{n \in \mathbb{N}}$ be a circuit family over a basis B. If C is U$_{\text{E}}$-uniform, then it is U$_{\text{D}}$-uniform as well.

For bounded fan-in circuit families, the following partial converse can be given:

Theorem 4.30. Let $C = (C_n)_{n \in \mathbb{N}}$ be a circuit family over B_0 of size s and depth d. Let $d(n) \geq \log s(n) \cdot \log \log s(n)$. If C is U_D-uniform, then C is U_E^*-uniform.

Proof. The proof relies on a divide-and-conquer method similar to that used to prove Theorem 2.35. Let M be a machine for $L_{DC}(C)$. We construct an alternating machine M' for $L_{EC}(C)$. An input $\langle y, \cdot, p, \cdot \rangle$ where $p = \epsilon$ can be decided easily looking at the definition of M. Given now an input $\langle y, \cdot, p, \cdot \rangle$ where $p \neq \epsilon$, we have to follow a path of length at most $\log s(n)$ in circuit C_n. For this it remains to verify a computation path of M of length at most $\log s(n)$. M' proceeds as follows: it guesses existentially the number of the configuration at the middle of the path, then branches universally and verifies recursively the two shorter paths. Paths of length 1 can be verified by simulating the machine for $L_{DC}(C)$.

On inputs of the form $\langle y, \cdot, \cdot, \cdot \rangle$ where $|y| = n$, this procedure needs $\log \log s(n)$ recursion steps. Each step needs time $\log s(n)$ (dominated by the time to write down the number of the gate at the middle of the current path). The final recursion step needs $O(\log s(n))$ steps for the simulation of M. Thus the overall time is $O(\log s(n) \cdot \log \log s(n)) = O(d(n))$. The space requirement is given by the space needed to record the gate numbers (which is $O(\log s(n))$) plus that needed by the subroutine calls to machine M. Since these calls need time $O(\log s(n))$, they cannot need more space, and thus the overall space requirement is given by $O(\log s(n))$. $\square$

Hence we see that for NC^2 and above, U_D-uniformity and U_L-uniformity coincide. For unbounded fan-in classes of the form $\mathrm{SIZE\text{-}DEPTH}_B(n^{O(1)}, 1)$ within NC^1 we now give the following converse to Observation 4.29.

Theorem 4.31. Let $C = (C_n)_{n \in \mathbb{N}}$ be a U_D-uniform circuit family over basis B of polynomial size and constant depth. Then there is an equivalent circuit family C' over $B \cup B_1$ of polynomial size and constant depth which is U_E-uniform.

Proof. We are given a machine M deciding $L_{DC}(C)$ in time $O(\log n)$ on inputs of the form $\langle y, \cdot, \cdot, \cdot \rangle$ where $|y| = n$. Then $L_{EC}(C)$ can be decided by an alternating machine M which works according to the divide-and-conquer method given in the proof of Theorem 4.30. The time bound is $O(\log n)$ and the alternation bound is constant, both due to the fact that we are only dealing here with constant-length paths.

We now construct a circuit family D simulating M using the method described in the proof of Theorem 4.7. This circuit family will have polynomial size and constant depth. Since we are dealing with only constant path lengths we claim that D is U_E-uniform. To verify that a tuple $\langle y, g, p, b \rangle$ is in $L_{EC}(D)$ is obvious for $p = \epsilon$. If $p \neq \epsilon$ then p can only be of constant length. In this case we have to simulate M, but M runs only for $O(\log n)$ steps.

From D we now construct for every triple $\langle g, p, b \rangle$ a circuit family $D^{\langle g, p, b \rangle}$, such that for each y, $|y| = n$, the following holds:

y is accepted by $D_n^{\langle g,p,b \rangle}$ $\iff$ $\langle y,g,p,b \rangle$ is accepted by $D_{|\langle y,g,p,b \rangle|}$

Certainly all these families share U_E-uniformity because $\mathcal{D}$ is U_E-uniform. By prefixing the names of all gates in $\mathcal{D}^{\langle g,p,b \rangle}$ with $\langle g,p,b \rangle$ we now get *one* uniformity machine for all families $\mathcal{D}^{\langle \cdot,\cdot,\cdot \rangle}$.

Finally we construct a circuit family $\mathcal{C}'$ simulating $\mathcal{C}$. The idea is that to obtain C_n' from C_n we will replace every wire connecting a gate g with its ith predecessor gate h by the circuit given in Fig. 4.1. Intuitively g is connected to copies of all gates $g_1, g_2, \ldots, g_s$ which exist in C_n, but making use of the uniformity circuits from the above we make sure that only the output of gate h will be a relevant predecessor to g. C_n' is now obtained by first replacing the wires connecting the output of C_n with its predecessors with the appropriate circuits as just described. Then we proceed with the bottom gates of these circuits (i.e., the gates $g_1, \ldots, g_s$ in the different subcircuits) and handle their input wires as before. Then we handle the input wires of the bottom gates of this new level, and so on. Altogether we will have as many levels in this construction as we have gates on a longest path in C_n. $\mathcal{C}'$ will be of polynomial size and constant depth. Uniformity of the construction follows from uniformity of the circuits for $L_{EC}(\mathcal{C})$ and the regular construction of Fig. 4.1. $\qquad \Box$

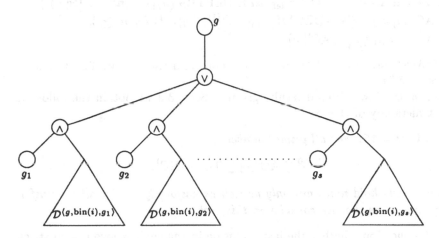

Fig. 4.1. U_E-uniform circuit replacing ith input wire of g

Uniformity issues are an important and non-trivial aspect of small-depth circuit classes as should be clear from this chapter so far. For classes above NC^2 we can safely use U_L-uniformity, and as we saw this condition collapses with all the stricter conditions, even with logtime-uniformity. For subclasses of NC^2 which are still superclasses of L (that is, SAC^1 and AC^1), logspace-uniformity is also acceptable as the characterizations given in the previous sections of this chapter showed.

For NC^1 and subclasses we have to use the more complicated U_E^*- or U_E-uniformity, which fortunately for constant-depth classes coincides with U_D-uniformity. That this is a good choice within NC^1 will become clear from the results we present below. These will show that under this uniformity condition the classes obtained are fairly robust. At the moment let us only note that by proceeding this way we obtain a result for the class AC^0 extending Theorem 4.7:

Corollary 4.32. $U_D\text{-}AC^0 = \text{ATIME-ALT}(\log n, 1)$.

Proof. See Exercise 4.17.   □

Remark 4.33. As in the case of machine characterizations of the other NC^i and AC^i classes (see Corollaries 4.4 and 4.8), the corollary above allows us to conclude that we may define $U_D\text{-}AC^0$ equivalently by restricting ourselves to circuits in input normal form.

4.5.2 A Hierarchy of Classes within NC^1

In Chaps. 1 and 3 we discussed subclasses of NC^1. Let us introduce the following shorthand notations:

Definition 4.34.   1. $TC^i =_{\text{def}} \text{SIZE-DEPTH}_{B_1 \cup \{MAJ\}} \left(n^{O(1)}, (\log n)^i \right)$.
2. $AC^i[m] =_{\text{def}} \text{SIZE-DEPTH}_{B_1(m)} \left(n^{O(1)}, (\log n)^i \right)$ for $m \geq 2$.
3. $ACC^i =_{\text{def}} \bigcup_{m \geq 2} AC^i[m]$.

("ACC" stands for "AC with counters", and the "T" in TC stands for "threshold.")

From the lower bound results given in Sect. 3.3 we obtain the following strict hierarchy of classes:

Corollary 4.35. *For all prime numbers p,*

$$AC^0 \subsetneq AC^0[p] \subsetneq TC^0 \subseteq NC^1.$$

These results hold non-uniformly as well as uniformly (under all the uniformity conditions that we consider in this book).

The question whether the last inclusion in the chain above is also strict, i. e., whether $TC^0 \stackrel{?}{=} NC^1$, is one of the main open problems in circuit complexity.

In the next two subsections we describe (some of) these classes from an algebraic and a logical point of view.

Let us recall the following notation: If A is a language, then $AC^0(A)$ is the class of all languages that can be reduced to A under constant-depth reductions (or equivalently: that can be recognized by polynomial-size constant-depth circuits that have gates for A besides the unbounded fan-in basis). Hence $TC^0 = AC^0(MAJ)$ and $AC^0[p] = AC^0(MOD_p)$.

4.5.3 Algebraic Characterizations

In this subsection we will first introduce a computation model related closely to that of Boolean circuits, the so-called *branching programs*. We will then obtain characterizations of L and NC^1 in terms of this model, which will finally lead us to an algebraic view of the class NC^1.

Definition 4.36. An n-input branching program is a tuple

$$P = (V, E, \beta, \alpha, v_0),$$

where

- (V, E) is a finite directed acyclic graph where each node has out-degree 0 or 2;
- $\beta \colon V \to \{0, 1, x_1, \ldots, x_n\}$ is a function giving the *type* of a node, if a node v has out-degree 0 then $\beta(v) \in \{0, 1\}$, otherwise $\beta(v) \in \{x_1, \ldots, x_n\}$;
- $\alpha \colon E \to \mathbb{N}$ is an injective function, inducing an ordering of the edges in P;
- $v_0 \in V$ is a particular node in V, the so-called *source* node.

P computes a function $f_P \colon \{0,1\}^n \to \{0,1\}$ as follows: Given an input $a_1 \cdots a_n \in \{0,1\}^n$, a pebble is placed on node $v_0 \in V$. A computation step now consists of the following:

> If the pebbled node is v, $\beta(v) = x_i$, and the successors of v are u_0 and u_1 where $\alpha(u_0) < \alpha(u_1)$ then the pebble is removed from v and placed on u_{a_i}, where the computation continues.

Repeat this step until the pebble is placed on a node v with out-degree 0. Now we set $f_P(a_1 \cdots a_n) = \beta(v)$ where v is the node which is pebbled when the computation stops.

The *size* of P is given by $|V|$ and the *length* of P is the length of a longest directed path in (V, E).

A *family of branching programs* is a sequence $\mathcal{P} = (P_n)_{n \in \mathbb{N}}$, where each P_n is an n-input branching program. The function computed by $\mathcal{P}$ is given by $f_{\mathcal{P}} \colon \{0,1\}^* \to \{0,1\}$, $f_{\mathcal{P}}(x) = f_{P_{|x|}}(x)$. Let $s, l \colon \mathbb{N} \to \mathbb{N}$. We say that $\mathcal{P}$ is of size s if, for every n, the size of P_n is bounded by $s(n)$, and $\mathcal{P}$ is of length l if, for every n, the length of P_n is bounded by $l(n)$.

Branching programs of polynomial size are of course of major interest. Therefore we define:

Definition 4.37. BP is the class of all languages $A \in \{0,1\}^*$ for which there is a family $\mathcal{P}$ of branching programs of polynomial size such that $f_{\mathcal{P}} = c_A$.

We want to relate BP to the complexity classes that we have encountered so far. A family of branching programs is an infinite object, and branching programs are a non-uniform computation model. Thus we will have to relate

BP to non-uniform complexity classes. The idea will be to use advice functions to encode families of branching programs (similarly to the simulation of circuits by Turing machines and vice versa in Chap. 2).

Let $\mathcal{P} = (P_n)_{n \in \mathbb{N}}$ be a family of branching programs. An admissible encoding scheme for $\mathcal{P}$ is given as follows: For every n, if $P_n = (V, E, \beta, \alpha, v_0)$ fix a numbering of V such that v_0 is numbered 0 and the highest number of a node in V is bounded by a polynomial in the size of C_n. Then a node $v \in V$ numbered by g is encoded by $\langle g, b \rangle$ if v is of out-degree 0 and $\beta(v) = b$, and by $\langle g, i, g_0, g_1 \rangle$ if v is of out-degree 2, $\beta(v) = x_i$, the successors of v are u_0 and u_1 with $\alpha(u_0) < \alpha(u_1)$, and g_0, g_1 are respectively the numbers of u_0, u_1. Finally fix an arbitrary order on the nodes of P_n. The encoding of P_n, denoted by $\overline{P_n}$ now, is the sequence of encodings of all $v \in V$ in that order.

Theorem 4.38. BP $=$ L/Poly.

Proof. ($\subseteq$): Let $\mathcal{P} = (P_n)_{n \in \mathbb{N}}$ be a family of polynomial-size branching programs computing the characteristic function of $A \in \{0, 1\}^*$. Fix an admissible encoding scheme and define $g(1^n)$ to be an encoding of P_n. It is clear that $g \in \text{Poly} = \text{F}\left(n^{O(1)}\right)$.

Define a Turing machine M to operate as follows: Given an input $\langle x, g(1^{|x|}) \rangle$, M simulates the computation of $P_{|x|}$ step by step. It uses one of its work tapes to hold the number of the current pebbled node. One step consists of scanning the advice to find the encoding of the current node and then updating the number of the pebbled node according to the input. This can clearly be done in logarithmic space.

($\supseteq$): Let M be a Turing machine working in logarithmic space, and let $g \in \text{Poly}$ be the advice function of M. P_n with input x, $|x| = n$ has to simulate the behavior of M on input $\langle x, g(1^n) \rangle$. The number of configurations of M on inputs of length $n + |g(1^n)|$ is bounded by a suitable polynomial p in n (see Theorem 2.9). Let us suppose that M has a unique accepting configuration K_f, and that if M reaches K_f further computation steps do not leave K_f; i.e., M accepts an input of length n if and only if after exactly $p(n)$ computation steps it is in configuration K_f. Let K_0 be the initial configuration of M.

We now define P_n. The nodes of P_n are all pairs $\langle K, t \rangle$, where K is a configuration of M for inputs of length $n + |g(1^n)|$, and $0 \le t \le p(n)$. The connections in P_n are given as follows:

- The source node of P_n is $\langle K_0, 0 \rangle$.
- If $t = p(n)$ then $\beta(\langle K, t \rangle) = 1$ if $K = K_f$, and $\beta(\langle K, t \rangle) = 0$ otherwise.
- Let $t < p(n)$ and the input head position in K be i, $i \le n$. Then $\beta(\langle K, t \rangle) = x_i$, and the successors of $\langle K, t \rangle$ will be $\langle K_0, t+1 \rangle$ and $\langle K_1, t+1 \rangle$ in this order, where K_0 is the successor configuration of K if the ith input symbol is 0 and K_1 is the successor configuration of K if the ith input symbol is 1.
- Let $t < p(n)$ and the input head position in K be i, $n < i \le n + |g(1^n)|$. Then $\beta(\langle K, t \rangle) = x_1$, and both successors of $\langle K, t \rangle$ will be $\langle K_0, t+1 \rangle$ where

K_0 is the successor of K and we assume the current input symbol is the ith symbol in the string $\langle 1^n, g(1^n) \rangle$.

Since by construction P_n simulates $p(n)$ steps of M, $f_{P_n}(x) = 1$ iff M accepts $\langle x, g(1^n) \rangle$. The size of P_n is clearly polynomial in n. $\square$

Of course, there is also a uniform version of Theorem 4.38. Let us say that a family $\mathcal{P} = (P_n)_{n \in \mathbb{N}}$ of branching programs of size s is logspace-uniform (or L-uniform for short) if there is an admissible encoding scheme such that the map $1^n \mapsto \overline{P_n}$ is in FDSPACE($\log s$). U_L-BP denotes the class of all languages which can be recognized by L-uniform branching programs of polynomial size.

Corollary 4.39. U_L-BP $=$ L.

Proof. Argue along the lines of the proof of Theorem 4.38 with the following modifications/additions: For the direction from left to right, we now let the simulating machine make a subroutine call to the uniformity machine instead of looking up the structure of the branching program in its input. For the direction from right to left, observe that the family constructed in the above proof is easily seen to be L-uniform. $\square$

Next we turn to a restricted form of branching programs, the so-called branching programs of *bounded width*.

Let $P = (V, E, \beta, \alpha, v_0)$ be a branching program. We say that P is *layered*, if the following holds: For all nodes v in V, all paths from v_0 to v have the same length. A *level* of P consists of all nodes that have the same distance to v_0. Observe that edges in E can then only connect nodes from neighboring levels.

Definition 4.40. A branching program P is of *width* k if it is layered and all levels in P have no more than k nodes. Let $h \colon \mathbb{N} \to \mathbb{N}$. A family of branching programs $\mathcal{P} = (P_n)_{n \in \mathbb{N}}$ is of *width* h if for all $n \geq 0$, the program P_n has width $h(n)$. $\mathcal{P}$ is of *bounded width* if $\mathcal{P}$ is of width h for a constant function h.

In the rest of this subsection we clarify the power of polynomial-size bounded-width branching programs.

Definition 4.41. BWBP is the class of all languages $A \in \{0,1\}^*$ for which there is a family $\mathcal{P}$ of branching programs of polynomial size and bounded width such that $f_{\mathcal{P}} = c_A$.

First let us show that BWBP coincides with a class of programs of an even more restrictive form.

Definition 4.42. Let M be a monoid. An M-*program* P over n variables is a sequence of triples $((\langle i_j, f_j, g_j \rangle))_{1 \leq j \leq \ell}$, where for every j, $i_j \in \{1, \ldots, n\}$ and $f_j, g_j \in M$. The triples $\langle i_j, f_j, g_j \rangle$ are called *instructions* of P. Given an input $a_1 \cdots a_n \in \{0,1\}^n$, such an instruction yields the *value*

$$\text{val}_{\langle i_j, f_j, g_j \rangle}(a_1 \cdots a_n) =_{\text{def}} \begin{cases} f_j, & \text{if } a_{i_j} = 1, \\ g_j, & \text{if } a_{i_j} = 0. \end{cases}$$

The *output* of B on input $a_1 \cdots a_n$ is defined to be

$$f_P(a_1 \cdots a_n) = \prod_{j=1}^{\ell} \text{val}_{\langle i_j, f_j, g_j \rangle}(a_1 \cdots a_n)$$

(where we use monoid multiplication). The *size* of P is ℓ.

Given a family $\mathcal{P} = (P_n)_{n \in \mathbb{N}}$, where P_n is an M-program over n variables, we define $f_{\mathcal{P}} \colon \{0,1\}^* \to M$ by $f_{\mathcal{P}}(x) =_{\text{def}} f_{P_{|x|}}(x)$. For $s \colon \mathbb{N} \to \mathbb{N}$, we say that $\mathcal{P}$ is of size s if P_n is of size $s(n)$ for all n.

Given a set $F \subseteq M$, the set $L(\mathcal{P}, F) =_{\text{def}} \left\{ x \in \{0,1\}^* \mid f_{\mathcal{P}}(x) \in F \right\}$ is the language *accepted* by B with respect to F. We say that A is accepted by $\mathcal{P}$ if there is some $F \subseteq M$ such that $A = L(\mathcal{P}, F)$.

Example 4.43. Let $M = S_2$ (see Appendix A8), and $F = \{(1,2)\}$ (here, $(1,2)$ denotes the transposition of two elements.) Define the family of M-programs $\mathcal{P} = (P_n)_{n \in \mathbb{N}}$, where each P_n consists of n instructions, and the i-th instruction in each program is $\langle i, (1,2), e \rangle$. Then $\mathcal{P}$ computes the parity function.

Example 4.44. Let $R \subseteq \{0,1\}^*$ be a regular language. It is well known from automata theory that then there is a finite monoid M, a subset $F \subseteq M$, and a homomorphism $h \colon \{0,1\} \to M$, such that for all $x \in \{0,1\}^*$, we have: $x \in R \iff h(x) \in F$. (Essentially, M can be chosen to be the transformation monoid of the minimal automaton for R [Eil76, Pin86].) Thus $R \in \text{BWBP}$.

Branching programs of bounded width and M-programs can simulate each other with a polynomial overhead in size, as we show next.

Lemma 4.45. Let $A \subseteq \{0,1\}^*$. There is a family of bounded-width branching programs $\mathcal{P}$ of polynomial size that computes c_A if and only if there is a finite monoid M, a set $F \subseteq M$, and an M-program $\mathcal{P}'$ of polynomial size such that $A = L(\mathcal{P}', F)$.

Proof. $(\Rightarrow)$: Given $\mathcal{P}$, first construct a family $\mathcal{P}''$ of bounded-width branching programs of polynomial size where within each level all nodes either query the same input bit or are nodes of fan-out 0 (see Exercise 4.19). Let k be the width of $\mathcal{P}''$. If we pick M to be the monoid of all functions from $\{1, \ldots, k\}$ into $\{1, \ldots, k\}$, then $\mathcal{P}''$ can be easily transformed into an M-program family $\mathcal{P}'$.

$(\Leftarrow)$: Given an M-program $\mathcal{P}'$, construct $\mathcal{P}$ as follows: Every level of a program in $\mathcal{P}$ will have one node per monoid element. There will be one such level for each instruction in the M-program. The connections between the levels are then constructed according to the multiplication of the corresponding instruction-elements. The details are left as an exercise (Exercise 4.20). The size of $\mathcal{P}$ is clearly polynomial since the size of $\mathcal{P}'$ is polynomial.   ◻

The following theorem characterizes the circuit class NC^1 in terms of branching programs.

Theorem 4.46 (Barrington). BWBP = NC^1.

Proof. ($\subseteq$): Let $A \in$ BWBP, $A = L(\mathcal{P}, F)$ for some M-program family $\mathcal{P} = (P_n)_{n \in \mathbb{N}}$ and $F \subseteq M$. We construct a circuit family $\mathcal{C} = (C_n)_{n \in \mathbb{N}}$.

First we observe that (under an encoding of the elements of M as words over $\{0, 1\}$) multiplication in M can be realized by constant-size circuits. The circuit C_n for $n \in \mathbb{N}$ then works as follows: First, at the bottom level the input word is transformed into a polynomial-length sequence of elements of M, according to the instructions in P_n. This sequence is then multiplied out in a tree-like fashion, making use of associativity of multiplication in M. This gives a binary tree of multiplications. At the top level membership in F is tested. All this results in a bounded fan-in circuit of logarithmic depth.

($\supseteq$): Let $A \in$ NC^1 be accepted by circuit family $\mathcal{C} = (C_n)_{n \in \mathbb{N}}$. We assume without loss of generality that every C_n is a circuit consisting of $\wedge$ and $\vee$ gates with a bottom level, where the inputs and their negations can be found.

We construct an M-program family for A, where we pick M to be any non-solvable group G (see Appendix A8). Let us suppose at the moment that G coincides with its commutator subgroup. The main idea in the construction is that the commutator (g, h) of two elements g, h from G has the "character" of an $\wedge$ gate: (g, h) is equal to the identity of G if and only if g or h is the identity, since the identity commutes with every group element. Hence if we associate the identity with the truth value *false* and all other group elements with *true*, then (g, h) is true if and only if both g and h are true.

Now let G be any non-solvable group. Let first us assume as above that the commutator subgroup of G is again G. Let $m = |G|$ be the order of G. Every $\sigma \in G$ not equal to the identity must have a representation as a product of at most m commutators in G.

We now prove by induction: Let C be a circuit as above of depth d. For every $\sigma \in G$, $\sigma \neq e$, there is a G-program P of length at most $(4m)^d$ such that for all inputs x, if $f_C(x) = 1$ then $f_P(x) = \sigma$, and if $f_C(x) = 0$ then $f_P(x) = e$.

If $d = 0$ then the output of C is equal to one of the inputs. In this case there is a one-instruction M-program with the properties claimed.

Now let C be of depth $d > 0$, and let $\sigma \neq e$. Assume first that the output gate of C is an $\wedge$ gate, and let A be the set accepted by C. Then $A = A_1 \cap A_2$ and there are depth $d - 1$ circuits C_1 and C_2 that accept respectively A_1 and A_2. Thus, by induction hypothesis, for every $\tau \in G$ there are programs P_1^τ and P_2^τ, both of length at most $(4m)^{d-1}$ such that for $i = 1, 2$, if an input x is accepted by C_i then P_i^τ outputs τ, otherwise P_i^τ outputs e. Let now $\sigma = (g_1, h_1) \circ (g_2, h_2) \circ \cdots \circ (g_l, h_l)$ for $l \leq m$ and $g_i, h_i \in G$ for $1 \leq i \leq l$. Define the program P to be the concatenation of the programs

$$P_1^{g_1} P_2^{h_1} P_1^{g_1^{-1}} P_2^{h_1^{-1}} P_1^{g_2} P_2^{h_2} P_1^{g_2^{-1}} P_2^{h_2^{-1}} \cdots P_1^{g_l} P_2^{h_l} P_1^{g_l^{-1}} P_2^{h_l^{-1}} \qquad (4.2)$$

Then if $x \in A_1 \cap A_2$, P will output σ, and if $x \notin A_1 \cap A_2$, P will output e. Thus P is a program that simulates C. The length of P is at most $4m(4m)^{d-1} = (4m)^d$.

As a second case we see what happens if the top gate of C is an $\vee$ gate. Let us look at the circuit C' that computes the complement of the output of C. This circuit has an $\wedge$ gate on top. Proceeding as in the first case we obtain a G-program P of length $(4m)^d$ that, on input x, outputs σ^{-1} if C' accepts x, and outputs e otherwise. Let $\langle i, g, h \rangle$ be the final instruction in P. Replace this instruction with $\langle i, g \circ \sigma, h \circ \sigma \rangle$. Then this new program outputs σ if C accepts x (and hence C' rejects x), and e otherwise.

If now G is not equal to its commutator subgroup, let G_1 be a subgroup of G that coincides with its commutator subgroup. Now proceed as above with G_1 instead of G. We thus obtain a G_1-program for A. Since $G_1 \subseteq G$ this is also a G-program for A.   □

The result of the theorem above can be formulated in another way. Given a monoid M, let the *word problem* for M, denoted by L_M, be the set of all sequences of elements of M (suitably encoded over the binary alphabet) that multiply out to the identity. Say that a language $B \subseteq \{0,1\}^*$ is complete for NC^1 under a reducibility $\mathcal{R}$, if $B \in \mathrm{NC}^1$ and for all languages $A \in \mathrm{NC}^1$, we have that A reduces to B under $\mathcal{R}$-reductions.

Corollary 4.47. *Let G be a non-solvable group. Then the word problem for G is complete for NC^1 under $\leq_{cd}$.*

Proof. Let $A \in \mathrm{NC}^1$ via circuit family $\mathcal{C} = (C_n)_{n \in \mathbb{N}}$. By Theorem 4.46 there is a family $\mathcal{P}$ of G-programs that simulates $\mathcal{C}$. The reducing circuit now looks as follows: Evaluating every instruction of $\mathcal{P}$, a polynomial-length sequence of elements of G is produced. This sequence is then input to a gate for the word problem for G.   □

Observe that the word problem for a finite monoid is a regular language. Thus we have:

Corollary 4.48. *A language A is in NC^1 if and only if A is reducible via constant-depth reductions to a regular language. Moreover there are regular languages which are complete for NC^1 under $\leq_{cd}$.*

We want to strengthen the result just given by proving that we can use even stricter reductions. For this, we first look at particular monoids, the so-called *free monoids*. Let Σ be an alphabet, and let $\circ$ denote concatenation of words over Σ. Then $(\Sigma^*, \circ)$ is the free monoid over Σ. A Σ-program over n variables is now an M-program over n variables where we pick as monoid M the free monoid over Σ. Such a program defines a mapping from $\{0,1\}^n$ to Σ^*. Given a family $\mathcal{P} = (P_n)_{n \in \mathbb{N}}$, where P_n is a Σ-program over n variables, this family thus defines a function $f_{\mathcal{P}} : \{0,1\}^* \to \Sigma^*$.

A *projection* is a function $f: \{0,1\}^* \to \{0,1\}^*$ for which there is a family $\mathcal{P}$ of $\{0,1\}$-programs such that $f = f_\mathcal{P}$. Say that A *projection-reduces* to B, in symbols: $A \leq_{\text{proj}} B$, if there is a projection f such that for all x, $x \in A$ if and only if $f(x) \in B$.

Now we can rephrase Theorem 4.46 as follows:

Corollary 4.49. *A language A is in* NC^1 *if and only if A is reducible via projection reductions to a regular language. Moreover there are regular languages which are complete for* NC^1 *under* $\leq_{\text{proj}}$.

Proof. ($\Rightarrow$): If A is in NC^1, then by Theorem 4.46 $A \in$ BWBP, hence there is a non-solvable finite monoid M and a family $\mathcal{P}$ of M-programs such that for all x, $x \in A$ if and only if $f_\mathcal{P}(x) = e$, where e is the identity of M. Now let the alphabet Σ be the set of all monoid elements and let R be the word problem for M, $R \subseteq \Sigma^*$. Look at $\mathcal{P}$ as a family of Σ-programs. Then $x \in A \iff f_\mathcal{P}(x) \in R$. Encoding R over the binary alphabet (e. g., using block encoding, where every letter $a \in \Sigma$ is encoded by a sequence of binary symbols, such that $|h(a)| = |h(b)|$ for all $a, b \in \Sigma$) yields a regular language R' to which A reduces via a projection.

If M is a non-solvable group then the obtained R' is complete for NC^1.

($\Leftarrow$): Let $R \subseteq \Sigma^*$, let $\mathcal{P}$ be a family of $\{0,1\}$-programs, such that for all inputs x, we have $x \in A \iff f_\mathcal{P}(x) \in R$. Then in fact A reduces to R via constant-depth reductions. The reducing circuit just has to evaluate (in parallel) all instructions of $\mathcal{P}$ and feed the sequence of outputs into an R gate. Hence $A \in$ NC^1 by Corollary 4.48. $\square$

Thus every language $A \in$ NC^1 reduces to a regular language, not only via constant-depth reduction (Corollary 4.48), but even in the very strict sense of the corollary just given, i. e., via a function computable by an M-program.

From the results above we conclude that NC^1 = L/Poly if and only if every polynomial-size branching program can be simulated by a polynomial-size branching program of bounded width.

Let us now turn to a uniform version of Barrington's Theorem. In the above we examined logspace-uniformity for branching programs. For the same reason that U$_L$-uniformity is not a good choice for the circuit class NC^1, we will not use it here for BWBP, since BWBP is (non-uniformly as well as for logspace-uniformity) a possibly strict subclass of L; hence the power of the uniformity machine is too strong for the class. What we want to do is examine uniformity notions which closely resemble U$_D$- and U$_E$-uniformity for circuits. For this purpose, we first define an encoding of a branching program by a so-called program language, similar to encodings of circuits via connection languages.

Definition 4.50. For a family of M-programs $\mathcal{P} = (P_n)_{n \in \mathbb{N}}$ and a fixed admissible encoding scheme, the *program language* $L_\mathrm{P}(\mathcal{P})$ is defined to be the language of all tuples

$$\langle y, k, i, f, g \rangle,$$

where in $P_{|y|}$ the kth instruction exists and is $\langle i, f, g \rangle$.

A measure for uniformity of a family $\mathcal{P}$ will be the complexity of its program language. As in the case of circuits, the classes ATIME($\log n$) and DTIME($\log n$) turn out to be good choices.

Definition 4.51. Let $\mathcal{P}$ be a family of M-programs of size s. Then we define:

1. $\mathcal{P}$ is U_D-uniform (or, logtime-uniform), if there is an admissible encoding scheme such that $L_P(\mathcal{P})$ can be recognized by a deterministic Turing machine running in time $O(\log s(n))$ on inputs of the form $\langle y, k, i, f, g \rangle$, where $|y| = n$.
2. $\mathcal{P}$ is U_B-uniform (or, ALOGTIME-uniform), if there is an admissible encoding scheme such that $L_P(\mathcal{P})$ can be recognized by an alternating Turing machine running in time $O(\log s(n))$ on inputs of the form $\langle y, k, i, f, g \rangle$, where $|y| = n$.

We will prefix the name of the class by U_D- or U_B- to indicate to which uniformity condition we are referring.

Theorem 4.52. *Let $A \subseteq \{0, 1\}^*$, and let G be any non-solvable group. The following statements are equivalent.*

1. $A \in$ ATIME($\log n$).
2. *There is a polynomial-size G-program family that accepts A, whose program language is in* ATIME($\log n$).
3. *There is a polynomial-size G-program family that accepts A, whose program language is in* DTIME($\log n$).

Proof. We first prove (2) $\Rightarrow$ (1). Let $A = L(\mathcal{P}, F)$ for some $F \subseteq G$, and suppose we are given an input x, $|x| = n$. Program P_n will produce a polynomial-length sequence of elements from G. We have to check that this sequence multiplies out to an element from F.

We define an alternating machine M as follows: M first existentially guesses an $f \in F$. Then M verifies that the sequence produced by P_n evaluates to f. For this, M existentially guesses an element $g \in G$ and then verifies universally that

– the first half of the sequence evaluates to g; and
– the second half of the sequence prefixed by g evaluates to f.

This can be implemented as a recursive procedure. For an input of length polynomial in n, a logarithmic number of recursion steps is necessary. The final recursion step consists of evaluating one instruction of P_n. This can be done in ATIME($\log n$) because the program language for $\mathcal{P}$ is in this class.

The implication (3) $\Rightarrow$ (2) is obvious. We now prove (1) $\Rightarrow$ (3).

Let $A \in \text{ATIME}(\log n)$ be accepted by the alternating machine M. Without loss of generality we assume that M is in input normal form. From M, a logarithmic-depth U_E-uniform circuit family $\mathcal{C} = (C_n)_{n \in \mathbb{N}}$ that accepts A can be obtained as in Theorem 2.48. The circuits in $\mathcal{C}$ immediately correspond to computation trees of M. From $\mathcal{C}$ we then construct a family $\mathcal{P} = (P_n)_{n \in \mathbb{N}}$ of G-programs as in Theorem 4.46. Let $|G| = m$.

It is sufficient to construct a machine that, given n in unary and k in binary, produces the kth instruction in P_n. Let the computation tree of M for input length n be a complete binary tree of depth t (where $t = O(\log n)$). This can be achieved by adding useless branches where necessary. First we want to determine from k the sequence of nondeterministic choices of M on the path that corresponds to the kth instruction. That is, our goal is to compute a binary string $\bar{s} = s_1 \cdots s_t$ such that bit s_i is the choice machine M made at its ith nondeterministic branch ($s_i = 0$ if M took the left branch, $s_i = 1$ if M took the right branch).

For simplicity let us suppose for the moment that every $\sigma \in G$ is the commutator of two other group elements. In this case the program P constructed in (4.2) is the concatenation of 4 shorter programs. We leave it to the reader to check that the construction in the proof of Theorem 4.46 is such that the non-deterministic choices of M are then given exactly by the odd-numbered bits in k, see Exercise 4.21.

Now in the general case we first make sure (by adding useless instructions, that is, instructions always evaluating to the identity) that all programs we construct according to (4.2) consist of the same number of shorter programs, and that this number is a power of 2, say, $4 \cdot 2^c$. Then every $(c+2)$th bit in the binary representation of k corresponds to a guess of M.

From this sequence $\bar{s}$ of choices we can immediately simulate the corresponding path of M in logarithmic time and determine the input bit that is queried.

To determine the elements from G that constitute the second and third component of the kth instruction we have to trace the construction of P_n in the proof of Theorem 4.46, while reading k bit for bit to determine in which case of the construction we are. This can in fact be done by a finite automaton with input k. The time is determined by the length of k, so this is again $O(\log n)$. We leave the cumbersome details, which require no new ideas, as an exercise (Exercise 4.21). $\square$

Corollary 4.53. $\text{ATIME}(\log n) = U_E\text{-}NC^1 = U_E^*\text{-}NC^1 = U_D\text{-}BWBP = U_B\text{-}BWBP$.

Corollary 4.54. *Let G be a non-solvable group. Then the word problem for G is complete for $\text{ATIME}(\log n)$ under $\leq_{cd}$, where the reducing circuit family is U_D-uniform.*

Proof. Completely analogous to the proof of Corollary 4.47. For any language $A \in \text{ATIME}(\log n)$ by Theorem 4.52 we obtain a family $\mathcal{P}$ of G-programs that accepts A. The reducing circuit, as before, looks as follows: Evaluating every instruction of $\mathcal{P}$, a polynomial-length sequence of elements of G is produced, which is input to a gate for the word problem for G. Uniformity is obvious from the uniformity of $\mathcal{P}$. □

We consider the results of this subsection as additional evidence that for NC^1, U_E^*-uniformity is the right choice.

4.5.4 A Logical Characterization

In this subsection we will show that the languages in AC^0 are exactly those languages that can be defined by first-order formulas. We start by introducing the logical framework. We will precisely define all logical concepts needed for the results presented, but since logical languages are not the central topic of this book, our exposition will be very short. For an elaborate and informative introduction to first-order logic, the reader may wish to consult [Sho67, Sch89, EFT94, Imm99].

It has become customary in the study of logical properties of languages to consider only non-empty words. (This reflects the logical tradition that all referred to structures are required to be non-empty.) For the rest of this subsection we therefore make the assumption that all languages do not contain the empty word ϵ.

Definition 4.55. First-order formulas are built using variables, numerical predicate symbols, input predicate symbols, and the *logical symbols* $\{\neg, \wedge, \vee, (,), \exists, \forall\}$. *Variables* are elements of the set $\{v_1, v_2, v_3, \dots\}$. All *numerical predicate symbols* are of the form R_i^r for $r \geq 0$ and $i \geq 1$. Here, r is called the *arity* of R_i^r. There are two input predicate symbols: Q_0 and Q_1.

The set of *first-order formulas* is the smallest set fulfilling the following properties:

(1) If x is a variable, $a \in \{0, 1\}$, then $Q_a(x)$ is a first-order formula.
(2) If $x_1, \dots, x_r$ are variables and R is a numerical predicate symbol of arity r, then $R(x_1, \dots, x_r)$ is a first-order formula.
(3) If ϕ and ψ are first-order formulas, then $\neg\phi$, $(\phi \vee \psi)$, and $(\phi \wedge \psi)$ are also first-order formulas.
(4) If ϕ is a first-order formula and x is a variable, then $\exists x\phi$ and $\forall x\phi$ are first-order formulas.

Formulas of type (1) and (2) are called *atomic* formulas.

Definition 4.56. Let ψ be a formula with a sub-formula of the form $\exists x\phi$ or $\forall x\phi$. Then all occurrences of the variable x in ϕ are called *bound* occurrences. All occurrences of x in ψ that are not bound (in ϕ or another sub-formula of ψ) are called *free*.

For example, in $\exists v_1 (R_1^2(v_1, v_2) \wedge \forall v_2 R_1^3(v_1, v_2, v_3))$ variable v_1 has two bound occurrences, variable v_2 has one bound and one free occurrence, and variable v_3 has one free occurrence. We will sometimes write $\phi(x_1, \ldots, x_k)$ to denote implicitly that $x_1, \ldots, x_k$ (and possibly other additional variables) occur free in ϕ. A formula without free occurrences of variables is said to be a *sentence*.

We want to use formulas to describe properties of words over the alphabet $\{0, 1\}$. The idea is that variables stand for positions in a word and $Q_a(x)$ expresses that the xth letter in the word is an a. For example, the formula

$$\exists v_1 Q_1(v_1)$$

is true for all words that have at least one letter 1. But how should we deal with free variables? Given a word $w \in \{0, 1\}^*$, how do we define the truth value of a formula $Q_1(v_1)$? We overcome this problem in the following way: "Inputs" to our formulas will be words $w \in \{0, 1\}^*$ plus an assignment of values $\{1, \ldots, |w|\}$ to the variables with free occurrences. Formally, we define:

Definition 4.57. Let V be a finite set of variables. A *V-structure* is a sequence

$$w = (a_1, U_1)(a_2, U_2) \cdots (a_n, U_n)$$

such that $U_i \cap U_j = \emptyset$ for $i \neq j$ and $\bigcup_{i=1}^n U_i = V$. Formally we look at w as a word over the alphabet $\{0, 1\} \times \mathcal{P}(V)$, and we say that n is the *length* of w. If $V = \emptyset$ then we write $a_1 \cdots a_n$ as a shorthand for $(a_1, \emptyset) \cdots (a_n, \emptyset)$.

Such a V-structure w determines first the input word to be $a_1 a_2 \cdots a_n$, and second an assignment to all variables $v \in V$ by setting $v \leftarrow i$ if $v \in U_i$. If $V = \{x_1, \ldots, x_k\}$ and $x_j \in U_{i_j}$ for $1 \leq j \leq k$, we will also use, for sake of readability, the notation $w = [a_1 \cdots a_m; x_1 \leftarrow i_1, \ldots, x_k \leftarrow i_k]$, but the reader should keep in mind that formally w is still a word over $\{0, 1\} \times \mathcal{P}(V)$.

Example 4.58. Consider the formula $\phi = Q_1(v_1)$. The set of variables free in ϕ is $V = \{v_1\}$.

1. Consider the input $w = (0, \emptyset)(1, \emptyset)(1, \{v_1\})(0, \emptyset)$, i. e., $w = [0110, v_1 \leftarrow 3]$. This fixes the actual input word to be 0110 and assigns value 3 to variable v_1. Hence intuitively w fulfills the property defined by ϕ.
2. Now consider $w = (0, \emptyset)(1, \emptyset)(1, \emptyset)(0, \{v_1\})$, i. e., $w = [0110, v_1 \leftarrow 4]$. Here ϕ does not hold for w.

Given now a formula ϕ with free variables from a set V and a V-structure w, we know how to interpret free occurrences of variables. But the question remains how to interpret numerical predicate symbols.

Definition 4.59. An *interpretation* I is a sequence of functions $I = (I^n)_{n \in \mathbb{N}}$ as follows: If R is a numerical predicate symbol of arity r, then for every $n \geq 1$,

$I^n(R)$ is a subset of $\{1,\ldots,n\}^r$. That is, I assigns to every numerical predicate symbol for every possible value of n an r-place relation over $\{1,\ldots,n\}$. The idea behind this is that, for structures of length n, the predicate symbol R will be interpreted by $I^n(R)$.

The *semantics* of first-order formulas can now be defined formally as follows.

Definition 4.60. Let I be an interpretation and ϕ be a first-order formula. For simplicity we require that no variable in ϕ has a bound and a free occurrence, and that no variable has bound occurrences in the scopes of two different quantifiers. (These requirements can of course always be achieved easily by renaming the variables.) Let V be the set of variables with free occurrences in ϕ and let w be a V-structure, $w = (a_1, U_1)(a_2, U_2) \cdots (a_n, U_n)$. We define that w *satisfies* ϕ (or, w is a *model* of ϕ) with respect to I, in symbols:

$$w \models_I \phi.$$

by induction on the structure of ϕ:

- If $\phi = Q_a(x)$ (for $a \in \{0,1\}$), then $w \models_I \phi$ if and only if there is an i, $1 \le i \le n$, such that $a_i = a$ and $x \in U_i$.
- If $\phi = R(x_1, \ldots, x_r)$ for a numerical predicate symbol R of arity r, and $x_j \in U_{i_j}$ for $1 \le j \le r$, then $w \models_I \phi$ if and only if $(i_1, \ldots, i_r) \in I^n(R)$.
- If $\phi = \neg\psi$, then $w \models_I \phi$ if and only if $w \not\models_I \psi$.
- If $\phi = \psi \vee \xi$, then $w \models_I \phi$ if and only if $w \models_I \psi$ or $w \models_I \xi$.
- If $\phi = \psi \wedge \xi$, then $w \models_I \phi$ if and only if $w \models_I \psi$ and $w \models_I \xi$.
- If $\phi = \exists x\psi$ for a variable x, then $w \models_I \phi$ if and only if there is some i, $1 \le i \le n$, such that

$$(a_1, U_1) \cdots (a_{i-1}, U_{i-1})(a_i, U_i \cup \{x\})(a_{i+1}, U_{i+1}) \cdots (a_n, U_n) \models_I \psi.$$

- If $\phi = \forall x\psi$ for a variable x, then $w \models_I \phi$ if and only if for all i, $1 \le i \le n$, we have

$$(a_1, U_1) \cdots (a_{i-1}, U_{i-1})(a_i, U_i \cup \{x\})(a_{i+1}, U_{i+1}) \cdots (a_n, U_n) \models_I \psi.$$

The *language defined by* ϕ is

$$L_I(\phi) =_{\text{def}} \{ (w \in \{0,1\} \times \mathcal{P}(V))^+ \mid w \models_I \phi \}.$$

If the interpretation I is understood from the context we omit index I and write $L(\phi)$ and $w \models \phi$.

Definition 4.61. Let $\mathcal{R}$ be a set of finite relations over N. We use the notation FO[$\mathcal{R}$] to refer to the class of all sets A such that $A = L_I(\phi)$ for a first-order sentence ϕ and an interpretation I, in which all numerical predicate symbols are interpreted by relations from $\mathcal{R}$ or the empty relation.

In the following we assume that the symbol R_1^2 is always interpreted as a linear order, i.e., for all interpretations I we have $I^n(R_1^2) = \{\, (l,m) \mid 1 \leq l < m \leq n \,\}$. We will simply write $x < y$ instead of $R_1^2(x,y)$. If no other relation (except the empty relation) is allowed, we use the notation FO[<].

Let us now consider as examples a few FO[<] sentences. For readability we will use $x, y, z, x_1, x_2, \ldots$ as the names of variables, and we will leave out parentheses that are unnecessary because of the associativity of $\wedge$ and $\vee$.

Example 4.62. $(\neg(x < y) \wedge \neg(y < x))$ expresses that in all possible models we must have $x = y$. Hence FO[<] can express equality. We will use the shorthand $x = y$ in the formulas below. Similarly, $>, \leq, \geq$ can be expressed.

Example 4.63. $\forall y (x \leq y)$ expresses that x is the first position in the word. The above formula will be abbreviated by first(x). Analogously, a formula last(x) is defined.

Example 4.64. $\exists x \, (\text{first}(x) \wedge Q_1(x))$ defines the language of all words that start with the letter 1.

Example 4.65. $\phi = \forall x \forall y \, (((\neg Q_1(x) \vee \neg Q_1(y)) \vee \exists z (x < z \wedge z < y)))$. Here, $L(\phi)$ is the set of all words with no consecutive appearances of the letter 1.

The formula $(\neg \psi \vee \xi)$ is more customarily written as $(\psi \to \xi)$, and we will use this notation further on. Furthermore, we abbreviate $((\phi \to \xi) \wedge (\xi \to \phi))$ by $(\phi \leftrightarrow \xi)$, and $\neg(\phi \leftrightarrow \xi)$ by $(\phi \oplus \xi)$.

Example 4.66. Let $k \in \mathbb{N}$, $k \geq 2$. The formula

$$\exists z_1 \exists z_2 \cdots \exists z_{k-1} \Big(x < z_1 \wedge z_1 < z_2 \wedge \cdots \wedge z_{k-2} < z_{k-1} \wedge z_{k-1} < y$$

$$\wedge \, \forall z \big((x \leq z \wedge z \leq y) \to$$

$$(z = x \vee z = z_1 \vee \cdots \vee z = z_{k-1} \vee z = y) \big) \Big)$$

expresses that position y is exactly k positions to the right of position x. This formula will be abbreviated by $\text{dist}_k(x,y)$. Analogously, formulas $\text{dist}_0(x,y)$ and $\text{dist}_1(x,y)$ can be constructed.

Example 4.67. Let $\phi(x)$ be a first-order formula with free variable x, and let $k \geq 1$. Then the formula $\psi = \exists y (\text{first}(y) \wedge \text{dist}_k(y,x) \wedge \phi(x))$ will be true for all inputs $(a_1, U_1)(a_2, U_2) \cdots (a_n, U_n)$, $n \geq k+1$, for which $(a_1, U_1) \cdots (a_{k+1}, U_{k+1} \cup \{x\}) \cdots (a_n, U_n) \models \phi$. Intuitively, ψ is ϕ with the value $k+1$ substituted for x. We also use the notation $\phi(k)$ for this formula.

Example 4.68. Let $k \in \mathbb{N}$. The formula $\exists x \exists y (\text{first}(x) \wedge \text{last}(y) \wedge \text{dist}_k(x,y))$ describes the set of all words of length exactly $k+1$.

We now come to the main result of this subsection. Let Arb denote the set of all finite relations over $\mathbb{N}$. Then FO[Arb] is the class of all sets definable by first-order sentences without restriction on the numerical predicates.

Theorem 4.69. $FO[Arb] = AC^0$.

Proof. ($\subseteq$): Fix an input length n. Let I be an arbitrary interpretation. For all first-order formulas ϕ with exactly k free variables $x_1, \ldots, x_k$, and for all $1 \leq m_1, \ldots, m_k \leq n$, we will construct a circuit $C_n^{\phi(m_1,\ldots,m_k)}$ with the following property: For every input $w \in \{0,1\}^n$, $w = a_1 \cdots a_n$, w is accepted by $C_n^{\phi(m_1,\ldots,m_k)}$ if and only if the $\{x_1, \ldots, x_k\}$-structure w is a model of ϕ, where $w = [a_1 \cdots a_n; x_1 \leftarrow m_1, \ldots, x_k \leftarrow m_k]$. That is, we construct a circuit for ϕ where we substitute the values $m_1, \ldots, m_k$ for the free variables.

We proceed by induction. A formula ϕ with k free variables $x_1, \ldots, x_k$, and natural numbers $m_1, \ldots, m_k$ are given.

- Let $\phi = \exists x \psi(x)$. If x does not occur free in ψ, then $C_n^{\phi(m_1,\ldots,m_k)} = C_n^{\psi(m_1,\ldots,m_k)}$. Otherwise, the free variables in ψ are $x_1, \ldots, x_k, x$. The circuit $C_n^{\phi(m_1,\ldots,m_k)}$ consists of an unbounded fan-in $\vee$ gate on top, whose inputs are all circuits $C_n^{\psi(m_1,\ldots,m_k,i)}$ for $1 \leq i \leq n$.
- If $\phi = \forall x \psi(x)$, we proceed as before with an $\wedge$ gate on top.
- If $\phi = \neg\psi$, $\phi = \psi \vee \xi$, or $\phi = \psi \wedge \xi$, then $C_n^{\phi(m_1,\ldots,m_k)}$ is immediate from the circuits $C_n^{\psi(m_1,\ldots,m_k)}$ and $C_n^{\xi(m_1,\ldots,m_k)}$.
- Let $\phi = Q_a(x)$ for $a \in i$. Then x must be x_i for an $i \in_r \{1, \ldots, k\}$, and hence $C_n^{\phi(m_1,\ldots,m_k)}$ is either the m_ith input gate (if $a = 1$) or the negation of the m_ith input gate (if $a = 0$).
- Let $\phi = R(y_1, \ldots, y_r)$ for some numerical predicate symbol R of arity r. Then the variables $y_1, \ldots, y_r$ must be among the $x_1, \ldots, x_k$; suppose $y_j = x_{i_j}$ for $1 \leq j \leq r$. Then $C_n^{\phi(m_1,\ldots,m_k)}$ is a circuit for the constant 1, if the tuple $(m_{i_1}, \ldots, m_{i_r})$ is in $I^n(R)$, or the constant 0 otherwise.

If now ϕ is a sentence then clearly this construction results in a for $L(\phi) \cap \{0,1\}^n$; and since the depth of C_n^{ϕ} does not depend on n, we see that $L(\phi) \in AC^0$.

($\supseteq$): Let $A \in AC^0$ be accepted by the family $\mathcal{C} = (C_n)_{n \in \mathbb{N}}$. We assume without loss of generality that every C_n is a tree and that for every gate v in $\mathcal{C}$, all paths from v to the input gates are of the same length. (See Exercise 4.35.)

Since $\mathcal{C}$ is of polynomial size there is a $k \in \mathbb{N}$ such that C_n has at most n^k gates. We will construct a formula ϕ and an assignment I such that for all inputs x, we have $x \in A \iff x \in L_I(\phi)$.

Thus now let $n \geq 1$. We can encode gates in C_n as k-tuples of numbers (i.e., input positions). That is, we identify k-tuples over $\{1, \ldots, n\}$ with the set $\{0, 1, \ldots, n^k - 1\}$. We next define the following relations over $\{1, \ldots, n\}$.

1. $\text{In}_n \subseteq \{1, \ldots, n\}^{k+1}$; $(i, m_1, \ldots, m_k) \in \text{In}_n$ if and only if $(m_1, \ldots, m_k)$ is the encoding of the ith circuit input gate x_i.
2. $\text{Out}_n \subseteq \{1, \ldots, n\}^k$; $(m_1, \ldots, m_k) \in \text{Out}_n$ if and only if $(m_1, \ldots, m_k)$ is the encoding of the circuit output gate.

3. Or$_n \subseteq \{1,\ldots,n\}^k$; $(m_1,\ldots,m_k) \in$ Or$_n$ if and only if $(m_1,\ldots,m_k)$ is the encoding of an $\vee$ gate in C_n.
4. And$_n \subseteq \{1,\ldots,n\}^k$; $(m_1,\ldots,m_k) \in$ And$_n$ if and only if $(m_1,\ldots,m_k)$ is the encoding of an $\wedge$ gate in C_n.
5. Neg$_n \subseteq \{1,\ldots,n\}^k$; $(m_1,\ldots,m_k) \in$ Neg$_n$ if and only if $(m_1,\ldots,m_k)$ is the encoding of a $\neg$ gate in C_n.
6. Pre$_n \subseteq \{1,\ldots,n\}^{2k}$; $(l_1,\ldots,l_k,m_1,\ldots,m_k) \in$ Pre$_n$ if and only if $(l_1,\ldots,l_k)$ is a predecessor of gate $(m_1,\ldots,m_k)$ in C_n.

These are finite relations over N. For simplicity, we use the names In, Out, Or, and so on not only for the above relations but also as numerical predicate symbols. The interpretation I will be such that I^n assigns In$_n$, Out$_n$, Or$_n$, And$_n$, Neg$_n$, and Pre$_n$ to the symbols In, Out, Or, And, Neg, and Pre, respectively. The different relations that appear as values of I^n are thus an encoding of the circuit C_n.

We will now construct for every d a formula Acc$_d(x_1,\ldots,x_k)$ with k free variables $x_1,\ldots,x_k$ such that the following holds: Let v be any gate in C_n on level d (that is, all paths from v to the circuit inputs have length d). Let v be encoded by $(m_1,\ldots,m_k)$, and let the input to the circuit be $a_1,\ldots,a_n \in \{0,1\}^n$. Then v evaluates to 1 on this input if and only if $[a_1 \cdots a_n; x_1 \leftarrow m_1,\ldots,x_k \leftarrow m_k] \models$ Acc$_d$.

We proceed by induction on d: If $d = 0$ then v must be an input gate x_i, $1 \leq i \leq n$. Thus we define:

$$\text{Acc}_0 = \exists x \big(\text{In}(x,x_1,\ldots,x_k) \wedge Q_1(x) \big).$$

Now let $d > 0$. We define Acc$_d = \Psi_1 \vee \Psi_2 \vee \Psi_3$, where the formulas Ψ_1, Ψ_2, Ψ_3 will handle the cases where v is an $\vee$, $\wedge$, or $\neg$ gate, respectively.

$$\Psi_1 = \Big(\text{Or}(x_1,\ldots,x_k) \wedge \exists y_1 \cdots \exists y_k \big(\text{Pre}(y_1,\ldots,y_k,x_1,\ldots,x_k)$$
$$\wedge \, \text{Acc}_{d-1}(y_1,\ldots,y_k) \big) \Big)$$

$$\Psi_2 = \Big(\text{And}(x_1,\ldots,x_k) \wedge \forall y_1 \cdots \forall y_k \big(\text{Pre}(y_1,\ldots,y_k,x_1,\ldots,x_k)$$
$$\rightarrow \text{Acc}_{d-1}(y_1,\ldots,y_k) \big) \Big)$$

$$\Psi_3 = \Big(\text{Neg}(x_1,\ldots,x_k) \wedge \exists y_1 \cdots \exists y_k \big(\text{Pre}(y_1,\ldots,y_k,x_1,\ldots,x_k)$$
$$\wedge \, \neg \text{Acc}_{d-1}(y_1,\ldots,y_k) \big) \Big)$$

Here Acc$_d(y_1,\ldots,y_k)$ is the formula that results from Acc$_d(x_1,\ldots,x_k)$ by renaming the free variables to $y_1,\ldots,y_k$. (If it is necessary to avoid conflicts during the above construction, we also rename other variables in different "copies" of the formulas Acc$_d$.)

Finally we construct a formula ϕ such that $L_I(\phi) = A$. Let d_0 be the depth of C_n, then obviously it is sufficient to set

$$\phi = \forall x_1 \cdots \forall x_k \big(\text{Out}(x_1,\ldots,x_k) \rightarrow \text{Acc}_{d_0}(x_1,\ldots,x_k) \big).$$

Observe that the constructed formulas ϕ and Acc_d are independent of the input length n, whereas the interpretations of the predicate symbols in Acc_d do depend on n.   □

Next let us turn to a uniform version of Theorem 4.69. In that theorem we allowed arbitrary numerical predicates. To prove $\text{AC}^0 \subseteq \text{FO[Arb]}$, these were used to encode the structure of the circuit family under consideration. For the other direction to allow arbitrary relations did not raise difficulties since we could use the non-uniformity of the circuit family to hardwire into the circuit all relevant facts about the relations actually used.

Now we want to characterize $\text{U}_\text{D}\text{-AC}^0$ by first-order formulas. Which relations should we allow? On one hand these relations have to be easy to compute (within $\text{U}_\text{D}\text{-AC}^0$). On the other hand, the relations should add enough power to FO[<] to be able to simulate circuits. (It is well known that $\text{FO[<]} \subsetneq \text{AC}^0$, see Exercises 4.32–4.34 and the bibliographic remarks on these exercises on p. 171.)

One central point deserves attention here: Circuits are able to look at particular bits in the binary representation of a number. We need this power in our logical language. When researchers tried to characterize Turing machine classes in a "machine independent" way (e. g. using algebraic operations or logical languages, as we do here), they encountered the same problem. Turing machines can operate with binary representations. Somehow we will have to simulate this in a logical framework. Certainly the following numerical predicate is useful for this purpose: For $i, j \geq 1$, $\text{bit}(i, j)$ holds if and only if bit no. j in the binary representation of i is a one. More precisely, we consider relations bit_n, where $\text{bit}_n \subseteq \{1, \ldots, n\}^2$, $(i, j) \in \text{bit}_n$ if the jth bit in the n-bit binary representation of i is on. (The rightmost bit here is the 1st bit, the leftmost bit is the nth bit.) If, besides linear order (and the empty relation), only these bit-relations are allowed in our interpretations, then we use the notation FO[<, bit]. We will show below that $\text{FO[<, bit]} = \text{U}_\text{D}\text{-AC}^0$, but first we prove some preliminary facts about the power of first-order logic enhanced with bit.

When in the following we say that first-order logic with bit can *express* some relation $R \subseteq \mathbb{N}^r$, then we mean that there is a formula ϕ^R (using only $<$ and bit) with r free variables $x_1, \ldots, x_r$ such that for all $m_1, \ldots, m_r, m \in \mathbb{N}$, $m_1, \ldots, m_r \leq m$, the following holds: $(m_1, \ldots, m_r) \in R$ if and only if $[w, x_1 \leftarrow m_1, \ldots, x_r \leftarrow m_r] \models \phi^R$ for all words w with $|w| \geq m$.

Lemma 4.70. *First-order logic with* bit *can express the relation* $\text{Plus} \subseteq \mathbb{N}^3$, *where* $(a, b, c) \in \text{Plus}$ *if and only if* $a + b = c$.

Proof. The definition is analogous to the definition of the constant-depth circuit for addition in Sect. 1.1. First define a formula $\text{Carry}(i)$ with four free variables a, b, c, i by

$$\text{Carry}(i) = \Big(\exists j \big(j < i \wedge \text{bit}(a,j) \wedge \text{bit}(b,j) \big)$$
$$\wedge \, \forall k \big((j < k \wedge k < i) \to (\text{bit}(a,k) \vee \text{bit}(b,k)) \big) \Big).$$

Now Plus is defined by the formula

$$\text{Plus}(a,b,c) = \forall i \Big(\text{bit}(c,i) \leftrightarrow \big((\text{bit}(a,i) \oplus \text{bit}(b,i)) \oplus \text{Carry}(i) \big) \Big).$$

$\square$

Lemma 4.71. *First-order logic with* bit *can express the relation* BSum $\subseteq$ $\mathbb{N}^2$, *where* $(a,b) \in$ BSum *if and only if the number of* 1s *in the binary representation of* a *is* b.

Proof. Let n be the input length. Let $a, b \leq n$. We want to express that b is equal to the number of 1's in the binary representation of a. Define $L = \ell(n)$ (see Appendix A2). Observe that for an input position x, $\ell(x)$ is the unique number l that satisfies the formula $\text{bit}(x,l) \wedge \forall i (i > l \to \neg\text{bit}(x,i))$. Define L' to be the smallest power of 2 for which $L \leq (L')^2$. Assume $(L')^2 \leq n$. (The finitely many exceptions to this inequality can be "hardwired" into the formula.)

Let the binary representation of a be $a_L \cdots a_1$. Define a number s whose binary representation consists of L' many blocks of length L' each, $\text{bin}(s) = \text{bin}_{L'}(s_1) \cdots \text{bin}_{L'}(s_{L'})$ for $1 \leq i \leq L'$, such that the following holds:

$$s_i =_{\text{def}} s_{i-1} + \sum_{j=(i-1)\cdot L'+1}^{i \cdot L'} a_j, \tag{4.3}$$

where additionally we set $s_0 =_{\text{def}} 0$. Then $s_{L'}$ is the sum of the bits of a.

Observe that since $(L')^2 \leq n$, variables in our logical language can hold values of s. Suppose that Ψ is a formula with one free variable s such that Ψ evaluates to true if and only if the value of s corresponds to a as defined in (4.3). Then a formula for BSum would look as follows:

$$\exists s \big(\Psi \wedge \text{``} b = s_{L'} \text{''} \big),$$

where "$b = s_{L'}$" can be easily expressed using the bit-predicate. Thus it remains to construct Ψ.

Fix an i, $1 \leq i \leq L'$. We want to express that s_i is correct. For this, define t^i to be a number whose binary representation again consists of L' many blocks of length L' each, $\text{bin}(t^i) = \text{bin}_{L'}(t^i_1) \cdots \text{bin}_{L'}(t^i_{L'})$ for $1 \leq j \leq L'$, such that the following holds:

$$t^i_j =_{\text{def}} \sum_{k=1}^{j} a_{(i-1)\cdot L'+k}. \tag{4.4}$$

So t_{j+1}^i is either t_j^i or $t_j^i + 1$, depending on one bit in a, and this can be easily expressed by a first-order formula. Let Φ be a formula with two free variables i and t that asserts that the variable t holds the correct value of t^i according to (4.4), i.e., in principle Φ looks like

$$\forall j \left(t_j^i = t_{j-1}^i + a_{(i-1)\cdot L'+j} \right).$$

(See Exercise 4.36.)

Hence $t_{L'}^i$ is the bit sum $\sum_{j=(i-1)\cdot L'+1}^{i\cdot L'} a_j$. Now the formula Ψ in principle looks as follows:

$$\forall i \left(1 \leq i \leq L' \to \exists t^i (\Phi \wedge \text{``}s_i = s_{i-1} + t_{L'}^i\text{''}) \right).$$

Addition can be defined in first-order logic with bit by Lemma 4.70.   □

Lemma 4.72. DLOGTIME $\subseteq$ FO[$<$, bit].

Proof. Let M be a Turing machine with k work tapes (including the index tape, which for the purpose of this proof is tape number 1) running in time $c \log n$ on inputs of length n. Assume without loss of generality that M never moves one of its tape heads to a position to the left of the position the head holds at the beginning of the computation. We will construct a formula Φ such that for all $w \in \{0,1\}^*$,

$$M \text{ accepts } w \iff w \models \Phi.$$

We will encode a computation of M as follows: Suppose the input length is n. Number all relevant tape cells by numbers $1, 2, \ldots, r$, where $r = O(\log n)$. Computation step number i of M will be encoded as $\langle q_t, a_{t,0}, a_{t,1}, \ldots, a_{t,k},$ $d_{t,1}, \ldots, d_{t,k} \rangle$, where q_t is the state of M before step t, $a_{t,0}$ is the input symbol at that position, whose number can be found on the index tape before step t, $a_{t,i}$ is the symbol which M writes on its ith work tape at step t, and $d_{t,i}$ is the move of the head of M on tape i, $d_{t,i} \in \{-1, 0, +1\}$ $(1 \leq i \leq k)$. Note that in general, M in the tth step does not have access to $a_{t,0}$, unless it is in its input query state (see Appendix A4). Thus our encodings of computations incorporate more than is necessary, but we will see below that proceeding in this way makes it easy to describe a computation in terms of a logical formula.

One computation step can be encoded with a constant number of bits, if we proceed as just described to encode the single steps. A sequence of $c \log n$ moves of M can thus be encoded with $O(\log n)$ many bits. Each variable stands for an input position, and thus can hold values with $\log n$ bits. Hence a computation of M can be encoded using the values of a constant number of variables, say, q variables $\bar{x} = \langle x_1, \ldots, x_q \rangle$. Making use of the bit-predicate, the individual values of $q_t, a_{t,0}, a_{t,1}, \ldots, a_{t,k}, d_{t,1}, \ldots, d_{t,k}$ can be recovered from $\bar{x}$. Our formula Φ will look like

$$\exists x_1 \cdots \exists x_q \Psi,$$

where Ψ checks that $\overline{x}$ encodes an accepting computation of M on input w.

To construct Ψ we first define two auxiliary formulas:

For every i, $1 \le i \le k$, let $P_i(t, p)$ express that before time step t, the head on the ith tape is on cell number p. To check this, one has to add all the values $d_{s,k}$ for $s < t$. This can easily be expressed using the formula BSum constructed in Lemma 4.71.

Next, for every i, $1 \le i \le k$, let $C_i(t, p, a)$ express that before time step t, the position in cell number p on tape i is the symbol a. Thus $C_i(t, p, a)$ holds if and only if $a_{s,i} = a$, where s is the maximal value less than t for which $P_i(s, p)$ holds. From this observation, a formula defining C_i can be constructed easily.

We can now check that $\overline{x}$ encodes an accepting computation of M as follows: Correctness of every $a_{t,0}$ can be checked by

$$\exists i \left(\text{"}i \text{ is the contents of } M\text{'s index tape at time } t\text{"} \wedge (a_{t,0} \leftrightarrow Q_1(i)) \right),$$

where the first sub-formula can be built using formula C_1 and the bit-predicate. The correctness of q_t and the $a_{t,1}, \ldots, a_{t,k}, d_{t,1}, \ldots, d_{t,k}$ has to be checked according to M's finite transition table. For this we need the value q_{t-1} (to be found in $\overline{x}$) and the symbols scanned by the work tape heads (which can be determined making use of the formulas P_i and C_i, namely, a is the current symbol on tape i if and only if for some position p, both $P_i(t, p)$ and $C_i(t, p, a)$ hold). $\qquad \square$

The following uniform version of Theorem 4.69 can now be given:

Theorem 4.73. $FO[<, \text{bit}] = U_D\text{-}AC^0$.

Proof. ($\subseteq$): The construction of the circuit family is exactly the same as in the proof of Theorem 4.69. We have to show that the resulting family is logtime-uniform.

We use a gate numbering as follows: The number p of a gate v will consist of a constant number of blocks of length $O(\log n)$ each. Such a number will be an encoding of the path from the circuit output gate v_0 to the current gate v, similar to that which we used in the definition of the extended connection language (see p. 123); i.e., if $p = \langle \text{bin}(i_1), \cdots, \text{bin}(i_{k-1}), \text{bin}(i_k) \rangle$ then v is the i_kth input of the i_{k-1}th input of the ... of the i_1th input of v_0. To compute connections in C_n is then easy: all that has to be done is to compare several of these blocks of logarithmic length.

To compute the type of a gate is easy for inner $\wedge$, $\vee$, and $\neg$ gates. To check an input gate, the number of the input bit that has to be read must be computed. First we have to determine the sequence of choices for the variables that were made on the path leading to that gate. This sequence can be recovered from the gate number. From this sequence we can then recover the number of the input bit. To evaluate an atomic sentence involving a numerical predicate (which in the circuit will appear as a constant 0 or constant 1 gate), again we first determine the sequence of choices for the variables that were made on the path leading to that gate. Then we have to evaluate one of the predicates $<$ and bit applied to input positions, i.e., numbers with a logarithmic number of bits in binary. Because of this restriction a logtime machine can evaluate atomic sentences and thus find out if the appropriate circuit is constant 1 or constant 0.

($\supseteq$): A language $A \in U_D\text{-}AC^0$ is given, accepted by family $C = (C_n)_{n \in \mathbb{N}}$. Then there is an admissible encoding scheme such that $L_{DC}(C)$ can be decided in logarithmic time. The largest number occurring as an encoding of a gate in C_n is polynomial in the size of C_n, hence polynomial in n. Therefore there exists a k such that gates can be encoded as k-tuples as in Theorem 4.69.

Observe that the direct connection language of C can now be decided in logarithmic time, and hence is in FO$[<, \text{bit}]$ (by Lemma 4.72). From a formula for $L_{DC}(C)$ we can construct formulas with the appropriate number of free variables which (for input length n) will express the relations In_n, Out_n, Or_n, And_n, Neg_n, and Pre_n. (See Exercise 4.37.)

Then we proceed as in the proof of Theorem 4.69 and construct the formula ϕ. In ϕ we finally replace the numerical predicates encoding the circuit family by the defining FO$[<, \text{bit}]$ formulas. This shows $A \in$ FO$[<, \text{bit}]$. $\square$

So we see that there is a close correspondence between AC^0 and first-order logic. Can similar results be proved for the other constant-depth classes defined in Sect. 4.5.2? For this we would need extensions of first-order logic. We will consider additional quantifiers besides the usual existential and universal quantifiers.

The modular counting quantifier $\exists^{\equiv p}$. Let $p \in \mathbb{N}$. Extend the syntax and semantics of first-order formulas as follows:

If ψ is a first-order formula and $y_1, \ldots, y_k$ are variables, then

$$\exists^{\equiv p} y_1, \ldots, y_k\, \psi$$

is a first-order formula.

Let $\phi = \exists^{\equiv p} y_1, \ldots, y_k \psi$ be a formula with free variables from V. Let w be a V-structure. Then $w \models \phi$ if and only if the number of all $V \cup \{y_1, \ldots, y_k\}$-structures w', which can be obtained from w by assigning arbitrary values to the variables $y_1, \ldots, y_k$ and for which $w' \models \psi$, is divisible by p.

The majority quantifier MAJ. Extend the syntax and semantics of first-order formulas as follows:

If ψ is a first-order formula and $y_1, \ldots, y_k$ are variables, then

$$\text{MAJ}\, y_1, \ldots, y_k\, \psi$$

is a first-order formula.

Let $\phi = \mathrm{MAJ}\, y_1, \ldots, y_k\, \psi$ be a formula with free variables from V. Let w be a V-structure. Then $w \models \phi$ if and only if for the majority of all $V \cup \{y_1, \ldots, y_k\}$-structures w', which can be obtained from w by assigning arbitrary values to the variables $y_1, \ldots, y_k$, we have $w' \models \psi$.

If $\mathcal{R}$ is a set of relations and Q is one of the quantifiers just defined, then the extension of FO[$\mathcal{R}$] by allowing Q-quantifiers is denoted by FO[$Q; \mathcal{R}$].

The following results can be given:

Corollary 4.74. *1.* FO[MAJ; Arb] = TC^0.
2. FO[$\exists^{\equiv p}$; Arb] = AC^0[p] *for any $p \in \mathbb{N}$.*

Proof. The proof is an easy generalization of the proof of Theorem 4.69. The directions from left to right are obvious. For the directions from right to left, we have to add an additional sub-formula Ψ_4 in the construction of Acc$_d$ which handles the case of a gate of type MOD$_p$ or MAJ.   □

The group quantifier Q_G. Let G be a finite group. Let $l \in \mathbb{N}$ such that $|G| \leq 2^l$. Fix an encoding of the elements of G by l bit strings. Let $k \in \mathbb{N}$. Extend the syntax and semantics of first-order formulas as follows:

If $\psi_1, \ldots, \psi_l$ are first-order formulas and $y_1, \ldots, y_k$ are variables, then

$$Q_G\, y_1, \ldots, y_k[\psi_1, \ldots, \psi_l]$$

is a first-order formula.

Let $\phi = Q_G\, y_1, \ldots, y_k[\psi_1, \ldots, \psi_l]$ be a formula with free variables from V. Let w be a V-structure of length n. Consider the order on $\{1, \ldots, n\}^k$ induced by the usual order on $\{1, \ldots, n\}$, i.e. $(0, \ldots, 0, 0) < (0, \ldots, 0, 1) < (0, \ldots, 0, 2) < \cdots < (0, \ldots, 1, 0) < (0, \ldots, 1, 1) < (0, \ldots, 1, 2) < \cdots$. Let w_i be the $V \cup \{y_1, \ldots, y_k\}$-structure obtained from w by assigning the ith vector of k-tuples over $\{1, \ldots, n\}$ (in the order specified above) as values to the variables $y_1, \ldots, y_k$. Let ϕ^w be the word of length $l \cdot n^k$ defined as $\phi^w =_{\mathrm{def}} s_1 s_2 \cdots s_{n^k}$, where for all i, $1 \leq i \leq n^k$, $s_i \in \{0, 1\}^l$, $s_i = s_{i,1} \cdots s_{i,l}$, and $s_{i,j} = 1$ if and only if $w_i \models \psi_j$ ($1 \leq j \leq l$). Now we define

$$w \models \phi \iff \phi^w \in L_G.$$

(Recall that L_G is the word problem for G.)

Intuitively, the following is happening here: We consider the l formulas $[\psi_1, \ldots, \psi_l]$. When we substitute values for the free variables $y_1, \ldots, y_k$, all these formulas evaluate to true or false. Thus, the l formulas define for a particular choice of values for $y_1, \ldots, y_k$ a string of l truth values (or, a length l binary string). Now we consider all possible assignments of values to the variables $y_1, \ldots, y_k$. Each such assignment gives a length l binary string as just explained, and ϕ^w is nothing other than the concatenation of these strings; thus ϕ^w is a binary string of length $l \cdot 2^k$. The formula ϕ evaluates to true if and only if this string encodes a sequence of elements from G which multiplies out to the identity.

It can be observed that the universal, existential, and modular counting quantifiers are special cases of group quantifiers (see Exercise 4.39).

Theorem 4.75. $FO[Q_G; Arb] = AC^0[L_G]$.

Proof. Again the proof is an extension of the proof of Theorem 4.69. The direction from left to right is easy. For the direction from right to left, we have to insert a sub-formula Ψ_4 in the definition of Acc_d, which handles the case of an L_G gate.

Let G be a finite group. Suppose the elements of G are encoded as binary strings of length l. Suppose without loss of generality that the encoding of the identity of G is the string 0^l.

Fix an input length n, and consider C_n. Encode gates in C_n as k tuples as in the proof of Theorem 4.69. As above, identify k-tuples over $\{1, \ldots, n\}$ with the set $\{0, 1, \ldots, n^k - 1\}$. We need additional relations here besides those used in the proof above: Suppose $G \subseteq \{1, \ldots, n\}^k$ is a relation, whose elements are all encodings of G gates in C_n. Second, an input to an L_G gate must be a sequence of gates of a length which is a multiple of l. This is because the predecessors to L_G are encodings of sequences of elements from G. Thus, the predecessor gates to G can be grouped into l-element sequences; call these "l-sequences." Each such l-sequence will produce an l bit string which is the encoding of one element from G. Furthermore the order of the predecessors of G is important here, different to the construction used to prove Theorem 4.69. Thus we define relations $Pre_i^G \subseteq \{1, \ldots, n\}^{3k}$ for $1 \le i \le l$, where $(y_1, \ldots, y_k, x_1, \ldots, x_k, z_1, \ldots, z_k) \in Pre_i^G$ if and only if in the $(y_1, \ldots, y_k)$th l-sequence of predecessors of gate $(x_1, \ldots, x_k)$, the ith element is the gate $(z_1, \ldots, z_k)$.

Remember that Acc_d will look as follows: $Acc_d = \Psi_1 \vee \Psi_2 \vee \Psi_3 \vee \Psi_4$. The free variables $x_1, \ldots, x_k$ of Acc_d are intended to be an encoding of the current gate. Formulas Ψ_1, Ψ_2, Ψ_3 are defined as before. Next we construct Ψ_4:

$$\Psi_4 = (G(x_1, \ldots, x_k) \wedge Q_G y_1, \ldots, y_k [\Theta_1, \ldots, \Theta_l]),$$

where the formula Θ_i (for $1 \le i \le l$) is defined as

$$\exists z_1 \cdots \exists z_k \left(Pre_i^G(y_1, \ldots, y_k, x_1, \ldots, x_k, z_1, \ldots, z_k) \wedge Acc_{d-1}(z_1, \ldots, z_k)\right).$$

In this way, the formulas $[\Theta_1, \ldots, \Theta_l]$ produce for given $y_1, \ldots, y_k$ the l-bit string corresponding to the $(y_1, \ldots, y_k)$th group element in the sequence which is input to the L_G gate. If no $(y_1, \ldots, y_k)$th group element exists, then $[\Theta_1, \ldots, \Theta_l]$ produce the string 0^l, which corresponds to the identity of G and thus may be inserted arbitrarily without affecting the result.

Hence we see that for an input word $w \in \{0, 1\}^n$, an L_G gate $(m_1, \ldots, m_k)$ at level d computes the value 1 if and only if $[w; x_1 \leftarrow m_1, \ldots, x_k \leftarrow m_k] \models Acc_d$. $\qquad \square$

Corollary 4.76. *Let G be a non-solvable group. Then* $\text{FO}[Q_G; \text{Arb}] = \text{NC}^1$.

Proof. Follows from Corollary 4.47 and Theorem 4.75.   □

Let us finish by mentioning that the characterizations of the classes TC^0, $\text{AC}^0[p]$, and NC^1 given above also hold uniformly (see Exercise 4.38):

Corollary 4.77. *1.* $\text{FO}[\text{MAJ}; <, \text{bit}] = \text{U}_\text{D}\text{-TC}^0$.
2. $\text{FO}[\exists^{\equiv p}; <, \text{bit}] = \text{U}_\text{D}\text{-AC}^0[p]$ *for any* $p \in \text{N}$.
3. $\text{FO}[Q_G; <, \text{bit}] = \text{ATIME}(\log n) = \text{U}_\text{E}^*\text{-NC}^1$ *for any non-solvable group* G.

4.6 P-Completeness

The class NC was defined to capture the notion of problems with fast parallel algorithms using a feasible amount of hardware. If a problem belongs to NC it can be parallelized efficiently. While (in the uniform case) NC $\subseteq$ P, it is not known if this inclusion is strict, i.e., if there are problems that can be solved sequentially in polynomial time but not be parallelized efficiently. However there is much evidence to suspect NC $\subsetneq$ P [GHR95, Chap. 5].

As the vast majority of researchers believe that NC $\neq$ P, it would therefore be interesting to know which problems in P are not in NC *under the assumption that* NC $\neq$ P. We will exhibit a few such problems in this section. These problems, the so-called P-complete problems, may be considered the "hardest problems in P."

In this section we assume that all our circuit families are uniform; for concreteness let us fix logspace-uniformity, a reasonable choice for the class NC as we argued above. We omit the prefix "U_L-" in the notation of the classes. (The results we present hold for other uniformity conditions as well.)

4.6.1 Reducibility Notions

Let us start with some definitions. What should be considered as a hardest problem in P? We want to say that B is such a problem if every other problem $A \in$ P can be solved efficiently with the help of B, i.e., with a subroutine for B. The idea of subroutines can be captured by the formal notion of *reducibility*, as has been made clear already, see Sects. 1.4 and 4.4. We have considered constant-depth reducibility there. In the context here, some other reducibility notions will be important.

In Sect. 1.4, we said that A is constant-depth reducible to B if there is a constant-depth circuit solving A with the help of gates for membership in B. Let us in this section use the notation $A \leq_{\text{cd}} B$ to denote that A is constant-depth reducible to B, *and* that the reduction can be performed by a uniform circuit family.

In Sect. 4.4 we considered NC^i circuits with gates for membership in arbitrary languages B. We will now pay particular attention to the case of NC^1.

Definition 4.78. For $A, B \subseteq \{0,1\}^*$, we say that A is NC^1 *Turing reducible* to B, in symbols: $A \leq_T^{NC^1} B$, if $A \in NC^1(B)$.

Thus A is NC^1 Turing reducible to B if there is an NC^1-circuit that can solve A with gates for B. That is, during the solution process for A we may use subcircuits for B at different stages, which may even depend on one another. (The term "Turing reduction" usually refers to this unrestricted use of subroutine calls.) The following stricter notion forbids this.

Definition 4.79. For $A, B \subseteq \{0,1\}^*$, we say that A is NC^1 many-one reducible to B (NC^1 reducible for short), in symbols: $A \leq_m^{NC^1} B$, if there is a function $f \in FNC^1$ such that for all $x \in \{0,1\}^*$, we have: $x \in A \iff f(x) \in B$.

Hence for A to be NC^1 reducible we require that A can be solved by NC^1 circuits with one call to a subroutine for B, and moreover this subroutine has to be called at the end. (Reductions of this kind are generally referred to as "many-one reductions.")

There is another kind of many-one reducibility defined strictly in terms of Turing machines, which we will also use.

Definition 4.80. For $A, B \subseteq \{0,1\}^*$, we say that A is logspace reducible to B, in symbols: $A \leq_m^{log} B$, if there is a function f computable on a deterministic Turing machine in space $O(\log n)$, such that for all $x \in \{0,1\}^*$, we have: $x \in A \iff f(x) \in B$.

Lemma 4.81. *Let $\leq^{\mathcal{R}}$ be one of the above reducibilities, i.e., $\leq^{\mathcal{R}} \in \{\leq_{cd}, \leq_m^{NC^1}, \leq_T^{NC^1}, \leq_m^{log}\}$. Then $\leq^{\mathcal{R}}$ is transitive.*

Proof. First, consider the case $\leq^{\mathcal{R}} \in \{\leq_{cd}, \leq_m^{NC^1}, \leq_T^{NC^1}\}$. Let $A \leq^{\mathcal{R}} B$ and $B \leq^{\mathcal{R}} C$. We then immediately obtain a reduction $A \leq^{\mathcal{R}} C$ by composing the circuits witnessing the first two reductions.

Let now $\leq^{\mathcal{R}} = \leq_m^{log}$. Let $A \leq_m^{log} B$ via f, i.e., for all $x \in \{0,1\}^*$, $x \in A \iff f(x) \in B$. Let $B \leq_m^{log} C$ via g. Let M_f and M_g be Turing machines operating in logarithmic space that compute respectively f and g. We now define a Turing machine M that computes $g \circ f$. On input x, we would like to run M_f first and then run M_g on the result produced by M_f. However this result may be too long to write down in logarithmic space. Hence we proceed as follows: On input x, start M_f until it produces its first output symbol. This must be the first symbol of the input of M_g. Then start M_g. Whenever M_g moves its head to the right, we continue in the simulation of M_f until the next output symbol is produced. When M_g moves its head to the left, say to input position i, we restart the complete simulation of M_f. We let

M_f run until it produces its ith output symbol (all previous output symbols are discarded). Then we go on with the simulation of M_g, intertwined with further calls to M_f as necessary, until finally M_g stops. The output of M_g is then the value $g(f(x))$.

What is the space needed by this procedure? Let $|x| = n$. The simulation of M_f certainly needs no more space than $O(\log n)$. The runtime of M_f is polynomial, hence $|f(x)|$ is polynomial in n, hence the space needed by the simulation of M_g is $O(\log n)$. Additionally we only have to store the value of the input head position of M_g, which is also $O(\log n)$.

Obviously, $x \in A \iff g(f(x)) \in C$. Hence machine M shows that $A \leq_m^{\log} C$.   □

Now we can say that a hardest problem in P is one to which all other problems are reducible.

Definition 4.82. Let $\leq^{\mathcal{R}} \in \{\leq_{cd}, \leq_m^{NC^1}, \leq_T^{NC^1}, \leq_m^{\log}\}$ and let $B \in \{0,1\}^*$. We say that B is P-complete under $\leq^{\mathcal{R}}$, if $B \in$ P and for all $A \in$ P we have: $A \leq^{\mathcal{R}} B$.

We will use "P-complete" as a shorthand for "P-complete under $\leq_m^{NC^1}$-reductions."

The P-complete problems now have the property we desired at the beginning of this section. They are only in NC if all problems in P are.

Theorem 4.83. *Let B be P-complete under $\leq^{\mathcal{R}}$; $\leq^{\mathcal{R}} \in \{\leq_{cd}, \leq_m^{NC^1}, \leq_T^{NC^1}, \leq_m^{\log}\}$. Then we have: $B \in$ NC if and only if NC = P.*

Proof. The direction from right to left is trivial, since a P-complete problem is always in P.

For the other direction, take any problem $A \in$ P. Then we know that $A \leq^{\mathcal{R}} B$. In all four cases of reducibilities, this clearly means that $A \in$ $NC^2(B)$. If now $B \in$ NC then there is some $i \in$ N such that $B \in NC^i$. Hence $A \in NC^2(NC^i) = NC^{i+1} \subseteq$ NC by Theorem 4.21, which had to be shown.   □

How do we prove that a certain problem B is P-complete? According to the definition we have to show that every problem A in P is reducible to B, which can be a hard task. Fortunately there is an easier way to achieve this.

Theorem 4.84. *Let B_0 be P-complete under $\leq^{\mathcal{R}}$; $\leq^{\mathcal{R}} \in \{\leq_{cd}, \leq_m^{NC^1}, \leq_T^{NC^1}, \leq_m^{\log}\}$. Let $B \in$ P. Then B is P-complete under $\leq^{\mathcal{R}}$ if and only if $B_0 \leq^{\mathcal{R}} B$.*

Proof. The direction from left to right is obvious, since $B_0 \in$ P and hence is reducible to B.

For the direction from right to left, let A be any problem in P. Then $A \leq^{\mathcal{R}} B_0$. But since $B_0 \leq^{\mathcal{R}} B$ and by the transitivity of $\leq^{\mathcal{R}}$ (Lemma 4.81), we get $A \leq^{\mathcal{R}} B$.   □

This means that we first have to identify one P-complete problem by showing that all sets from P are reducible to it. (This is a so-called *generic proof*, since it has to show how we can reduce a generic problem from P to the complete set.) Then we can go on to prove more completeness results according to Theorem 4.84 by using this problem.

While problems in NC are regarded as efficiently parallelizable, P-complete problems (under any of the above defined reducibilities) are not in NC (unless NC = P, contrary to widespread belief). Hence P-complete problems are not feasibly parallelizable (where "feasible" means: in NC). This, however, does not neglect the possibility that, even if a problem is P-complete, there might be a parallel algorithm solving it in time $\sqrt{n}$, say, with a polynomial number of processors, which is still an enormous speed-up compared to linear (or even worse) sequential runtime.

4.6.2 Circuit Value Problems

We now show that, given an encoding of a Boolean circuit and an input, to compute the outcome of the circuit is P-complete. Since we know of no other complete problem so far, we have to give a generic proof.

Formally, we consider the following problem:

> *Problem:*   Circuit Value Problem (CVP)
> *Input:*   an encoding $\overline{C}$ of a circuit C over $\mathcal{B}_0$ with n inputs, and a word $x \in \{0,1\}^n$
> *Question:*   is x accepted by C?

Lemma 4.85. *CVP is P-complete under $\leq_m^{\log}$-reductions.*

Proof. That CVP is in P follows from Theorem 2.15.

Let now $A \in$ P. By Lemma 2.6 we can assume that A is accepted by an oblivious Turing machine running in polynomial time. In Lemma 2.5 we showed how every oblivious polynomial-time Turing machine can be simulated by a polynomial-size circuit. Now define a function f as follows: $f(x)$ is the encoding of the circuit constructed in this lemma for inputs of length $|x|$, together with input x. Then clearly $x \in A$ if and only if $f(x) \in$ CVP.

We showed in Theorem 2.27 that the head positions are computable in logarithmic space. Together with the regular structure of the construction of Lemma 2.5, this shows that f is computable in logarithmic space.   ∎

What we want to show now is that CVP is P-complete under NC^1 reductions and even stricter reductions. Therefore we give another construction of a circuit from a Turing machine which is easier to compute. The obstacle in the proof of Lemma 4.85 is that we only know how to compute the head positions in logarithmic space. We now present a reduction using a larger, but more regular, circuit such that the encoding can be generated with less resources.

Theorem 4.86. CVP *is* P-*complete.*

Proof. CVP is in P as above.

Let now $A \in$ P. Let M be a deterministic Turing machine that accepts A. Let p be a polynomial bounding the runtime of M.

Fix an input length n, and let $x \in \{0, 1\}^n$. Then x is in A if and only if there are configurations $K_0, K_1, K_2, \ldots, K_{p(n)}$ such that K_0 is the initial configuration of M on input x, K_{i+1} is obtained from K_i by one step of M for $0 \le i \le p(n) - 1$, and finally $K_{p(n)}$ is an accepting configuration. Every configuration of M on inputs of length n can be encoded as a binary word of length $q(n)$, where q is another polynomial.

We now define a circuit C_n^M as follows: C_n^M will have (among others) gates $g_{i,j}$ for $0 \le i \le p(n)$ and $1 \le j \le q(n)$. C_n^M will consist of $p(n) + 1$ layers. In layer i we will have all gates $g_{i,j}$. Given now an input $x \in \{0, 1\}^n$, the gates $g_{i,1}, \ldots, g_{i,q(n)}$ will encode the ith configuration in the computation of M on input x. For this, we have to make sure that

1. $g_{0,1}, \ldots, g_{0,q(n)}$ encode M's initial configuration;
2. $g_{i+1,1}, \ldots, g_{i+1,q(n)}$ encode a configuration which is obtained in one step of M from the configuration encoded by $g_{i,1}, \ldots, g_{i,q(n)}$ $(0 \le i \le p(n)-1)$;

This can easily be achieved by a polynomial-size circuit. For (1) we essentially have to copy the input x into the configuration. For (2) we have to compute the bits $g_{i+1,j}$ according to M's transition function. Observe that every bit $g_{i+1,j}$ depends only on a constant number of the bits $g_{i,1}, \ldots, g_{i,q(n)}$. Finally, we let C_n^M accept its input if and only if $g_{p(n),1}, \ldots, g_{p(n),q(n)}$ encode an accepting configuration of M.

Now define a function f as follows: $f(x)$ is the encoding of the circuit constructed above for inputs of length $|x|$, together with input x. Then certainly $x \in A$ if and only if $f(x) \in$ CVP.

Thus, the size of C_n^M is certainly polynomial in n. The structure of C_n^M is very regular. It consists of layers which look similar, and every layer is built using small, constant-size subcircuits for the transition function of M. From this one can conclude that the transformation f can be computed in U_L-FNC^1. We leave the details to the reader (see Exercise 4.42). □

Using the method of the proof, it can be shown that the reduction can even be performed in constant-depth, see Exercise 4.43.

In the remainder of this subsection, we will show that the above problem remains P-complete for several restrictions on the shape of the circuits that are allowed.

We first consider the case where negations are forbidden.

Problem:	Monotone Circuit Value Problem (MCVP)
Input:	an encoding $\overline{C}$ of a circuit C over $\{\vee^2, \wedge^2\}$ with n inputs, and a word $x \in \{0, 1\}^n$
Question:	is x accepted by C?

Theorem 4.87. MCVP *is* P-*complete.*

Proof. MCVP is in P since CVP is. We now reduce CVP to MCVP. Then the completeness follows from Theorem 4.84.

Let C be a circuit over $\mathcal{B}_0$ with n inputs, and let $x \in \{0,1\}^n$. Define a circuit C' over $\{\vee^2, \wedge^2\}$ with $2n$ inputs as follows: If g is a gate in C then C' will have gates g^0 and g^1. If g is an $\wedge$ gate with predecessors g_l and g_r, then g^1 will be an $\wedge$ gate with predecessor g_l^1 and g_r^1, and g^0 will be an $\vee$ gate with predecessors g_l^0 and g_r^0. An $\vee$ gate is handled dually. If g is an input gate x_i then g^1 will be the input gate x_{2i-1} and g^0 will be the input gate x_{2i}. Now if g is an $\neg$ gate with predecessor g_l, then g^1 will be an $\wedge$ gate whose both predecessors are g_l^0, and g^0 will be an $\vee$ gate whose both predecessor are g_l^1. If g is the output gate of C then g^1 will be the output gate of C'. Finally define an input x' as follows: If $x = a_1 \cdots a_n$ then $x' = a_1\overline{a_1} \cdots a_n\overline{a_n}$.

It is obvious from the construction that for a gate $g \in C$, g^1 computes the same value as g while g^0 computes the negation of that value. Hence we see that x is accepted by C if and only if x' is accepted by C'. Certainly the transformation from (C, x) to (C', x') can be computed in logarithmic depth. $\qquad\qquad\Box$

Define the functions NOR and NAND by $\mathrm{NOR}(x,y) =_{\mathrm{def}} \neg(x \vee y)$ and $\mathrm{NAND}(x,y) =_{\mathrm{def}} \neg(x \wedge y)$.

Problem: NOR Circuit Value Problem (NOR-CVP)
Input: an encoding $\overline{C}$ of a circuit C with n inputs having only NOR gates, and a word $x \in \{0,1\}^n$
Question: is x accepted by C?

The problem NAND-CVP is defined analogously.

Theorem 4.88. NOR-CVP *and* NAND-CVP *are* P-*complete.*

Proof. Follows immediately from the fact that {NAND} and {NOR} are complete bases, that is, they can be used to simulate the basis $\mathcal{B}_0$. $\qquad\Box$

4.6.3 Graph Problems

Next we consider a few problems from graph theory. In the reductions below we restrict the proof to arguing membership in P and the construction of a transformation from a known complete problem. We leave it to the reader to check that the transformation can be carried out by a function from FNC^1.

An *alternating graph* is a triple $G = (V, W, E)$, where $(V \cup W, E)$ is a directed graph. Nodes in V are called existential, while nodes in W are called universal. Given a node $s \in V \cup W$, the set of all nodes *reachable* from s, $\mathrm{Reach}_G(s)$, is defined as the smallest set satisfying:

– $s \in \mathrm{Reach}_G(s)$;

– if $v \in V$ and there is a $w \in \mathrm{Reach}_G(s)$ such that $(w,v) \in E$, then $v \in \mathrm{Reach}_G(s)$;

– if $v \in W$ and for all nodes w such that $(w,v) \in E$ we have $w \in \mathrm{Reach}_G(s)$, then $v \in \mathrm{Reach}_G(s)$.

Let us consider the following problem:

> *Problem:*   Alternating Graph Accessibility Problem (AGAP)
> *Input:*   an alternating graph G and two nodes $s, t \in G$
> *Question:*   is $t \in \mathrm{Reach}_G(s)$?

Theorem 4.89. AGAP *is P-complete.*

Proof. AGAP is in P by Proposition 2.37, since it can be solved by an alternating Turing machine in logarithmic space.

We now reduce MCVP to AGAP: Let C be a circuit over $\{\vee^2, \wedge^2\}$ with n inputs and $x \in \{0,1\}^n$. Define the graph G as follows: G has two existential vertices 0 and 1. Moreover for every gate $g \in C$ there is a vertex $g \in G$. If g is an input gate in C then g is existential in G. If in C g is connected to an input which is set to 0 under x then there is an edge in G from 0 to g, otherwise there is an edge from 1 to g. If g is an $\vee$ or $\wedge$ gate then its predecessors in C again become g's predecessors in G. If g is an $\vee$ gate in C then g is existential in G. If g is an $\wedge$ gate in C then g is universal in G.

Now the following holds: x is accepted by C if and only if the node in G that corresponds to the output gate of C, is in $\mathrm{Reach}_G(1)$. $\qquad\Box$

Let $G = (V, E)$ be an undirected graph. An *independent set* in G is a set of nodes $I \subseteq V$ such that no two elements from I are connected by an edge in G. An independent set I is *maximal* if there is no independent set I' such that $I \subsetneq I'$.

Suppose now that $V = \{v_1, \ldots, v_n\}$ and there is an order on V, without loss of generality $v_1 < v_2 < \cdots < v_n$. Any maximal independent set I can now be ordered as a sequence of vertices, or in other words, a maximal independent set is a word over the alphabet V. Using lexicographic order we can now define an order on maximal independent sets. The smallest maximal independent set in this order is called the *lexicographically first maximal independent set*.

> *Problem:*   Lexicographically First Maximal Independent Set
>   (LFMIS)
> *Input:*   an undirected graph G with an order on the vertices; a
>   vertex $v \in G$
> *Question:*   is v in the lexicographically first maximal independent
>   set of G?

Theorem 4.90. LMFIS *is P-complete.*

Proof. We first show that LMFIS is in P. We are given an undirected graph $G = (V, E)$, $V = \{v_1, \ldots, v_n\}$ and $v_1 < v_2 < \cdots < v_n$. The algorithm in Fig. 4.2 computes the lexicographically first maximal independent set of G and obviously runs in polynomial time.

---

input $G = (V, E)$, $V = \{v_1, \ldots, v_n\}$, $v_1 < v_2 < \cdots < v_n$;
output lexicographically first maximal independent set I;

   $I := \emptyset$;
   for $i := 1$ **to** n **do**
      if v_i is not connected to any vertex in I **then**
         $I := I \cup \{v_i\}$

Fig. 4.2. Algorithm to compute lexicographically first maximal independent sets

We now reduce NOR-CVP to LMFIS. Let C be a circuit with n inputs, which has only NOR gates. Let $x \in \{0,1\}^n$ be an input to C. Order the gates of C topologically. Let $g_1, \ldots, g_n$ be the gates of C in that order with g_n being the output gate. Define a graph G as follows: G consists of the same nodes as C with the same connections (but as undirected edges). Additionally, G has a vertex 0 that is adjacent to all input gates x_i that are set to 0 under the current input word x. The order of nodes in G is defined to be $0 < g_1 < \cdots < g_n$.

Now the following holds: Let I be the lexicographically first maximal independent set of G. Vertex 0 will certainly be in I. A vertex g_i is included in I if and only if in C it evaluates to true. Thus circuit C accepts the input x if and only if g_n is in the lexicographically first maximal independent set of G. □

4.6.4 Problems from Logic, Algebra, and Language Theory

We close this section by proving the P-completeness of some problems from different areas. Let us start with two questions from propositional logic.

   Problem: Satisfiability of Horn Formulas (HORN-SAT)
   Input: a Horn formula Φ
   Question: is Φ satisfiable?

Theorem 4.91. HORN-SAT *is P-complete.*

Proof. Observe that a clause with at most one unnegated literal can be written as an implication, for example $\neg X_1 \vee \cdots \vee \neg X_k \vee Y$ is the same as $(X_1 \wedge \cdots \wedge X_k) \Rightarrow Y$. The algorithm given in Fig. 4.3 determines whether a Horn formula Φ in variables $X_1, \ldots, X_n$ is satisfiable. Here, T is an n-element array of Boolean values; T represents the assignment in which X_i is assigned true if and only if $T[i] = 1$ $(1 \leq i \leq n)$. The algorithm clearly has polynomial runtime. Correctness of the algorithm is argued in any standard introduction to complexity, e. g., [Pap94].

> **input** Horn-formula Φ with variables $X_1, \ldots, X_n$;
>
> $T := 0$;
> **for** all clauses $\{X_i\}$ in Φ **do**
> $\quad T[i] := 1$;
> **while** there is a clause $(X_{i_1} \wedge \cdots \wedge X_{i_k}) \Rightarrow X_j$ in Φ
> $\quad$ such that $T[i_1] = \cdots = T[i_k] = 1 \neq T[j]$ **do**
> $\quad\quad T[j] := 1$;
> **if** T satisfies Φ **then** output "satisfiable"
> $\quad\quad\quad\quad\quad$ **else** output "not satisfiable"

Fig. 4.3. Algorithm to test the satisfiability of Horn formulas

We now show that the complement of AGAP can be reduced to HORN-SAT. Since the complement of the AGAP problem is P-complete (see Exercise 4.44), the result follows by Theorem 4.84.

Let G be an alternating graph, and $s, t \in G$. For each node $v \in G$ introduce a propositional variable X_v. Construct the formula Φ as follows: Initially we start with the empty formula and then add more and more clauses to Φ.

- Add a clause $\{X_s\}$ to Φ.
- If v is an existential node in G, and (w, v) is an edge in G, then add the clause $X_w \Rightarrow X_v$ to Φ.
- If v is a universal node in G, and $w_1, w_2, \ldots, w_k$ are all predecessors of v in G, then add the clause $(X_{w_1} \wedge X_{w_2} \wedge \cdots \wedge X_{w_k}) \Rightarrow X_v$ to Φ.
- Add a clause $\neg X_t$ to Φ.

The following holds: In any satisfying assignment of Φ, all variables X_v for which $v \in \text{Reach}_G(s)$ must get the value true. Thus, Φ is not satisfiable if and only if $t \in \text{Reach}_G(s)$. $\qquad\Box$

Say that a unit clause is a clause consisting of exactly one literal. The empty clause is the clause consisting of no literals. Let C and C' be clauses. The clause D can be obtained from C and C' by **unit resolution**, if one of C and C' is a unit clause, say C', and D is obtained from C by removing the literal $\neg X$ if $C' = \{X\}$, and by removing X if $C' = \{\neg X\}$.

> *Problem:* Unit Resolution (UNIT)
> *Input:* a set of clauses Φ
> *Question:* can the empty clause be deduced from Φ by unit resolution?

Theorem 4.92. UNIT *is P-complete.*

Proof. Membership in P is given by Exercise 4.45. We now reduce CVP to UNIT. Let C be a circuit with n inputs and $x \in \{0,1\}^n$. We represent the gates of C by certain sets of clauses. Φ will then be the union of all these. As variables in Φ we use the gates of C.

If g is an $\wedge$ gate in C with predecessors g_l and g_r, then we represent g by $g \Leftrightarrow g_l \wedge g_r$. This formula is equivalent to the conjunction $\{(g \Rightarrow g_l) \wedge (g \Rightarrow g_r) \wedge (g_l \wedge g_r \Rightarrow g)\}$. This conjunction corresponds to the following set of clauses: $\{(\neg g \vee g_l), (\neg g \vee g_r), (\neg g_l \vee \neg g_r \vee g)\}$. If g is an $\vee$ gate in C with predecessors g_l and g_r, then we represent g analogously by $\{(\neg g_l \vee g), (\neg g_r \vee g), (\neg g \vee g_l \vee g_r)\}$. If g is a $\neg$ gate with predecessor g_l, we represent g by $\{(g \vee g_l), (\neg g \vee \neg g_l)\}$. If g is the input gate x_i, then we represent g by $\{g\}$, if the ith bit in x is on, and by $\{\neg g\}$ otherwise. Finally, if g_o is the output gate of C, then we add a clause $\{\neg g_o\}$ to Φ.

Then a clause $\{g\}$ for a gate g in C can be derived by unit resolution if and only if g evaluates to true in C; a clause $\{\neg g\}$ for a gate g in C can be derived by unit resolution if and only if g evaluates to false in C. Thus the empty clause can be derived if and only if the clause $\{g_o\}$ can be derived which is true if and only if C accepts x. $\square$

Next we turn to abstract algebra. Let X be a finite set and $\circ$ be a binary operation on X, i.e., $\circ$ is a total function $\circ\colon X \times X \to X$. If $S \subseteq X$, then the *closure* of S under $\circ$ is defined as the smallest set T such that

$- S \subseteq T$;
$-$ if $u, v \in T$ then $u \circ v \in T$.

We consider the following problem:

  Problem: Generability (GEN)
  Input: a finite set X, a binary operation $\circ$ on X, a subset $S \subseteq X$, and $t \in X$
  Question: is t in the closure of X under $\circ$?

Theorem 4.93. GEN *is P-complete.*

Proof. Let X be a finite set with binary operation $\circ$, and let $S \subseteq X$. The algorithm in Fig. 4.4 computes the closure of S under $\circ$. The runtime is clearly polynomial.

> **input** finite set X with operation $\circ$, $S \subseteq X$;
> **output** closure T of S under $\circ$;
>
> $T := S$;
>   **while** there are $u, v \in T$ such that $u \circ v \notin T$ **do**
> $T := T \cup \{u \circ v\}$

Fig. 4.4. Algorithm to compute the closure of set S

Next we reduce UNIT to GEN. Let Φ be a set of clauses. Define X to be the set of all subclauses of clauses in Φ, plus a special element 0. Let S be the set of all clauses in Φ. For clauses C, D, let $C \circ D$ be the unit resolvent of C and D, if it exists; otherwise let $C \circ D = 0$. Moreover, for all clauses C, we set $C \circ 0 = 0 \circ C = 0$. Let w be the element in X representing the empty clause.

Observe that the empty clause can be derived from Φ by unit resolution if and only if w is in the closure of S with respect to $\circ$.   ❑

As our last problem, we consider the membership problem for context-free languages.

> *Problem:*   Context-free Grammar Membership (CFGMEM)
> *Input:*   a context-free grammar G, and a word w
> *Question:* can w be generated by G?

Theorem 4.94. CFGMEM *is* P-*complete.*

Proof. Containment in P is given by the well-known Cocke-Younger-Kasami algorithm, see [HU79]. We next reduce GEN to CFGMEM.

Now let X be a finite set with binary operation $\circ$, and let $S \subseteq X$ and $t \in X$. We construct the grammar G as follows: The set of nonterminals is X, the set of terminals is arbitrary; for concreteness let us fix the set $\{\circ\}$. The start symbol is t. The production rules are all rules of the form $x \to yz$ for which $y \circ z = x$, and all rules $x \to \epsilon$ for $x \in S$. Then it is easy to verify that t is in the closure of S under $\circ$ if and only if the grammar G generates the empty word ϵ.   ❑

In the last three subsections we have considered many problems whose importance for all areas of computer science, especially algorithm design, is evident. We have established reductions among them as summarized in Fig. 4.5. Since the circuit value problem, the lowest problem in the figure, is P-complete, this shows that all these problems are P-complete, which implies that for none of them can we hope to find a parallel algorithm with poly-logarithmic runtime using a reasonable amount of hardware. In other words, under the assumption NC $\neq$ P they are all not contained in NC.

4.7 Overview of the Classes of the NC Hierarchy

In Fig. 4.6 we present the inclusion structure of the classes of the NC hierarchy as well as typical problems in these classes. Most problems are complete for one of the subclasses of the hierarchy. For lucidity of presentation we omit uniformity specifications, which were discussed in detail in the different sections of the present chapter.

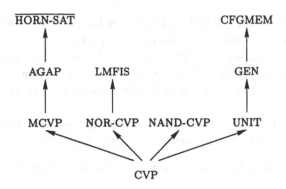

Fig. 4.5. Summary of reductions. An arrow from A to B means that A is reducible to B.

We add references for problems not studied in the text of this chapter. The class DET, which is mentioned in the figure, is defined and examined in Exercise 4.15. Most of the problems from the figure are covered in detail in the textbook [GHR95].

Bibliographic Remarks

The definition of NC first appears explicitly in [Coo79], but the class was studied earlier by Nicholas Pippenger [Pip79]. The characterization of NC^i on alternating Turing machines (Corollary 4.3) is from [Ruz81]. The characterization of AC^i (Theorem 4.7) is attributed to [Cook and Ruzzo, 1983, unpublished] in [Coo85].

Semi-unbounded fan-in circuits were considered in [Ven86, Ven91]. In these references, equivalence of semi-unbounded fan-in circuit families with so-called semi-unbounded alternating Turing machines was proved, see Exercise 4.3. The equivalence of SAC^1 with polynomial tree-size on alternating machines (Theorem 4.12) is also from there. However this latter result builds heavily on a previous theorem proved in [Ruz80], which shows how polynomial tree-size can be simulated time-efficiently (see Exercise 4.5); this can be used to obtain the inclusion ASPACE-TREESIZE($\log n, n^{O(1)}$) $\subseteq SAC^1$. Ruzzo also proved that polynomial tree-size characterizes the class LOGCFL of all problems reducible to a context-free language under logspace reductions (see Exercises 4.8–4.10). Combining this with the results from Sect. 4.3 proves LOGCFL $\subseteq NC^2$. Many more characterizations of complexity classes using semi-unbounded fan-in circuits were obtained by Vinay, see [Vin91]. The closure of the classes SAC^i ($i \geq 1$) under complementation (Theorem 4.14) was given in [BCD$^+$89]. The technique of inductive counting which was used to obtain this result, was earlier introduced independently in [Imm88] and

P	different circuit value problems, HORN-SAT, lexicographically first maximal independent set, linear programming [DLR79, Kha79], maximum flow [LW90] (*all these problems are complete for* P)
UI	
NC	maximal independent set [KW85], maximal matching [Lev80]
UI	
⋮	
UI	
NC³	
UI	
AC²	
UI	
SAC²	
UI	
NC²	determinant, iterated matrix product matrix inverse [Coo85] (*all these problems are complete for* DET)
UI	
AC¹	shortest path in weighted undirected graphs [Coo85]
UI	
LOGCFL = SAC¹	context-free languages [BLM93] (*complete*)
UI	
NL	directed graph accessibility [Sav70] (*complete*)
UI	
L	graph accessibility for directed forests [Sav70] (*complete*)
UI	
NC¹	regular languages, Boolean formula value problem [Bus87] (*complete*)
UI	
TC⁰	majority, iterated addition, multiplication, sorting; in P-uniform TC⁰: division, iterated multiplication (*complete*)
UI	⟵ strict inclusion if p is prime
AC⁰[p]	mod$_p$ (*complete*)
UI	
AC⁰	addition

Fig. 4.6. The classes of the NC hierarchy

[Sze87]. That SAC^0 is not closed under complementation (Theorem 4.16) was proved in [Ven91].

Circuits with threshold gates were first examined in [HMP$^+$87]. The class TC^0 was defined in this paper.

In the Bibliographic Remarks for Chap. 2 we mentioned that the question whether time and space for deterministic Turing machines correspond simultaneously to size and depth for circuits remains open so far. For the class NC this means that it is not clear if NC = SC, where SC =$_{def}$ DTIME-SPACE($n^{O(1)}, (\log n)^{O(1)}$). (SC stands for "Steve's class" in recognition of the contribution of [Coo79]; the name was coined by N. Pippenger.) Cook, who showed that deterministic context-free languages are in SC, conjectured that there are context-free languages outside NC. As we saw in this chapter this conjecture does not hold – even LOGCFL $\subseteq$ NC^2; nevertheless the majority of complexity theorists still believe SC $\neq$ NC. Pippenger [Pip79] showed that NC corresponds simultaneously to polynomial time and polylogarithmic number of head reversals on deterministic Turing machines.

The decomposability of the NC hierarchy was examined in [Wil90]. NC circuits with oracle gates had already been considered in [Coo85] to define a notion of NC^1 reducibility (see also Exercise 4.15), and in [Wil87]. The result that if two classes in the NC hierarchy are equal then the whole hierarchy collapses is also from [Wil90], though it was proved earlier in [J. Chen, 1987, unpublished manuscript].

Uniformity issues for classes within NC^1 were examined in [BIS90]. It should be remarked that their definition of "DLOGTIME-uniform NC^1" is *not* the same as our definition of U$_D$-NC^1. Their notion of DLOGTIME-uniformity refers to a certain formula language for NC^1 circuits which has to be acceptable in logarithmic time (see Exercise 4.26). The class "DLOGTIME-uniform NC^1" in their sense coincides with ATIME($\log n$), and hence is equal to our U$_E$-NC^1. It is not clear whether this also holds for U$_D$-NC^1. While clearly ATIME($\log n$) $\subseteq$ U$_D$-NC^1, the converse inclusion is not known. It seems conceivable that U$_D$-NC^1 could be a little larger [David A. Mix Barrington, 1998, personal communication].

The computation model of branching programs was introduced in [Lee59]. Equivalence of polynomial-size branching programs and non-uniform logspace (Theorem 4.38) has been known since the mid-1960s [Cobham, 1966, unpublished; Pudlak and Žak, 1983, unpublished]. A complexity theory of branching programs was developed in the monograph [Mei86]. More types of branching programs, corresponding to different complexity classes for Turing machines, were examined there.

The collapse of bounded-width branching programs (Theorem 4.46) with the class NC^1 was given in [Bar89], see also [Weg87, Chap. 14.5]. The uniform version (Theorem 4.52) is from [Bar89, BIS90]. The examination of the fine structure of NC^1 from an algebraic point of view was continued in [BT88]. There the M-program model was examined for different varieties of monoids.

Say that a monoid M is *solvable*, if every subgroup of M is solvable. Say that M is *aperiodic*, if it contains no non-trivial subgroups. In the same way as non-solvable monoids characterize NC^1, as we showed in this chapter, solvable monoids characterize ACC^0 and aperiodic monoids characterize AC^0. This shows that a converse of Theorem 4.47 would be very interesting. If we knew that a word problem for a group can be complete for NC^1 only if the group is non-solvable, then we would have shown $ACC \neq NC^1$, a major breakthrough in circuit complexity. But as remarked in Chap. 3, even $AC[6] \stackrel{?}{=} NP$ is open.

The M-program model was extended to other algebraic structures (not necessarily monoids) in [BLM93]. More precisely, programs over *groupoids* were examined in that paper, and it was shown that in this way, the complexity class $LOGCFL = SAC^1$ is obtained. (A groupoid is an algebraic structure with a binary operation that possesses an identity element. In contrast to monoids, associativity is not required.)

The part of finite model theory that studies the connection between expressibility power of certain logics and formal languages and computational complexity theory was initiated in the 1960s independently by [Büc62] and [Tra61] who gave characterizations of the class of regular languages. In 1974 Fagin obtained a characterization of NP [Fag74]. Since then the area has developed in a flourishing way. For a recent state of the art, see [EF95, Imm99]. The relation between constant-depth circuits and first-order logic (Theorem 4.69) is from [Imm87]. (An independent proof appeared in [GL84].) The uniform version of this relation was presented (via the detour of using PRAMS) in [Imm89b].

The generalization to non-standard quantifiers was given in [BIS90]. (For further developments, see [BI94].) The paper [BIS90] not only considers group quantifiers but more generally what they called *monoidal quantifiers*. Furthermore, building on the above-mentioned characterization of LOGCFL using groupoids, [LMSV99] examined *groupoidal quantifiers*. All these quantifiers are special cases of so-called *Lindström quantifiers*, introduced in [Lin66], see also [EF95, Chap. 10]. Lindström quantifiers also play a role in other fields of computational complexity theory, see, e. g., [Ste92, Got97, Vol98a].

Clearly the question arises as to what is so particular about the bit predicate. This predicate appeared in the paper [Jon75] (under the name σ, see Exercise 4.41), and was used later a number of times to obtain different machine-independent characterizations of complexity classes (sometimes disguised by allowing other bit-operations, e. g. shift and integer division by 2). The question of what are the properties of bit that explain its importance is examined in [Lin92]. However, it should be remarked that allowing integer addition and multiplication is already enough to simulate bit: it can be shown that exponentiation can be defined from addition and multiplication (see, e. g., [HP93, 301] and [Smo91, 192]). From this it follows that $FO[<, bit] = FO[+, \times]$ as pointed out by [Lindell, 1994, e-mail communication]. A self-contained presentation of the result can be found in [Imm99, Sect. 1.2.1].

The first P-complete problem (under logspace-reduction) was exhibited in [Lad75]. NC^1 reductions were introduced in [Coo85]. The theory then developed very quickly. An early pioneering paper is [JL76]. Most of the reductions we have given in this chapter were proved there. A survey can be found in [Tor93]. The theory of P-completeness is systematically developed and studied in the textbook [GHR95]. This book parallels the classical reference [GJ79] for questions on NP-completeness.

An interesting question is: if $NC \neq P$, are there "intermediate" problems which are neither P-complete nor in NC? The existence of such problems (under the assumption $NC \neq P$) was proved in [Ser90, Vol91, RV97]. However no natural examples are known so far, though there are several polynomial-time solvable problems for which it is not clear whether they are in NC or whether they are P-complete. These are all candidates for intermediate problems. Noteworthy examples are the lexicographically first maximal matching problem and the circuit value problem for so-called comparator circuits [MS92]. More examples can be found in [GHR95, Appendix B].

Exercises

4.1. Show that the inclusion chain $SAC^i \subseteq AC^i \subseteq NC^{i+1} \subseteq SAC^{i+1}$ holds for U_P-, U_L-, U_E^*-, and U_E-uniformity.

4.2. Let $i \geq 1$.

(1) Let $C = (C_n)_{n \in \mathbb{N}}$ be an AC^i circuit family. Prove: C is U_L-uniform if and only if $L_{DC}(C) \in DSPACE(\log n)$.
(2) Prove that the same result holds in the case that C is an SAC^i circuit family.

Hint: Generalize the proof of Lemma 2.25.

4.3. An alternating Turing machine M is *semi-unbounded* if there is a $c \in \mathbb{N}$ such that, for all x accepted by M, there is an accepting computation subtree T of M on input x with the following property: Along any path in T connecting two configuration nodes which are of existential type, the number of non-existential nodes in between is bounded by c.

Show that a language A belongs to U_L-SAC^1 if and only if there is a semi-unbounded alternating Turing machine that accepts A, which runs in logarithmic space and makes a logarithmic number of alternations.

Hint: Follow the argument of the proof of Theorem 4.12.

4.4. Let T be a tree with z nodes, $z > 3$, where each node in T has at most two successors. Then there is a node v in T such that, if we let z_1 be the number of descendants of v including v and z_2 be the number of nodes in T

which are not proper descendants of v (that is, again including v itself), then $\frac{z}{12} \leq z_1, z_2 \leq \frac{11z}{12}$.

4.5. Let $s, t: \mathrm{N} \to \mathrm{N}$, $s(n) \geq \log n$. Prove: ASPACE-TREESIZE$(s(n), t(n))$ $\subseteq$ ATIME-SPACE$(s(n) \cdot \log t(n), s(n))$.

4.6. Prove: $\mathrm{U_L}$-SAC^i = ASPACE-TREESIZE$(\log n, 2^{(\log n)^i})$.

4.7. Give a characterization of uniform SAC^0 in terms of alternating Turing machines. What uniformity conditions are reasonable?

4.8. Let CFL denote the class of all context-free languages. Show that CFL $\subseteq$ ASPACE-TREESIZE$(\log n, n^2)$.

4.9. (1) Show that if $A \leq_{\log} B$ and $B \in$ ASPACE-TREESIZE$(\log n, n^{O(1)})$, then $A \in$ ASPACE-TREESIZE$(\log n, n^{O(1)})$.
(2) Let LOGCFL $=_{\mathrm{def}} \{ A \mid$ there is a set $B \in$ CFL such that $A \leq_m^{\log} B \}$. Show that LOGCFL $\subseteq$ ASPACE-TREESIZE$(\log n, n^{O(1)})$.

4.10.** Show that ASPACE-TREESIZE $\subseteq$ LOGCFL.

4.11. Prove Corollary 4.17.

4.12. Prove that Corollary 4.25 also holds in the case of $\mathrm{U_E}$-uniformity.

4.13. Prove Corollary 4.27.

4.14. Let $i \geq 1$. Prove:

(1) If NC^i = AC^i, then for all $j > i$, NC^j = AC^j.
(2) If AC^i = NC^{i+1}, then for all $j \geq i$, AC^j = NC^{j+1}.

4.15.** Consider the following problems:

Problem: INTDET
Input: an $n \times n$ matrix A with n bit integer entries
Output: the determinant of A

Problem: MATPOW
Input: an $n \times n$ matrix A with n bit integer entries, a number $k \leq n$
Output: the k-th power of A

Problem: ITMATPROD
Input: n matrices of dimension $n \times n$ with n bit integer entries
Output: the product of the input matrices

(1) Let $f, g: \{0,1\}^* \to \{0,1\}^*$ be polynomially length-bounded. We say that f is NC^1 reducible to g; in symbols: $f \leq_T^{NC^1} g$, if $f \in FNC^1(\mathrm{bits}(g))$. Say that f and g are NC^1-equivalent, if both $f \leq_m^{NC^1} g$ and $g \leq_m^{NC^1} f$.

Prove that MATPOW, ITMATPROD, and INTDET are pairwise NC^1-equivalent.

(2) Define DET to be the class of all functions that are NC^1 reducible to INTDET. Recall our convention that prefixing a circuit class with "F" means that we consider functions instead of sets.

Show that DET $\subseteq FNC^2$ non-uniformly as well as under all uniformity conditions we considered.

4.16. Let $A \subseteq \{0,1\}^*$ such that c_A is symmetric. Prove:

(1) $A \in TC^0$.
(2) If $A \in P$ then $A \in U_P\text{-}TC^0$.
(3) If $A \in L$ then $A \in U_L\text{-}TC^0$.

4.17. Prove Corollary 4.32.

4.18.* Let $A \in \{0,1\}^*$. Prove:

(1) $A \in \mathrm{DSPACE}(\log s)/F(O(s))$ if and only if there is a family of branching programs $\mathcal{P}$ of size $O(s)$ that accepts A.
(2) Let s be space-constructible. Show that $A \in \mathrm{DSPACE}(s)$ if and only if there is a family of L-uniform branching programs $\mathcal{P}$ of size $2^{O(s)}$ that accepts A.

4.19. Show that every family $\mathcal{P}$ of bounded-width branching programs of size s can be simulated by a family $\mathcal{P}'$ of bounded-width branching programs of size $s^{O(1)}$ with the additional property that for every program in $\mathcal{P}'$ all nodes within the same layer have the same type under the β-function.

4.20. Complete the proof of Lemma 4.45.

4.21. Fill in the details in the proof of the implication (1) $\Rightarrow$ (3) of Theorem 4.52.

4.22. Let D be the set of constant-depth degrees (see Exercise 1.13). For $c, d \in D$, let $c < d$ denote that $c \leq d$ but not $d \leq c$. Let $0 =_{\mathrm{def}} [\emptyset]_{\equiv_{cd}}$ and $1 =_{\mathrm{def}} [L_{S_5}]_{\equiv_{cd}}$, where L_{S_5} denotes the word problem for group S_5. Show that there are constant-depth degrees $d_1, d_2, d_3, \ldots$ such that for all $i, j \geq 1$, $i \neq j$, we have: $0 < d_i < 1$ and $d_i \neq d_j$.

4.23.* Say that a layered circuit C (see p. 90) is of bounded width if there is a $k \in \mathbb{N}$ such that for every level l in C the number of gates of level l is at most k. Prove:

(1) A bounded fan-in circuit family of depth d and width k ($d: \mathbb{N} \to \mathbb{N}$, $k \in \mathbb{N}$) can be simulated by a branching program family of length $O(d)$ and width $O(2^k)$.

(2) A branching program family of width k and length l ($k \in \mathbb{N}$, $l: \mathbb{N} \to \mathbb{N}$) can be simulated by a bounded fan-in circuit family of width $O(\log k)$ and depth $O(l)$.

(3) NC^1 is the class of languages that can be accepted by polynomial-size circuit families of bounded width.

4.24. Consider projection reducibility defined on p. 133.

(1) Show that the relation $\leq_{\mathrm{proj}}$ is transitive.
(2) Show that $\leq_{\mathrm{proj}} \subseteq \leq_{\mathrm{cd}}$, i.e., if A and B are languages such that $A \leq_{\mathrm{proj}} B$, then $A \leq_{\mathrm{cd}} B$.

4.25. Let $A \in \mathrm{ATIME}(\log n)$ and let G be a non-solvable group. Show that A is constant-depth reducible to the word problem for G, where the reducing circuit family is U_D-uniform.

4.26.** The set of *Boolean formulas* with n inputs is the smallest class of strings over the alphabet $\Sigma_F =_{\mathrm{def}} \{0, 1, x, \neg, \vee, \wedge, (,), \flat\}$, satisfying the following properties:

– 0 and 1 are Boolean formulas.
– $x\mathrm{bin}(i)$ is a Boolean formula, where $1 \leq i \leq n$.
– If α and β are Boolean formulas, then so are $\neg\alpha$, $(\alpha \vee \beta)$ and $(\alpha \wedge \beta)$.
– If α is a Boolean formula, $\alpha = vw$ for $v, w \in \Sigma_F^*$, then $v\flat w$ is a Boolean formula. (Hence we may insert blank symbols into formulas at arbitrary positions.)

Given an input $a_1 \cdots a_n$ and an n-input formula α, the value of $\alpha(a_1 \cdots a_n)$ is defined in the obvious way. If $\alpha(a_1 \cdots a_n) = 1$ we say that the input is accepted by α. A formula family is a sequence of formulas, one for each input length. Given a formula family, the formula language is defined to be the set of all triples

$$\langle y, i, c \rangle,$$

where the ith character in the formula for inputs of length $|y|$ is the letter $c \in \Sigma_F$.

Thus, a formula is nothing else than a bounded fan-in circuit of fan-out 1, but in the description of the formula we are allowed to insert blanks at arbitrary positions.

Define DLOGTIME-uniform NC^1 to be the class of all languages A such that there is a formula family of polynomial length and logarithmic depth that accepts A, for which the formula language is in $\mathrm{DTIME}(\log n)$. (Here, the length of a formula is simply its length as a string over Σ_F, and the depth of a formula is defined to be the depth of the corresponding circuit.)

Prove: DLOGTIME-uniform NC^1 equals $\mathrm{ATIME}(\log n)$.

4.27.* Prove: A language A is in U_P-NC if and only if there is a logspace-bounded alternating Turing machine M accepting A, such that M accesses its input only during the first $(\log n)^{O(1)}$ computation steps.
Hint: Transfer the proof of Theorems 2.48 and 2.49 into the context of ptime-uniformity.

4.28. Let A be a language over a one-letter alphabet, i.e., $A \subseteq 0^*$. Show that if $A \in P$ then $A \in U_P$-NC.

4.29.** Prove: U_L-NC $= U_P$-NC if and only if all languages in P over a one-letter alphabet are actually contained in U_L-NC, i.e., $P \cap \mathcal{P}(0^*) \subseteq U_L$-NC.
Hint: Let U be a set complete for DTIME(2^n), and define $T(U) =_{\text{def}} \{ 0^i \mid \text{bin}(i) \in U \}$. Show that every set in U_P-NC is $\leq_T^{NC}$-reducible to $T(U)$.

4.30. Construct a formula $\phi \in \text{FO}[<]$ with two free variables x and y such that in all possible models, the value of y will be one greater than the value of x.

4.31. Which languages are defined by the following formulas? (We use the shorthands defined in Sect. 4.5.4.)

(1) $\exists v_1 (\forall v_2 (v_1 \geq v_2) \wedge Q_a(v_1))$.
(2) $\forall v_1 (\forall v_2 (v_1 \geq v_2) \wedge Q_a(v_1))$.
(3) $\exists v_1 (\forall v_2 (v_1 \geq v_2) \rightarrow Q_a(v_1))$.
(4) $\exists v_1 (\forall v_2 (v_1 \geq v_2) \wedge Q_a(v_1) \wedge \exists v_3 (v_3 < v_1))$.

4.32. Show that all languages in FO[<] are regular.

4.33. Show that all star-free regular sets over $\{0, 1\}$ are in FO[<]. (The class of *star-free* regular languages is defined to be the smallest class of sets fulfilling the following properties: The empty set is star-free. All sets consisting of one word are star-free. If L and L' are star-free, then $\{0,1\}^+ \setminus L$, $L \cup L'$, and $L \cdot L'$, the set of all concatenations of one word from L and one word from L', are star-free.)

4.34.** Show that all languages in FO[<] are star-free.

4.35. (1) Show that for every AC^0 circuit family $\mathcal{D}$ there is an equivalent AC^0 circuit family $\mathcal{C} = (C_n)_{n \in \mathbb{N}}$ which has the property that every C_n is a tree, and that for every gate $v \in C_n$ all paths from v to the circuit inputs are of the same length.
(2) Show that the result under (1) still holds in the context of U_D-uniformity, i.e., if $\mathcal{D}$ is U_D-uniform then $\mathcal{C}$ is also U_D-uniform.
(3) Show that the above also holds for $AC^0[p]$-circuits, TC^0-circuits, and $AC^0(L_M)$-circuits, where M is a monoid.

4.36. To complete the proof of Lemma 4.71, give detailed constructions of the formulas Φ and Ψ defined there. In particular, explain how to "implement" the universal quantifiers $(\forall i, \forall j)$ that range over blocks of bits.

4.37. Let $\mathcal{C}$ be a circuit family such that $L_{\mathrm{DC}}(\mathcal{C}) \in \mathrm{FO}[<, \mathrm{bit}]$. Show that there are formulas that define the relations In, Out, Or, And, Neg, and Pre. For example, in the case of the relation In, show that there is a formula ϕ^{In} with $k+1$ free variables $x_0, x_1, \ldots, x_k$ such that if v is a gate in C_n that is encoded as $(m_1, \ldots, m_k)$, then v is the input gate x_i if and only if $[a_1 \cdots a_n; x_0 \leftarrow i, x_1 \leftarrow m_1, \ldots, x_k \leftarrow m_k] \models \phi^{\mathrm{In}}$ for arbitrary $a_1, \ldots, a_n \in \{0, 1\}$.

4.38. Prove Corollary 4.77.

4.39. Show that there is a finite group G_e such that for every first-order formula ϕ with free variables from $V \cup \{x\}$ there are first-order formulas $\phi_1, \ldots, \phi_l$ such that for all V-structures w,

$$w \models \exists x \phi \iff w \models Q_{G_e} x[\phi_1, \ldots, \phi_l].$$

Similarly, prove the existence of groups G_u and G_p in the case of a universal and an $\exists^{=p}$ quantifier.

Is this result also valid for majority quantifiers?

4.40. Suppose that in the proof of Theorem 4.75 the identity of G is not encoded as 0^l but as $a_1 \cdots a_l$. Modify the construction of formulas $\Theta_1, \ldots, \Theta_l$ to work in this new setting.

4.41.[*] Let σ be a three-place predicate such that $\sigma(x, i, a)$ holds for $x, i \in \{0, 1\}^*$, $a \in \{0, 1\}$, if and only if a is the ith symbol of x.

The class of *log-bounded rudimentary predicates*, denoted by $\mathrm{RUD}_{\log}$, is the smallest class of predicates such that the following holds:

(1) $\sigma(x, i, a)$ and $\neg\sigma(x, i, a)$ are in $\mathrm{RUD}_{\log}$.
(2) For every $d \in \mathbb{N}$, the predicates $C_d(x, u, v, w)$ and $C_d^c(x, u, v, w)$ are in $\mathrm{RUD}_{\log}$, where $C_d(x, u, v, w)$ holds iff $uv = w$ and $|uvw| \leq d\log|x|$, and $C_d^c(x, u, v, w)$ holds iff $uv \neq w$ and $|uvw| \leq d\log|x|$.
(3) If $Q(x, y_1, \ldots, y_m)$ and $R(x, y_1, \ldots, y_m)$ are predicates in $\mathrm{RUD}_{\log}$, then the predicates $Q(x, y_1, \ldots, y_m) \wedge R(x, y_1, \ldots, y_m)$ and $Q(x, y_1, \ldots, y_m) \vee R(x, y_1, \ldots, y_m)$ are in $\mathrm{RUD}_{\log}$. ("Closure under positive Boolean operations")
(4) If $Q(x, y_1, \ldots, y_t)$ is a predicate in $\mathrm{RUD}_{\log}$ and $\xi_1, \ldots, \xi_t$ are strings over the alphabet $\{0, 1\} \cup \{y_1, \ldots, y_m\}$, then $R(x, y_1, \ldots, y_m)$ is a predicate in $\mathrm{RUD}_{\log}$, where $R(x, y_1, \ldots, y_m)$ holds iff $Q(x, \xi_1, \ldots, \xi_t)$. ("Closure under explicit transformation")
(5) If $Q(x, y_1, \ldots, y_{m+1})$ is a predicate in $\mathrm{RUD}_{\log}$ and $d \in \mathbb{N}$, then the predicates $R_d(x, y_1, \ldots, y_m)$ and $S_d(x, y_1, \ldots, y_m)$ are in $\mathrm{RUD}_{\log}$, where

$$R_d(x, y_1, \ldots, y_m) \iff \exists z \big(|z| \leq d\log|x| \wedge Q(x, y_1, \ldots, y_m, z)\big)$$
$$S_d(x, y_1, \ldots, y_m) \iff \forall z \big(|z| \leq d\log|x| \implies Q(x, y_1, \ldots, y_m, z)\big)$$

("Closure under log-bounded existential and universal quantification")

Prove that for every language $A \subseteq \{0,1\}^*$, the following holds: $A \in$ U_D-AC^0 if and only if A as a predicate (i.e., the predicate defined by $A(x) \iff x \in A$) is in $RUD_{\log}$.

4.42. Prove that the function f defined in the proof of Theorem 4.86 is in U_L-FNC^1.

4.43.* Prove that CVP is P-complete under constant-depth reductions.

4.44. Show that if A is P-complete, then the complement of A, $\overline{A}$, is P-complete as well.

4.45. Show that UNIT $\in$ P.

4.46. Prove: If P $=$ NC, then there is an $i \geq 0$ such that NC $=$ NC^i.

Notes on the Exercises

4.4–4.5. See [Ruz80].

4.3. This result is from [Ven91].

4.8–4.10. The equivalence LOGCFL $=$ ASPACE-TREESIZE$(\log n, n^{O(1)})$, which is established in this sequence of exercises, was proved in [Ruz80]. The proof of the inclusion ASPACE-TREESIZE $\subseteq$ LOGCFL (Exercise 4.10) heavily depends on a characterization of LOGCFL on so-called *auxiliary pushdown automata* given in [Sud78]. Ruzzo then proved how to simulate these devices by tree-size bounded alternating Turing machines.

4.6. See [Ven86, Ven91].

4.15. NC^1 reducibility and the class DET were defined in [Coo85]. The results from the exercises are established there, with the exception of INTDET $\leq_m^{NC^1}$ MATPOW which is from [Ber84].

4.21. See [BIS90, Proposition 6.4].

4.23. This exercise is due to [Bar89]. Barrington looked at the symmetric group S_5 which is not solvable and thus first obtained an S_5-program for any language in NC^1, which can then be transformed into a width 5 branching program using the method of Lemma 4.45. Refining the statement of the exercise (and using a certain normal form for bounded-width circuits) he even proved that NC^1 can be done by polynomial-size circuits of width 4.

4.26. Formula families were introduced and defined in [Bus87], but a different uniformity condition was used (the formula language was required to be in ATIME$(\log n)$).

The class DLOGTIME-uniform NC^1 was defined and examined in [BIS90]. The equality of this class with ATIME$(\log n)$ is also given in this paper. (See the discussion about uniformity for NC^1 in the Bibliographic Remarks of this chapter on p. 162.)

4.27–4.29. P-uniform NC (PUNC) was examined in [All89]. The results in the exercises are proved there.

4.32–4.34. The equality of the class of star-free regular sets and the logically defined class FO[<] was proved by [MP71]. The proof can also be found in [Str94, EF95].

4.41.   The class of space-bounded rudimentary predicates for arbitrary bounds s (not necessarily logarithmic) was defined in [Jon75]. This exercise is from [AG91], see also [Reg93].

4.45.   An algorithm can be found in [JL76].

5. Arithmetic Circuits

Arithmetic circuits are circuits whose gates compute operations over a semi-ring instead of the usual Boolean connectives (more precisely: operations over a semi-ring other than the Boolean semi-ring). The theory of arithmetic circuits, a very rich field with strong upper and lower bounds on the complexity of practical computational problems, serves as a theoretical basis for computer algebra and algebraic computations in symbolic computation software packages.

We will only briefly explore the field and study, following some relevant definitions, a few examples for arithmetic circuits. We refer the interested reader to the surveys cited in the Bibliographic Remarks for this chapter. However we will present an extensive discussion about a connection between a particular type of arithmetic circuit and the Boolean circuits that we have studied so far.

5.1 Complexity Measures and Reductions

We start by introducing the computation model of arithmetic circuits.

Let $\mathcal{R} = (R, +, \times, 0, 1)$ be a semi-ring. A function over $\mathcal{R}$ is a function $f: R^m \to R$ for some $m \in \mathbb{N}$. Here m is called the arity of f. A family of functions over $\mathcal{R}$ is a sequence $f = (f^n)_{n \in \mathbb{N}}$, where f^n is a function over $\mathcal{R}$ of arity n.

Definition 5.1. An *arithmetic basis* is given by a pair $(\mathcal{R}, B)$, where $\mathcal{R} = (R, +, \times, 0, 1)$ is a semi-ring and B is a finite set, whose elements are functions over R, or families of functions over R.

We now want to define arithmetic circuits over basis $(\mathcal{R}, B)$. These will be defined analogously to Boolean circuits we considered so far, but instead of gates computing some functions over the Boolean semi-ring, we now allow gates computing functions from B, i.e., functions over $\mathcal{R}$. In most of the applications below, $\mathcal{R}$ will be a ring or a field and B will consist of (a subset of) the ring or field operations plus sometimes the constants 0 or 1.

Definition 5.2. Let $(\mathcal{R}, B)$ be an arithmetic basis. An *arithmetic circuit* over $(\mathcal{R}, B)$ (or, an $\mathcal{R}$-circuit over B) with n inputs and m outputs is a tuple

$$C = (V, E, \alpha, \beta, \omega),$$

where (V, E) is a finite directed acyclic graph, $\alpha \colon E \to \mathbf{N}$ is an injective function, $\beta \colon V \to B \cup \{x_1, \ldots, x_n\}$, and $\omega \colon V \to \{y_1, \ldots, y_m\} \cup \{*\}$, such that the following conditions hold:

1. If $v \in V$ has in-degree 0, then $\beta(v) \in \{x_1, \ldots, x_n\}$ or $\beta(v)$ is a 0-ary function from B.
2. If $v \in V$ has in-degree $k > 0$, then $\beta(v)$ is a k-ary function from B or a family of functions from B.
3. For every i, $1 \leq i \leq n$, there is at most one node $v \in V$ such that $\beta(v) = x_i$.
4. For every i, $1 \leq i \leq m$, there is exactly one node $v \in V$ such that $\omega(v) = y_i$.

For every $v \in V$, we define a function $\mathrm{val}_v \colon R^n \to R$ as follows. Let $a_1, \ldots, a_n \in R$.

1. If $v \in V$ has fan-in 0 and $\beta(v) = x_i$ for some i, $1 \leq i \leq n$, then $\mathrm{val}_v(a_1, \ldots, a_n) =_{\mathrm{def}} a_i$. If $v \in V$ has fan-in 0 and $\beta(v) = a$ for some element $a \in B$, then a must be a 0-ary (i.e., constant) function and we define $\mathrm{val}_v(a_1, \ldots, a_n) =_{\mathrm{def}} a$.
2. Let $v \in V$ have fan-in $k > 0$, and let $v_1, \ldots, v_k$ be the gates that are predecessors of v ordered in such a way that $\alpha(v_1) < \cdots < \alpha(v_k)$. Let $\beta(v) = f \in B$. If f is a k-ary function over $\mathcal{R}$, then let

$$\mathrm{val}_v(a_1, \ldots, a_n) =_{\mathrm{def}} f\big(\mathrm{val}_{v_1}(a_1, \ldots, a_n), \ldots, \mathrm{val}_{v_k}(a_1, \ldots, a_n)\big).$$

Otherwise f must be a family of functions over $\mathcal{R}$, $f = (f^n)_{n \in \mathbf{N}}$. In this case, we define

$$\mathrm{val}_v(a_1, \ldots, a_n) =_{\mathrm{def}} f^k\big(\mathrm{val}_{v_1}(a_1, \ldots, a_n), \ldots, \mathrm{val}_{v_k}(a_1, \ldots, a_n)\big).$$

For $1 \leq i \leq m$, let v_i be the unique gate $v_i \in V$ with $\omega(v_i) = y_i$. Then the *function computed by C*, $f_C \colon R^n \to R^m$, is for all $a_1, \ldots, a_n \in R$ given by

$$f_C(a_1, \ldots, a_n) =_{\mathrm{def}} \big(\mathrm{val}_{v_1}(a_1, \ldots, a_n), \ldots, \mathrm{val}_{v_m}(a_1, \ldots, a_n)\big).$$

A family of arithmetic circuits over $(\mathcal{R}, B)$ (or, a family of $\mathcal{R}$-circuits over B) is a sequence $C = (C_n)_{n \in \mathbf{N}}$, where every C_n is an arithmetic circuit over $(\mathcal{R}, B)$ with n inputs. Let f^n be the function computed by C_n. Then we say that C computes the family of functions $f_C \colon R^* \to R^*$, defined by $f_C = (f^n)_{n \in \mathbf{N}}$.

We define the size and depth of arithmetic circuits as the number of non-input gates in V and the length of a longest directed path in (V, E), respectively, exactly as in the case of Boolean circuits.

Let us consider a few examples. We start with problems from linear algebra.

Problem: MATPROD
Input: two $n \times n$ square matrices A, B over $\mathcal{R} = (R, +, \times, 0, 1)$
Output: their product $C = A \cdot B$

Theorem 5.3. *MATPROD can be computed by $\mathcal{R}$-circuits over $\{+, \times\}$ of depth $O(\log n)$ and size $O(n^3)$.*

Proof. Let $A = (a_{i,j})_{\substack{1 \leq i \leq n \\ 1 \leq j \leq n}}$, and $B = (b_{i,j})_{\substack{1 \leq i \leq n \\ 1 \leq j \leq n}}$. Then $C = (c_{i,j})_{\substack{1 \leq i \leq n \\ 1 \leq j \leq n}}$, where $c_{i,j} = \sum_{k=1}^{n} a_{i,k} \cdot b_{k,j}$ for $1 \leq i, j \leq n$. The circuit computing MATPROD is the following: First compute in one stage all products $a_{i,k} \cdot b_{k,j}$ for $1 \leq i, j, k \leq n$; then compute in a second stage all values $c_{i,j}$.

The first stage needs n^3 size and constant depth, the second stage needs $n^2(n-1)$ size and logarithmic depth. □

On a more formal level, what we have just done is define a problem MATPROD$_\mathcal{R}$ for every semi-ring $\mathcal{R}$, and then construct circuits over $(\mathcal{R}, \{+, \times\})$ that solve this problem. We will, however, continue as above and not explicitly mention the underlying algebraic structure as a parameter in the definition of the problems we will study. It will always be clear how our obtained results can be translated into mathematically more exact statements.

In the arithmetic circuits we construct next we want to re-use the circuit for the matrix product. Formally, we again employ the concept of *reductions*.

Definition 5.4. Let $\mathcal{R} = (R, +, \times, 0, 1)$ be a semi-ring and f, g be families of functions over $\mathcal{R}$. Let $(\mathcal{R}, B)$ be an arithmetic basis.

1. We say that f log-depth $(\mathcal{R}, B)$-reduces to g; in symbols: $f \leq_{\log}^{(\mathcal{R}, B)} g$, if f can be computed by a family of $\mathcal{R}$-circuits over $B \cup \{g\}$ of polynomial size and logarithmic depth.
2. We say that f constant-depth $(\mathcal{R}, B)$-reduces to g; in symbols: $f \leq_{cd}^{(\mathcal{R}, B)} g$, if f can be computed by a family of $\mathcal{R}$-circuits over $B \cup \{g\}$ of polynomial size and constant depth.

Lemma 5.5. *Let $\mathcal{R} = (R, +, \times, 0, 1)$ be a semi-ring and suppose that g can be computed by $\mathcal{R}$-circuits over B of size s and depth d.*

1. *Let $f \leq_{\log}^{(\mathcal{R}, B')} g$; then there is a family of $\mathcal{R}$-circuits over $B \cup B'$ computing f, which is of size $s(n)^{O(1)}$ and depth $O\big(s(n) \cdot \log n\big)$.*
2. *Let $f \leq_{cd}^{(\mathcal{R}, B')} g$; then there is a family of $\mathcal{R}$-circuits over $B \cup B'$ computing f, which is of size $s(n)^{O(1)}$ and depth $O\big(s(n)\big)$.*

Proof. A circuit for f is obtained by replacing in the reducing circuit all gates for some g^k with the appropriate $\mathcal{R}$-circuits computing g^k. □

In Def. 5.4 we only considered families of functions over $\mathcal{R}$, i.e., functions $f, g: \mathcal{R}^* \to \mathcal{R}$. As we did in Chap. 1 for the Boolean case, the concept of reductions above can be generalized to the case that f, g are functions from $\mathcal{R}^* \to \mathcal{R}^*$, such that Lemma 5.5 still holds. We leave the formal details as an exercise (Exercise 5.1).

We now consider the problem of iterated matrix multiplication.

> *Problem:*   ITMATPROD
> *Input:* $n \times n$ square matrices $A_1, \ldots, A_n$ over $\mathcal{R} = (R, +, \times, 0, 1)$
> *Output:*   their product

Theorem 5.6. ITMATPROD $\leq_{\log}^{(\mathcal{R}, \emptyset)}$ MATPROD.

Proof. Because matrix multiplication is associative, we can easily compute the product of n matrices by a binary tree of depth $\lceil \log n \rceil$ of gates multiplying two matrices.   ☐

Corollary 5.7. ITMATPROD *can be computed by* $\mathcal{R}$*-circuits over* $\{+, \times\}$ *of depth* $O((\log n)^2)$ *and polynomial size.*

Proof. Follows immediately from Lemma 5.5 and Theorem 5.6.   ☐

Next we turn to some problems involving polynomials. If in the following, $\mathcal{R} = (R, +, \times, 0, 1)$ is a ring, and $f \in \mathcal{R}[x]$, $f(x) = a_0 + a_1 x + a_2 x^2 + \cdots + a_n x^n$, then we will represent f by $(a_0, a_1, \ldots, a_n)$; in symbols: $f = \mathrm{Pol}(a_0, a_1, \ldots, a_n)$.

> *Problem:*   POLYPROD
> *Input:*   elements $a_0, \ldots, a_n, b_0, \ldots, b_n$ from a ring $\mathcal{R}$
> *Output:* $c_0, \ldots, c_{2n}$ such that $\mathrm{Pol}(a_0, \ldots, a_n) \cdot \mathrm{Pol}(b_0, \ldots, b_n) = \mathrm{Pol}(c_0, \ldots, c_{2n})$

Theorem 5.8. POLYPROD *can be computed by* $\mathcal{R}$*-circuits over* $\{+, \times\}$ *of depth* $O(\log n)$ *and size* $O(n^3)$.

Proof. Let $f, g \in \mathcal{R}[x]$, $f(x) = a_0 + a_1 x + a_2 x^2 + \cdots + a_n x^n$ and $g(x) = b_0 + b_1 x + b_2 x^2 + \cdots + b_n x^n$, then $(f \cdot g)(x) = c_0 + c_1 x + c_2 x^2 + \cdots + c_{2n} x^{2n}$, where for $0 \leq k \leq 2n$, we have

$$c_k = \sum_{\substack{0 \leq i, j \leq n \\ i+j=k}} a_i \cdot b_j.$$

Thus the obvious circuit for POLYPROD in a first step computes all products $a_i \cdot b_j$ for $0 \leq i, j \leq n$, and in a second step all the c_k.

This circuit has size $O(n^2)$ and depth $O(\log n)$.   ☐

Problem: ITPOLYPROD
Input: polynomials $f_1, \ldots, f_n \in \mathcal{R}[x]$, each given by n coefficients
Output: their product, given by n^2 coefficients

Theorem 5.9. ITPOLYPROD $\leq_{\log}^{(\mathcal{R}, \emptyset)}$ POLYPROD.

Proof. Analogous to the proof of Theorem 5.6. $\qquad\qquad\square$

Corollary 5.10. ITPOLYPROD *can be computed by* $\mathcal{R}$-*circuits over* $\{+, \times\}$ *of depth* $O((\log n)^2)$ *and polynomial size.*

Finally, let us consider the problem of inverting a polynomial over a field $\mathcal{F} = (F, +, -, \times, /, 0, 1)$. Let $f_1, f_2, g \in \mathcal{F}[x]$. We say that f_1 and f_2 are *congruent* modulo g; in symbols: $f_1 \equiv f_2 \pmod{g}$, if there is an $h \in \mathcal{F}[x]$ such that $f_1 - f_2 = g \cdot h$. We say that f is invertible modulo x^{n+1} if there is some $g \in \mathcal{F}[x]$ such that $fg \equiv 1 \pmod{x^{n+1}}$. In this case g is called a *(modular) inverse* of f. The following lemma states a necessary and sufficient condition for a polynomial to be invertible.

Lemma 5.11. *Let* $\mathcal{F} = (F, +, -, \times, /, 0, 1)$ *be a field, and let* $f \in \mathcal{F}[x]$, $f(x) = a_0 + a_1 x + a_2 x^2 + \cdots + a_n x^n$. *Let* $m \in \mathbb{N}$. *Then* f *is invertible modulo* x^{m+1} *if and only if* $a_0 \neq 0$.

Proof. ($\Rightarrow$): Let f be invertible modulo x^{m+1}. Let g be the inverse of f modulo x^{m+1}, then there is a polynomial h such that $f(x)g(x) - 1 = h(x)x^{m+1}$. In particular this means that $f(0)g(0) = 1$, i.e., $a_0 = f(0) \neq 0$.

($\Leftarrow$): Let $b = a_0^{-1}$. Define the polynomial h as $h(x) =_{\text{def}} (a_0 - f(x)) \cdot b$, and let $g(x) =_{\text{def}} b \cdot \sum_{i=0}^{m} (h(x))^i$.

We claim that g is an inverse of f modulo x^{m+1}. Observe that $h(x) \equiv 0 \pmod{x}$, which implies $(h(x))^{m+1} \equiv 0 \mod x^{m+1}$. Moreover we have $f(x) = (1 - h(x)) \cdot b^{-1} = (1 - h(x)) \cdot a_0$. Thus we conclude

$$f(x) \cdot g(x) = (1 - h(x)) a_0 \cdot b \sum_{i=0}^{m} (h(x))^i$$

$$= 1 - (h(x))^{m+1}$$

$$\equiv 1 \pmod{x^{m+1}}.$$

$\qquad\qquad\square$

Observe that this proof shows that if f is of degree n and $f(0) \neq 0$, then there is an inverse of f modulo x^{m+1} of degree $m \cdot n$.

We now define the following problem:

Problem: POLYINV

Input: elements $a_0, \ldots, a_n$ from a field $\mathcal{F} = (F, +, -, \times, /, 0, 1)$

Output: $b_0, \ldots, b_{n^2}$ such that $\mathrm{Pol}(b_0, \ldots, b_{n^2})$ is an inverse of $\mathrm{Pol}(a_0, \ldots, a_n)$ modulo x^{n+1}, if such a polynomial exists; the constant 0 polynomial otherwise

Theorem 5.12. POLYINV $\leq_{\mathrm{cd}}^{(\mathcal{F}, \{+, -, \times, /, 0, 1\})}$ ITPOLYPROD.

Proof. The function g defined in the proof of Lemma 5.11 is an inverse of f modulo x^{m+1}. Thus the algorithm given in Fig. 5.1 solves the problem. An

> **input** $f \in \mathcal{F}[x]$, $f(x) = a_0 + a_1 x + a_2 x^2 + \cdots + a_n x^n$;
> **output** inverse g of f modulo x^{n+1};
>
> > **if** $a_0 = 0$ **then** let g be the constant 0 polynomial
> > **else begin**
> > > $b := a_0^{-1}$;
> > > $h := (a_0 - f(x)) \cdot b$;
> > > **for all** i, $0 \leq i \leq n$ **do** $f_i := h^i$;
> > > $g := b \cdot \sum_{i=0}^{n} f_i$.
> > **end**

Fig. 5.1. Algorithm for polynomial inversion

arithmetic circuit computing the different terms in the algorithm is straightforward, if we are allowed to use gates for iterated polynomial multiplication (see also Exercise 5.2). This shows that POLYINV $\leq_{\mathrm{cd}}^{\mathcal{R}}$ ITPOLYPROD. $\square$

Corollary 5.13. POLYINV *can be computed by circuits over the arithmetic basis* $(\mathcal{F}, \{+, -, \times, /, 0, 1\})$ *of depth* $O((\log n)^2)$ *and polynomial size.*

Proof. Follows from the above theorem and Lemma 5.5. $\square$

5.2 Evaluating Arithmetic Circuits by Straight-Line Programs

In this section we will present a very efficient way of evaluating small-depth arithmetic circuits by so-called *straight-line programs*. In a sense we will prove an arithmetization of Barrington's Theorem (Theorem 4.46).

Definition 5.14. Let $\mathcal{R} = (R, +, \times, 0, 1)$ be a ring. An n-input straight-line program over $\mathcal{R}$ using m registers is a sequence of instructions $P = (s_t)_{1 \leq t \leq l}$, where each instruction s_t is of one of the following forms:

- $R_j \leftarrow R_j + c \cdot R_i$,
- $R_j \leftarrow R_j - c \cdot R_i$,
- $R_j \leftarrow R_j + x_k \cdot R_i$,

- $R_j \leftarrow R_j - x_k \cdot R_i$,

where $i, j \in \{1, \ldots, m\}$, $i \neq j$, $c \in R$, $k \in \{1, \ldots, n\}$. The *length* of P is defined to be l.

Such a program computes a function $f_P \colon \{0,1\}^n \to \{0,1\}$ as follows: Let $x = x_1 \cdots x_n \in \{0,1\}^n$. At the beginning of the computations, register R_1 has as contents the value 1 and all other registers have contents 0. Then the instructions $s_1; s_2; \ldots; s_l$ are performed in this order, changing the contents of the registers in the natural way. Finally, the contents of register R_1 is the result of the computation, i.e., the value $f_P(x)$.

It is clear that every arithmetic circuit can be transformed into a straight-line program that simulates the circuit. The instructions just have to compute the values of the gates of the circuit in any topological order. How many registers do we need for this procedure? The naive way needs essentially as many registers as we have gates in the circuit. A cleverer approach would evaluate the gates according to a depth-first search of the circuit. Then the number of registers is equal to the depth of the circuit.

We now show that remarkably every arithmetic circuit can be evaluated using only three registers. As in the case of Boolean circuits, we define an arithmetic formula to be an arithmetic circuit where every gate has fan-out at most one. Obviously every arithmetic circuit can be transformed into an equivalent arithmetic formula of the same depth.

Theorem 5.15. *Let R be a ring, and let C be an arithmetic formula over the basis $(R, \{+, \times, 0, 1, -1\})$ computing the function $f \colon \{0,1\}^n \to \{0,1\}$. Let d be the depth of C. Then there is an n-input straight line program P over R of length $4^d + 2$ such that $f = f_C$. P uses only 3 registers.*

Proof. Let P be a straight-line program and let $f \colon \{0,1\}^n \to \{0,1\}$ be a Boolean function. Let $i, j \in \{1, 2, 3\}$, $i \neq j$. We say that P *offsets* R_j by $R_i \cdot f(x_1, \ldots, x_n)$, if the execution of P changes the values of the registers as follows: The value of R_j is incremented by $R_i \cdot f(x_1, \ldots, x_n)$ (taking the value of R_i before execution of P), and the values of the other registers R_k, $k \in \{1, 2, 3\} \setminus \{j\}$, after the execution of P are the same as before the execution of P. We say that P offsets R_j by $-R_i \cdot f(x_1, \ldots, x_n)$, if the execution of P decrements the value of R_j by $R_i \cdot f(x_1, \ldots, x_n)$ and does not change the values of all R_k, $k \in \{1, 2, 3\} \setminus \{j\}$.

Let C be an arithmetic formula of depth d, and let $i, j \in \{1, 2, 3\}$, $i \neq j$. We prove by induction on d that there is a straight-line program P which offsets R_j by $R_i \cdot f_C(x_1, \ldots, x_n)$, and a straight-line program Q which offsets R_j by $-R_i \cdot f_C(x_1, \ldots, x_n)$.

$d = 0$: If C is a constant $c \in R$, then P consists of the statement $R_j \leftarrow R_j + c \cdot R_i$ and Q consists of the statement $R_j \leftarrow R_j - c \cdot R_i$. If C is an input variable x_k, then P consists of the statement $R_j \leftarrow R_j + x_k \cdot R_i$ and Q consists of the statement $R_j \leftarrow R_j - x_k \cdot R_i$.

$d \to d+1$: Let C be an arithmetic formula of depth $d+1$. Let the top gate of C be a $+$ gate, and let D and E be the subformulas whose roots are the two predecessors of C's top gate, i.e., $f_C(x_1, \ldots, x_n) = f_D(x_1, \ldots, x_n) + f_E(x_1, \ldots, x_n)$.

The program P that offsets R_j by $R_i \cdot f_C(x_1, \ldots, x_n)$ looks as follows:
1. offset R_j by $R_i \cdot f_D(x_1, \ldots, x_n)$;
2. offset R_j by $R_i \cdot f_E(x_1, \ldots, x_n)$.

The program Q that offsets R_j by $-R_i \cdot f_C(x_1, \ldots, x_n)$ is
1. offset R_j by $-R_i \cdot f_D(x_1, \ldots, x_n)$;
2. offset R_j by $-R_i \cdot f_E(x_1, \ldots, x_n)$.

Formulas D and E are of depth at most d, hence by induction hypothesis, the required subprograms exist.

Let now the top gate of C be a $\times$ gate with subformulas D and E, i.e., $f_C(x_1, \ldots, x_n) = f_D(x_1, \ldots, x_n) \times f_E(x_1, \ldots, x_n)$.

Let $k \in \{1, 2, 3\}$, $k \neq i$, $k \neq j$. The program P that offsets R_j by $R_i \cdot f_C(x_1, \ldots, x_n)$ looks as follows.
1. offset R_j by $-R_k \cdot f_E(x_1, \ldots, x_n)$;
2. offset R_k by $R_i \cdot f_D(x_1, \ldots, x_n)$;
3. offset R_j by $R_k \cdot f_E(x_1, \ldots, x_n)$;
4. offset R_k by $-R_i \cdot f_D(x_1, \ldots, x_n)$.

It is clear that after the execution of P the values of R_i and R_k are the same as before. If r_i, r_j, r_k denote the values of R_i, R_j, R_k before the execution of P, then the value of R_j after the execution of P is

$$
r_j - r_k \cdot f_E(x_1, \ldots, x_n) + (r_k + r_i \cdot f_D(x_1, \ldots, x_n)) \cdot f_E(x_1, \ldots, x_n)
$$
$$
= r_j + r_i \cdot f_D(x_1, \ldots, x_n) \cdot f_E(x_1, \ldots, x_n);
$$

hence P works as required.

The program Q that offsets R_j by $-R_i \cdot f_C(x_1, \ldots, x_n)$ looks as follows.
1. offset R_j by $R_k \cdot f_E(x_1, \ldots, x_n)$;
2. offset R_k by $R_i \cdot f_D(x_1, \ldots, x_n)$;
3. offset R_j by $-R_k \cdot f_E(x_1, \ldots, x_n)$;
4. offset R_k by $-R_i \cdot f_D(x_1, \ldots, x_n)$.

Again, it is clear that after the execution of P the values of R_i and R_k are the same as before. Let r_i, r_j, r_k be the values of R_i, R_j, R_k before the execution of P. Then the value of R_j after the execution of P is

$$
r_j + r_k \cdot f_E(x_1, \ldots, x_n) - (r_k + r_i \cdot f_D(x_1, \ldots, x_n)) \cdot f_E(x_1, \ldots, x_n)
$$
$$
= r_j - r_i \cdot f_D(x_1, \ldots, x_n) \cdot f_E(x_1, \ldots, x_n);
$$

hence Q works as required.

Let now C be an arithmetic formula with n inputs. Then the following program computes f_C:

1. $R_2 \leftarrow R_2 + 1 \cdot R_1$;

2. $R_1 \leftarrow R_1 - 1 \cdot R_2$;
3. offset R_1 by $R_2 \cdot f_C(x_1, \ldots, x_n)$.

It is clear that after the first two instructions, the value of R_2 is 1 and the values of R_1 and R_3 are 0. Hence in the end, R_1 will hold the value $f_C(x_1, \ldots, x_n)$. Clearly the length of this program is bounded by $4^d + 2$, where d is the depth of C. $\qquad\square$

5.3 Counting Problems in Boolean Circuits

In this section we will consider arithmetic circuits over the semi-ring $\mathbb{N}$ of the naturals and the rings $\mathbb{Z}$ of the integers. When we restrict the inputs of our arithmetic circuits to be either 0 or 1, we will get an interesting connection with Boolean circuits. Let us first define those classes which will be of main interest for the rest of this chapter.

Definition 5.16. Let $\mathcal{R} = (R, +, \times, 0, 1)$ be a semi-ring. A *counting arithmetic circuit* (or *counting circuit* for short) over $(\mathcal{R}, B)$ with n inputs $x_1, \ldots, x_n$ is an arithmetic circuit C over $(\mathcal{R}, B)$ with $2n$ inputs $x_1, \ldots, x_{2n}$ and one output. Such a circuit computes a function $f_C \colon \{0,1\}^n \to R$ as follows: $f(a_1 \cdots a_n)$ is the value C computes when the first n of its inputs are set to $a_1, \ldots, a_n$ and the second n of its inputs are set to the negations of these bits, i.e., to $\overline{a_1}, \ldots, \overline{a_n}$.

If $\mathcal{C} = (C_n)_{n \in \mathbb{N}}$ is a family of counting arithmetic circuits, where C_n computes the function $f^n \colon \{0,1\}^n \to R$, then the function computed by $\mathcal{C}$, $f_C \colon \{0,1\}^* \to R$, is given by $f_C(x) =_{\text{def}} f^{|x|}(x)$.

Why we call these circuits *counting circuits* will become clear soon.

We start by considering the semi-ring $\mathbb{N}$ of the natural numbers, and the operations $+$ and $\times$ as circuit gates. However we will not restrict ourselves to fan-in 2. By $+^n$ and $\times^n$ we will denote addition and multiplication of n numbers in $\mathbb{N}$. 0^0 and 1^0 denote the 0-ary constants 0 and 1.

Definition 5.17. 1. A function $f \colon \{0,1\}^* \to \mathbb{N}$ is in the class $\#\text{SAC}^1$, if there is a family $\mathcal{C}$ of counting arithmetic circuits over the basis $(\mathbb{N}, \{(+^n)_{n \in \mathbb{N}}, \times^2, 0^0, 1^0\})$ of polynomial size and logarithmic depth computing f.
  2. A function $f \colon \{0,1\}^* \to \mathbb{N}$ is in the class $\#\text{NC}^1$, if there is a family $\mathcal{C}$ of counting arithmetic circuits over $(\mathbb{N}, \{+^2, \times^2, 0^0, 1^0\})$ of polynomial size and logarithmic depth computing f.
  3. A function $f \colon \{0,1\}^* \to \mathbb{N}$ is in the class $\#\text{AC}^0$, if there is a family $\mathcal{C}$ of counting arithmetic circuits over $(\mathbb{N}, \{(+^n)_{n \in \mathbb{N}}, (\times^n)_{n \in \mathbb{N}}, 0^0, 1^0\})$ of polynomial size and constant depth computing f.

In cases 1 and 3 we additionally require that the maximal fan-in of a gate in a circuit is bounded by a polynomial in the input length.

The restriction about the fan-in in unbounded fan-in circuits is simply to prevent the unnatural situation that a polynomial-size circuit has an exponential number of wires (without the additional stipulation, this would be possible since our circuits are allowed to have multiple wires between gates).

The choice of naming for the three classes just defined is motivated by the fact that their resources (depth, size, fan-in) resemble those of the Boolean circuit classes SAC^1, NC^1, and AC^0. However, there is an even closer connection as we show next.

Recall that in SAC^1 circuits we allowed negations but only adjacent to input gates. For the rest of this chapter, we also assume that all NC^1 and AC^0 circuits have this property, i.e., we assume the *convention that all Boolean circuits we consider below are in input normal form*. (By Corollary 4.4 and Remark 4.33 this is no restriction.)

For circuits of this form, we defined the notion of an accepting subtree above (see Def. 4.15). If C is a Boolean circuit and x is an input of C, then let $\#C(x)$ denote the number of accepting subtrees of C with input x.

Lemma 5.18. *Let C be a Boolean circuit of size s and depth d with n inputs. Then there is a counting N-circuit D over $\{+, \times\}$ of size s and depth d with n inputs such that for all $x \in \{0, 1\}^n$, we have:*

$$\#C(x) = f_D(x).$$

The fan-in allowed for $+$ and $\times$ gates is the same as the fan-in of the $\vee$ and $\wedge$ gates (respectively) in C.

Proof. D is obtained from C by replacing every $\vee$ gate by a $+$ gate and every $\wedge$ gate by a $\times$ gate. The correctness of this construction follows from a simple induction. □

Lemma 5.19. *Let C be a counting N-circuit over $\{+, \times\}$ of size s and depth d with n inputs. Then there is a Boolean circuit D over $\{\neg, \vee, \wedge\}$ of size $s+n$ and depth d with n inputs such that for all $x \in \{0, 1\}^n$, we have:*

$$f_C(x) = \#D(x).$$

The fan-in allowed for $\vee$ and $\wedge$ gates is the same as the fan-in of the $+$ and $\times$ gates (respectively) in C.

Proof. Analogous to the proof of the above lemma. □

Both lemmas immediately yield the following theorem:

Theorem 5.20. *Let $\mathcal{K}$ be one of the classes AC^0, NC^1, or SAC^1, and let $f : \{0, 1\}^* \to \mathbb{N}$. Then $f \in \#\mathcal{K}$ if and only if there is a $\mathcal{K}$ circuit family $C = (C_n)_{n \in \mathbb{N}}$ such that for all $x \in \{0, 1\}^*$, $f(x) = \#C_{|x|}(x)$.*

So we see that the arithmetic classes $\#SAC^1$, $\#NC^1$, and $\#AC^0$ have a nice characterization in terms of Boolean circuits. In the following subsections, where we address these classes in turn, we will make extensive use of this connection.

Remark 5.21. For the characterization of the classes above in terms of accepting subtrees, it is essential that in Def. 4.15 we first unwind the given circuit into a tree.

More precisely, given a circuit C in input normal form and an input x, define an *accepting subcircuit* of C on x as a subcircuit H of C with the following properties:

- H contains the output gate.
- For every $\wedge$ gate v in H, all the predecessors of v are in H.
- For every $\vee$ gate v in H, exactly one predecessor of v is in H.
- All gates in H evaluate to 1 on input x.

The only difference from the definition of accepting subtrees is exactly the point that here we do not start by unwinding the graph into a tree. For a circuit C and an input x, let $\#_cC(x)$ be the number of accepting subcircuits of C on x.

As an example, consider the circuit in Fig. 5.2. For input $x_1 = x_2 = x_3 = 1$, this circuit has four accepting subtrees but only two accepting subcircuits.

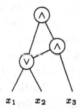

Fig. 5.2. Subtrees vs. subcircuits

We now consider the problems of counting the number of accepting subtrees and subcircuits of a given circuit C on input x. The first problem relates to evaluation of arithmetic circuits, as we have just seen; note that a polynomial-size circuit can evaluate to a number that is double-exponential in the input length. Thus even the time needed to write down the result is exponential, which somehow hides the true complexity of the problem. We therefore only consider the problem of evaluating such a circuit modulo a number exponential in the input length. More formally, we define:

Problem: AST

Input: circuit C over $\{\wedge, \vee\}$, an input $x \in \{0,1\}^*$, a number k in unary

> *Output:*　$\#C(x) \bmod 2^k$

> *Problem:*　ASC
> *Input:*　circuit C over $\{\wedge, \vee\}$, an input $x \in \{0,1\}^*$, a number k in unary
> *Output:*　$\#_c C(x) \bmod 2^k$

From the above relation to arithmetic circuits, it is clear that AST is in P (see Exercise 5.4). For ASC however, this is not known. In fact, this problem is complete for the class #P, a class presumably much harder than P (see Bibliographic Remarks, p. 208; a definition of #P will be given in Sect. 5.4).

Before we examine the classes $\#\mathrm{SAC}^1$, $\#\mathrm{NC}^1$, and $\#\mathrm{AC}^0$ defined above in more detail, let us consider another point. These classes are classes over the semi-ring N, and they are defined using resource bounds motivated from the consideration of Boolean circuit classes. What if we use the same resources but consider the ring Z instead of N? Formally, we extend the arithmetic bases over which we compute by additionally allowing the constant -1^0 (the 0-ary constant -1 function).

Definition 5.22.　1.　A function $f\colon \{0,1\}^* \to \mathbb{Z}$ is in the class Gap-SAC^1, if there is a family $\mathcal{C}$ of counting arithmetic circuits over the basis $(\mathbb{Z}, \{(+^n)_{n\in\mathbb{N}}, \times^2, 0^0, 1^0, -1^0\})$ of polynomial size and logarithmic depth computing f.
   2.　A function $f\colon \{0,1\}^* \to \mathbb{Z}$ is in the class Gap-NC^1, if there is a family $\mathcal{C}$ of counting arithmetic circuits over $(\mathbb{Z}, \{+^2, \times^2, 0^0, 1^0, -1^0\})$ of polynomial size and logarithmic depth computing f.
   3.　A function $f\colon \{0,1\}^* \to \mathbb{Z}$ is in the class Gap-AC^0, if there is a family $\mathcal{C}$ of counting arithmetic circuits over $(\mathbb{Z}, \{(+^n)_{n\in\mathbb{N}}, (\times^n)_{n\in\mathbb{N}}, 0^0, 1^0, -1^0\})$ of polynomial size and constant depth computing f.
   In cases 1 and 3 we additionally require that the maximal fan-in of a gate in a circuit is bounded by a polynomial in the input length.

These latter classes again have a characterization in terms of counting accepting subtrees, as we will show below.

5.3.1 Counting in SAC^1

In this subsection we will give a characterization of the class $\#\mathrm{SAC}^1$ in terms of the *algebraic degree* of an arithmetic circuit.

Definition 5.23. Let $\mathcal{R} = (R, +, \times, 0, 1)$ be a semi-ring and let C be an $\mathcal{R}$-circuit over $\{(+^n)_{n\in\mathbb{N}}, (\times^n)_{n\in\mathbb{N}}, 0^0, 1^0\}$. We define inductively for every node v in C its *algebraic degree* (*degree* for short):

1.　If v is a node of in-degree 0 then v has degree 1.

2. If v is a $+$ node then v's degree is the maximum of the degree of v's predecessors in C.
3. If v is a $\times$ node then v's degree is the sum of the degrees of v's predecessors.

Finally, we say that the degree of C is the maximum of the degree of all nodes in C.

Suppose $C = (C_n)_{n \in \mathbb{N}}$ is a family of $\mathcal{R}$-circuits of depth d with fan-in 2 $\times$ gates, then clearly the degree of C_n is bounded by $2^{d(n)}$. Thus we conclude in particular:

Lemma 5.24. *If $f \in \#SAC^1$ then there is a family of arithmetic circuits over basis $(\mathbb{N}, \{+^2, \times^2, 0^0, 1^0\})$ of polynomial size and polynomial degree that computes f.*

We now want to show that the converse relationship also holds, if the semi-ring $\mathcal{R}$ is commutative.

Theorem 5.25. *Let $\mathcal{R}$ be a commutative semi-ring. Then every family $C = (C_n)_{n \in \mathbb{N}}$ of arithmetic circuits of polynomial size and polynomial degree over $(\mathcal{R}, \{+^2, \times^2, 0^0, 1^0\})$ can be simulated by a family of circuits of polynomial size and logarithmic depth over $(\mathcal{R}, \{(+^n)_{n \in \mathbb{N}}, \times^2, 0^0, 1^0\})$.*

Proof. Let $C = (C_n)_{n \in \mathbb{N}}$ be a family of $\mathcal{R}$-circuits of polynomial size. Let $k \in \mathbb{N}$ be such that the arithmetic degree of C_n is bounded by n^k.

We now transform C into an equivalent circuit family $C' = (C'_n)_{n \in \mathbb{N}}$. Circuit C'_n will have gates $\langle g, i \rangle$ where g is a gate in C_n and $0 \leq i \leq n^k$. The idea is that every gate $\langle g, i \rangle$ will have degree i in C'_n; that is we can recover the degree of a node from its label.

To achieve this we use the following construction, which besides the above has a number of other convenient properties which we will use in the remainder of the proof:

1. We produce from C a circuit family C^* with the following two properties: First, there are no consecutive $+$ gates in any C_n^*. (This can be achieved by inserting a $\times$ gate between adjacent $+$ gates and setting the other input of the $\times$ gate to the constant 1.) Second, there is no $+$ gate g in C_n^* which has two input wires connecting it to the same predecessor. (If g is a $+$ gate which, say, has gate h twice as predecessor, we insert a gate computing $2 \times h$ and make this a predecessor of g.)
2. If g is a $+$ gate in C_n^*, then $\langle g, i \rangle$ is a $+$ gate in C'_n whose predecessors are all gates $\langle h, i \rangle$, where h is a predecessor of g in C_n^*.
3. Let g be a $\times$ gate in C_n^* and let g_l and g_r be the predecessors of g. Now $\langle g, i \rangle$ for $i > 1$ will be a $+$ gate in C'_n with $i - 1$ predecessors. Each of these predecessors is a $\times$ gate, and the predecessors of the jth of these are the gates $\langle g_l, j \rangle$ and $\langle g_r, i - j \rangle$ $(1 \leq j \leq i - 1)$. The relative order

of $\langle g_l, j \rangle$ and $\langle g_r, i - j \rangle$ is fixed in such a way that the degree of the left (i. e., first) predecessor of the $\times$ gate is not higher than the degree of the right (second) predecessor. This rearrangement of the predecessors of a $\times$ gate is possible since we assume that $\mathcal{R}$ is commutative.

4. Each gate $\langle g, 1 \rangle$, where g is not of in-degree 0 in C_n^*, is constant 0 in C_n'.
5. Let g be an in-degree 0 gate in C_n^*. Then $\langle g, 1 \rangle$ is an in-degree 0 gate in C_n' associated with the same type (input gate or constant) as g. Moreover $\langle g, i \rangle$ for $i > 1$ is a root of a subcircuit of degree i computing the constant 0 function.
6. If g is the output gate of C_n^* then the output of C_n' is a $+$ gate whose predecessors are all gates $\langle g, i \rangle$.

Thus C_n' simulates C_n. (See Exercise 5.5.) Observe that C' is still of polynomial size and degree.

Say that gates g and h in an arithmetic circuit are $+$-adjacent, if there is a path from g to h which in between visits only $+$ gates. By the construction of family C', the following property is obvious. Let g and h be any two nodes in C_n' which are $+$-adjacent. Then all paths from g to h with only $+$ gates in between have length at most 3.

We now define a nondeterministic procedure search which given a node g performs a particular depth-first search in the subcircuit with root g as defined in Fig. 5.3. Procedure search is nondeterministic (a predecessor of

```
procedure search(g: gate);
begin if g is a + gate then
          begin nondeterministically pick a predecessor h of g;
                search(h)
          end
      else if g is a × gate then
          begin let g₁, g₂ (in this order) be the predecessors of g;
                push g₂;
                search(g₁)
          end
      else { g must be an in-degree 0 node }
          begin if the stack is empty then halt;
                otherwise, let h be the top element of the stack;
                pop;
                search(h)
          end
end;
```

Fig. 5.3. Procedure search

a $+$ node is picked nondeterministically). Each definite sequence of these choices leads to an actual depth-first search, which we will refer to as an *exploration* of g. The last in-degree 0 node visited during an exploration of g will be called the *terminal node* of that exploration.

The *exploration height* of g is the maximum stack height of any exploration of g, i.e., the maximum stack height during executions of search(g) for any sequence of nondeterministic choices. The only way the stack height can increase is if we encounter a $\times$ gate g. Since the degree of the first predecessor of g which we search then is at most half the degree of g (by step 3 in the construction of C') we see that each time the stack height increases the degree of the node under consideration decreases by a factor of $\frac{1}{2}$; hence the exploration height of any node is at most logarithmic in the input length.

Let g be a node in C'_n and fix an input x of length n. When we evaluate g by multiplying out the expression given by the subcircuit with root g using the distributive laws, we get a sum of products. Since every exploration of g nondeterministically makes a choice for a sum, we see that such an exploration corresponds to a product in the above term. Hence the value computed by g, $\mathrm{val}_g(x)$, is the sum over all explorations e of g of the product of the values of the leaves encountered on e.

Let l be a gate in C'_n of in-degree 0. Then we define $\mathrm{val}_g^l(x)$ to be the sum over all explorations e of g, which have terminal node l, of the product of the values of the leaves encountered on e. If no exploration has terminal node l then we set $\mathrm{val}_g^l(x) =_{\mathrm{def}} 0$. Thus $\mathrm{val}_g(x)$ is the sum of all $\mathrm{val}_g^l(x)$ over all leaves l,

$$\mathrm{val}_g(x) = \sum_{\text{leaf } l} \mathrm{val}_g^l(x). \tag{5.1}$$

Now if h is an arbitrary gate in C'_n (not necessarily of in-degree 0), we look at the circuit we obtain from C'_n by replacing h with a leaf with value 1, and then let $\mathrm{val}_g^h(x)$ be the sum over all explorations e of g in this modified circuit, which have terminal node h, of the product of the values of the leaves encountered on e. Again, if no exploration has terminal node h we set $\mathrm{val}_g^h(x) =_{\mathrm{def}} 0$.

We now show how a #SAC^1 circuit can compute the value $\mathrm{val}_g^h(x)$ for any gates g, h in C'_n.

A. If g is an in-degree 0 node, then $\mathrm{val}_g^g(x)$ is either a constant or can be determined immediately from the input x; for $h \neq g$ we have $\mathrm{val}_g^h(x) = 0$.

B. If g is a $+$ gate, then

$$\mathrm{val}_g^h(x) = \sum_{\text{predecessor } g' \text{ of } g} \mathrm{val}_{g'}^h(x).$$

C. If g is a $\times$ gate, g and h are $+$-adjacent, and h is of in-degree 0, we proceed as follows: Let g_1, g_2 in this order be the predecessors of g. Since h can only be the terminal node of an exploration of g if it appears beneath the second predecessor g_2 of g, we have $\mathrm{val}_g^h(x) = 0$ if h does not appear beneath g_2. Otherwise we have

$$\text{val}_g^h(x) = h \times \text{val}_{g_1}(x)$$
$$= h \times \sum_{\text{leaf } l} \text{val}_{g_1}^l(x),$$

To argue correctness of this equality, we first claim that if g and h are $+$-adjacent, then there is no path from g to h via g_2 which contains a $\times$ gate. This holds since there is a path from g to h containing only $+$ gates, hence by construction of C'_n the degree of h must be equal to the degree of g_2. If there were another path from g_2 to h containing a $\times$ gate, then the degree of h would have to be smaller than the degree of g_2, again by construction of C'_n. This is a contradiction, hence this second path cannot exist. Second we observe that the $+$ path from g to h is unique (recall that a $+$ gate never has two input wires to the same predecessor gate).

Now the correctness of the above formula follows, since any exploration e with terminal node h must nondeterministically guess the nodes on the (unique) path from g to h. Hence the value $\text{val}_g^h(x)$ must be the value of this path (which is h) times the result of the evaluation of the other predecessor g_1 of g.

D. If g is a $\times$ gate, g and h are $+$-adjacent, and h is not of in-degree 0, then

$$\text{val}_g^h(x) = \sum_{\text{leaf } l} \text{val}_{g_1}^l(x),$$

where g_1 is the first predecessors of g. Correctness follows as in the previous case by observing that if h is not a leaf then we have to treat it by definition as a leaf with value 1.

E. The final case remaining is that of g being a $\times$ gate where g and h are not $+$-adjacent.

If there is an exploration e with terminal node h, then look at the path p leading from g to h. Exploration e must for every $+$ gate g' on p pick the one predecessor of g' which is also an element of p. For every $\times$ node g' on p, since h is the terminal node of e it must be visited during the exploration of the second predecessor of g', hence the right predecessor of g' must be on p. Hence we conclude that there is a unique sequence of gates

$$g = g_0, g_1, \ldots, g_m = h,$$

such that for $0 \le i < m$, gate g_i is a $\times$ gate and g_i is $+$-adjacent to g_{i+1}, where the path witnessing this includes the second predecessor of g_i. Uniqueness follows from the same considerations as in construction case C. For a gate g let $\deg(g)$ denote its degree in C'_n, and let $\deg_2(g)$ be the degree of the second predecessor of g in C'_n. Let $d =_{\text{def}} \frac{\deg(g) + \deg(h)}{2}$. Then there is exactly one i, $0 \le i \le m$, such that $\deg(g_i) \ge d > \deg_2(g_i)$. The product of all leaves encountered on exploration e is the product of all leaves encountered before visiting g_i times the product of those leaves encountered after visiting g_i. Hence altogether we have

$$\text{val}_g^h(x) = \sum_{\substack{g' \text{ such that} \\ \deg(g') \geq d > \deg_2(g')}} \text{val}_g^{g'}(x) \times \text{val}_{g'}^h(x).$$

Our circuit C_n'' now looks as follows: It computes all terms val_g^h for gates g, h in C_n' according to cases A–E above. Then it computes all terms val_g using (5.1). Clearly the size of this circuit is polynomial in n. To analyze the depth, we observe the following: If g is a $\times$ gate and g and h are $+$-adjacent (cases C and D in the construction), then the subcircuit computing $\text{val}_g^h(x)$ contains only gates for $\text{val}_g^{h'}(x)$ where the exploration height of g' is at least one less than the exploration height of g. If g is a $\times$ gate and g and h are not $+$-adjacent (case E above), then the subcircuit computing val_g^h contains only gates computing $\text{val}_{g'}^{h'}$ where $\deg(g') - \deg(h')$ is at most half of $\deg(g) - \deg(h)$. Since between any two $\times$ gates in C_n' there are at most two $+$ gates, the above considerations imply that the depth of C_n'' is logarithmic.   □

Corollary 5.26. *A function f is in $\#\text{SAC}^1$ if and only if there is a family of arithmetic circuits over basis $(\mathbb{N}, \{(+^n)_{n \in \mathbb{N}}, \times^2, 0^0, 1^0\})$ of polynomial size and polynomial degree that computes f.*

The above result also holds uniformly.

Theorem 5.27. *A function f is in $U_L\text{-}\#\text{SAC}^1$ if and only if there is a logspace-uniform family of $\mathbb{N}$-circuits over $\{(+^n)_{n \in \mathbb{N}}, \times^2, 0^0, 1^0\}$ of polynomial size and polynomial degree that computes f.*

Proof. We have to show that the constructions of the proof of Theorem 5.25 can be made uniform. This is easily seen for the construction of circuit families C^* and C'. The final circuit family C'' is logspace-uniform because of the following observations: The constructions in cases A and B are clearly logspace-uniform. Whether g and h are $+$-adjacent can be checked in logarithmic space since the paths of $+$ gates witnessing this fact are of length at most 3. Hence the subcircuits constructed in cases C and D are also logspace-uniform. For case E observe that the condition $\deg(g') \geq d > \deg_2(g')$ can be checked in logarithmic space due to the fact that the degree of a gate can be recovered from its label; hence this part of the construction is also logspace-uniform.   □

Let us now turn to the class Gap-SAC^1. As shown in Theorem 5.20 there is a nice relationship between evaluation of arithmetic circuits over $\mathbb{N}$ and counting accepting subtrees in Boolean circuits. We show next that Gap-SAC^1 can be characterized as the class of all functions which can be written as the difference of two $\#\text{SAC}^1$ functions. By Theorem 5.20 this also indirectly relates the class Gap-SAC^1 to counting subtrees in circuits.

Let $\mathcal{F}$ and $\mathcal{G}$ be classes of functions. Then we define $\mathcal{F} - \mathcal{G} =_{\text{def}} \{ f - g \mid f \in \mathcal{F}, g \in \mathcal{G} \}$.

Theorem 5.28. Gap-SAC^1 = #SAC^1 − #SAC^1 *and* U$_L$-Gap-SAC^1 = U$_L$-#SAC^1 − U$_L$-#SAC^1.

Proof. The direction from right to left is clear.

For the other direction, let $f \in$ Gap-SAC^1 via the Z-circuit family $\mathcal{C} = (C_n)_{n \in \mathbb{N}}$. Fix an input length n and consider C_n. We first construct a circuit C'_n over the semi-ring N of the naturals as follows: For every gate g from C_n there will be two gates g_P and g_N in C'_n such that for all inputs x of length n, we have $\text{val}_g(x) = \text{val}_{g_P}(x) - \text{val}_{g_N}(x)$.

The construction is by induction on the depth of a gate g: If g is an input gate or a constant $(0, 1, -1)$, the gates g_N and g_P are clear. If g is a $+$ gate with predecessors $g^1, \ldots, g^s$, and $g_P^1, \ldots, g_P^s, g_N^1, \ldots, g_N^s$ are gates in C'_n such that for all x and all i, $1 \leq i \leq s$, $\text{val}_{g^i}(x) = \text{val}_{g_P^i}(x) - \text{val}_{g_N^i}(x)$, then we define g_P to be the sum of all gates $g_P^1, \ldots, g_P^s$ and g_N to be the sum of all gates $g_N^1, \ldots, g_N^s$. If g is a $\times$ gate with predecessors g^1 and g^2, and $g_P^1, g_P^2, g_N^1, g_N^2$ are gates in C'_n such that for all x and $i = 1, 2$, $\text{val}_{g^i}(x) = \text{val}_{g_P^i}(x) - \text{val}_{g_N^i}(x)$, then we define g_P to be the sum of $g_P^1 \times g_P^2$ and $g_N^1 \times g_N^2$, and g_N to be the sum of $g_P^1 \times g_N^2$ and $g_N^1 \times g_P^2$.

It is now obvious how to use C'_n to define two N-circuits A_n and B_n such that $f_C(x) = f_A(x) - f_B(x)$.

Clearly this construction is logspace-uniform if $\mathcal{C}$ is logspace-uniform. □

Is there an upper bound for #SAC^1 and Gap-SAC^1 in terms of the Boolean circuit classes we considered before? The answer is yes:

Theorem 5.29. #SAC$^1 \subseteq$ FTC^1.

Proof. See Exercise 5.6. □

An analogous inclusion holds for Gap-SAC^1 under an appropriate encoding of the integers, e. g., two's complement representation. However it is not known if this can be improved, e. g., if #SAC$^1 \subseteq$ FAC^1 is open.

5.3.2 Counting in NC^1

We now turn to the problem of counting accepting subtrees in NC^1 circuits. We want to examine a characterization of the class #NC^1, based on a particular way we developed previously for looking at NC^1 computations.

Recall that in Sect. 4.5.3 we defined Σ-programs (where Σ is an alphabet), and proved that a language A is in NC^1 if and only if there is a regular language $R \subseteq \Sigma^*$ and a family $\mathcal{P}$ of Σ-programs, such that for all inputs x, we have $x \in A \iff f_\mathcal{P}(x) \in R$ (Corollary 4.49). In this subsection we will use this result to establish a connection between counting subtrees in NC^1 circuits and counting accepting paths in finite automata.

Let Σ be an alphabet. A nondeterministic finite automaton over Σ is a tuple $M = (Q, \delta, q_0, F)$, where

- Q is a finite set, the set of *states*;
- $\delta \colon Q \times \Sigma \to \mathcal{P}(Q)$ is the *transition function*; we explicitly remark that δ need not be a total function;
- $q_0 \in Q$ is the *initial state*; and
- $F \subseteq Q$ is the set of *final states*.

Given an input word $x \in \Sigma^*$, $x = a_1 \cdots a_n$, an *accepting path* for x in M is a sequence $q_1, \ldots, q_n$ of states such that the following holds:

1. For $i = 1, \ldots, n$, $\delta(q_{i-1}, a_i) \ni q_i$.
2. $q_n \in F$.

Let $\#M(x)$ be the number of accepting paths for x in M. An input word x is accepted by M if $\#M(x) > 0$.

We now define two classes of functions #BWBP and Gap-BWBP. (BWBP, as in Sect. 4.5.3, stands for "bounded width branching programs.")

Definition 5.30. The class #BWBP consists of all functions $f \colon \{0,1\}^* \to \mathbb{N}$, for which there exist an alphabet Σ, a family $\mathcal{P}$ of Σ-programs, and a nondeterministic finite automaton M over Σ such that for all $x \in \{0,1\}^*$, we have $f(x) = \#M\big(f_{\mathcal{P}}(x)\big)$. Furthermore, let the class Gap-BWBP be defined as Gap-BWBP = #BWBP − #BWBP.

We want to compare #BWBP and Gap-BWBP in the following with $\#NC^1$ and Gap-NC^1. As a convenient tool we introduce another notion of computation of functions. The reader should not be concerned that we are introducing too many complexity classes: We will soon prove that the two classes defined below are identical to #BWBP and Gap-BWBP, respectively; they are just a convenient way of coping with the above classes.

Definition 5.31. Let $\mathcal{R} = (R, +, \times, 0, 1)$ be a semi-ring. Let $k \in \mathbb{N}$, let Δ be a finite set of $k \times k$ matrices, let v_0 be a $1 \times k$ row vector and let v_f be a $k \times 1$ column vector, where all elements are drawn from $\mathcal{R}$. A (Δ, v_0, v_f)-*program* P *over n variables* is a sequence of triples $\big(\langle i_j, A_j, B_j \rangle\big)_{1 \leq j \leq \ell}$, where for every j, $i_j \in \{1, \ldots, n\}$ and $A_j, B_j \in \Delta$. The triples $\langle i_j, A_j, B_j \rangle$ are called *instructions* of P. Given an input $a_1 \cdots a_n \in \{0,1\}^n$, such an instruction yields the *value*

$$\mathrm{val}_{\langle i_j, A_j, B_j \rangle}(a_1 \cdots a_n) =_{\mathrm{def}} \begin{cases} A_j, & \text{if } a_{i_j} = 1, \\ B_j, & \text{if } a_{i_j} = 0. \end{cases}$$

The *output* of P on input $a_1 \cdots a_n$ is defined to be

$$f_P(a_1 \cdots a_n) = v_0 \times \left(\prod_{j=1}^{\ell} \mathrm{val}_{\langle i_j, A_j, B_j \rangle}(a_1 \cdots a_n) \right) \times v_f$$

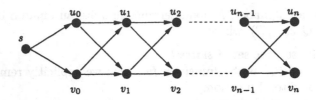

Fig. 5.4. Graph D_n

(where $\times$ is matrix multiplication over $\mathcal{R}$).

The *size* of P is ℓ.

Given a family $\mathcal{P} = (P_n)_{n \in \mathbb{N}}$, where every P_n is an (Δ, v_0, v_f)-program over n variables, we define $f_{\mathcal{P}} : \{0,1\}^* \to \mathcal{R}$ by $f_{\mathcal{P}}(x) =_{\text{def}} f_{P_{|x|}}(x)$. For $s : \mathbb{N} \to \mathbb{N}$, we say that $\mathcal{P}$ is of size s if P_n is of size $s(n)$ for all n.

Finally, define a *matrix program* over $\mathcal{R}$ to be a family of (Δ, v_0, v_f)-programs, for some Δ, v_i, v_f as above.

Of particular interest, of course, will be the semi-ring of the natural numbers and the ring of the integers.

Definition 5.32. 1. BPM($\mathbb{N}$) is the class of all functions $f : \{0,1\}^* \to \mathbb{N}$, for which there exists a matrix program $\mathcal{P}$ over $\mathbb{N}$ such that $f = f_{\mathcal{P}}$.
  2. BPM($\mathbb{Z}$) is the class of all functions $f : \{0,1\}^* \to \mathbb{Z}$, for which there exists a matrix program $\mathcal{P}$ over $\mathbb{Z}$ such that $f = f_{\mathcal{P}}$.

BPM stands for "branching program over matrices."

Before we relate #BWBP, BPM($\mathbb{N}$), and #NC^1, we prove a graph-theoretical result, which will be useful later. A directed acyclic graph G is *layered* if G is connected and there is a node s such that for all nodes $v \in G$, all paths from s to v are of the same length. Layer i in G (for $i \in \mathbb{N}$) is defined to be the set of all nodes $v \in G$ such that the length of paths from s to v is exactly i.

For the definition of function ℓ, see Appendix A2.

Lemma 5.33. *For $m \in \mathbb{N}$, there is a layered directed acyclic graph $H(m)$ with two particular nodes $s, t \in H(m)$ such that the following holds:*

1. *There are exactly m paths from s to t.*
2. *$H(m)$ consists of $\ell(m) + 2$ different layers.*
3. *Every layer in $H(m)$ consists of at most 3 nodes.*

Proof. Let $n \in \mathbb{N}$. As a first step, we define the graph $D_n = (V_n, E_n)$ by $V_n =_{\text{def}} \{s, u_0, u_1, \ldots, u_n, v_0, v_1, \ldots, v_n\}$ and $E_n =_{\text{def}} \{(s, u_0), (s, v_0)\} \cup \{(u_i, u_{i+1}) \mid 0 \le i < n\} \cup \{(v_i, v_{i+1}) \mid 0 \le i < n\} \cup \{(u_i, v_{i+1}) \mid 0 \le i < n\} \cup \{(v_i, u_{i+1}) \mid 0 \le i < n\}$; see Fig. 5.4. Then, obviously, there are exactly 2^i paths from s to u_i, and 2^i paths from s to v_i, for all $0 \le i \le n$.

Now let $m \in \mathbb{N}$, $l = \ell(m)$. Let the binary representation of m be $m = m_{l-1} \cdots m_1 m_0$. Define the graph $H(m) = (V(m), E(m))$ by $V(m) =_{\text{def}}$ $V_{l-1} \cup \{w_1, \ldots, w_l\}$ and $E(m) =_{\text{def}} E_{l-1} \cup \{ (w_i, w_{i+1}) \mid 1 \leq i < l \} \cup$ $\{ (v_i, w_{i+1}) \mid 0 \leq i < l \wedge m_i = 1 \}$. As an example consider the graph $H(42)$ in Fig. 5.5; the binary representation of 42 is 101010.

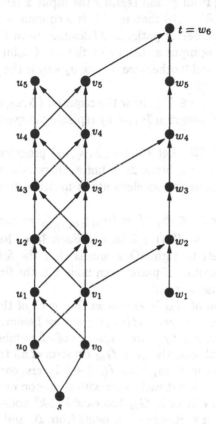

Fig. 5.5. Graph $H(42)$, having exactly 42 paths from s to t

If we now set $t =_{\text{def}} w_l$, then it is easy to see that $H(m)$ has exactly m paths from s to t.

Obviously $H(m)$ is layered and every layer consists of at most 3 nodes. The number of layers is $\ell(m) + 2$.   □

The classes of functions introduced in this section now can be related as follows:

Theorem 5.34. $\#\text{BWBP} = \text{BPM}(\mathbb{N}) \subseteq \#\text{NC}^1$.

Proof. $\#\text{BWBP} \subseteq \text{BPM}(\mathbb{N})$: Let $f \in \#\text{BWBP}$ by the Σ-program $\mathcal{P}$ and the finite automaton $M = (Q, \delta, q_1, F)$ over Σ. Let $Q = \{q_1, \ldots, q_s\}$. For every

$a \in \Sigma$, we define its transition matrix $M_a = \left((m_a)_{ij}\right)_{1 \le i,j \le s}$, where $(m_a)_{ij}$ is 1 if $\delta(q_i, a) \ni a_j$; $M_{ij} = 0$ otherwise. Given an input word $x = a_1 a_2 \cdots a_l$, let $M_x = \left((m_x)_{ij}\right)_{1 \le i,j \le s}$, $M_x = M_{a_1} \times M_{a_2} \times \cdots \times M_{a_l}$. Then it is easy to see that for all $i, j \le s$, $(m_x)_{ij}$ is the number of paths on which q_j can be reached starting from q_i and reading the input x (see Exercise 5.7). If we define $v_0 =_{\text{def}} (1, 0, 0, \ldots, 0)$ then $v_0 \cdot M_x$ is a column vector v_x over $\mathbb{N}$ whose i-th entry is the number of paths in M leading from the initial state q_1 to state q_i while reading input x. Let v_f be the $n \times 1$ column vector whose i-th entry is 1 if $q_i \in F$, and 0 otherwise. Hence $v_x \times v_f$ is the number of accepting paths of M for input x.

Let $\Delta =_{\text{def}} \{ M_a \mid a \in \Sigma \}$. Now the required (Δ, v_0, v_f)-program is easily obtained from the Σ-program $\mathcal{P}$ just by replacing in every instruction a letter $a \in \Sigma$ by $M_a \in \Delta$.

$\text{BPM}(\mathbb{N}) \subseteq \#\text{BWBP}$: Let $\mathcal{P}$ be a (Δ, v_0, v_f)-program, where all elements of Δ are $k \times k$ matrices. Since Δ is finite, there is a number $m \in \mathbb{N}$ such that all numbers appearing as elements in matrices from Δ are taken from $\{0, 1, \ldots, m\}$.

For each matrix $M \in \Delta$, $M = (m_{ij})_{1 \le i,j \le k}$, we now construct a graph G_M, consisting of $l = \ell(m) + 2$ layers. Each layer has $3k^2$ nodes, which we number from left to right. Our aim is that the following holds for all $1 \le i, j \le k$: The number of paths from node i in the first layer to node j in the last layer is equal to m_{ij}.

The construction of G_M is as follows: For each of the k^2 possible matrix entries m_{ij}, we take the graph $H(m_{ij})$ given by Lemma 5.33, where we add leading zeroes to the binary representations of all numbers m_{ij} to produce a binary word of length exactly $\ell(m)$. G_M consists of all these graphs arranged in parallel. Each of the $H(m_{ij})$ has $\ell(m) + 2$ layers, each layer consisting of at most 3 nodes. We fill out each layer with singleton nodes if necessary, such that in the end every layer in G_M has exactly $3k^2$ nodes.

If $M_1, \ldots, M_s$ is a sequence of elements from Δ, and $\widehat{M} = M_1 \times \cdots \times M_s$, then the graph $\widehat{G}$, obtained from the graphs $G_{M_1}, \ldots, G_{M_s}$ by identifying the first layer of $G_{M_{i+1}}$ with the last layer of G_{M_i} (for $1 \le i < s$) has the following property: For all $1 \le i, j \le n$, the number of paths leading from node i in the first layer of $\widehat{G}$ to node j in the last layer is equal to the element at row i and column j in $\widehat{M}$.

We now define a nondeterministic finite automaton N, having state set $Q = \{q_1, q_2, \ldots, q_{3k^2}\}$, working over the alphabet $\Sigma =_{\text{def}} \{ f \mid f : \{1, \ldots, 3k^2\} \to \{1, \ldots, 3k^2\} \}$. Thus, every letter $a \in \Sigma$ induces a transformation on $3k^2$ nodes, and hence, every graph G_M corresponds to a sequence of $l - 1$ letters.

Finally it only remains to deal with the vectors v_0 and v_f. If $v_0 = (a_1, \ldots, a_k)$, we add to N an initial state q_0 which has a_i paths to state q_i $(1 \le i \le k)$. Let the length of all these paths be l_0. All transitions in the automaton constituting these paths are defined to read the letter a_0. If

$v_f = (b_1, \ldots, b_k)^{\mathrm{T}}$, then we introduce k final states, the ith of which can be reached from state q_i on exactly b_i different paths ($1 \leq i \leq k$). Let the length of all these paths be l_f. All transitions in the automaton constituting these paths are defined to read the letter a_f. The constructions corresponding to the vectors v_0 and v_f are analogous to the constructions of the G_M above.

Our Σ-program now looks as follows: First it produces l_0 many letters a_0, then it behaves as the given program $\mathcal{P}$, simulating the output of one matrix from Δ by outputting the corresponding $l-1$ letters from Σ. Finally l_f many letters a_f are produced.

$\mathrm{BPM}(\mathbb{N}) \subseteq \#\mathrm{NC}^1$: We use the techniques from Sect. 5.1 to construct an arithmetic circuit for matrix multiplication. In particular, Corollary 5.7 showed how iterated matrix multiplication over $\mathbb{N}$ can be done by $\mathbb{N}$-circuits over $\{+, \times\}$ of depth $O((\log n)^2)$. Here we only deal with constant-size matrices, hence the depth reduces to $O(\log n)$ and the claim follows.   □

Of course one would like to know whether $\#\mathrm{NC}^1 \subseteq \#\mathrm{BWBP}$. Unfortunately this question is open. However when we consider the corresponding gap classes, equality is known, as we will prove next.

Theorem 5.35. Gap-BWBP = BPM($\mathbb{Z}$) = Gap-NC^1.

Proof. BPM($\mathbb{Z}$) $\subseteq$ Gap-BWBP: For $x \in \mathbb{Z}$, define the 2×2 matrix M_x as follows: If $x \geq 0$ then

$$M_x =_{\mathrm{def}} \begin{pmatrix} x & 0 \\ 0 & x \end{pmatrix}.$$

If $x < 0$ then

$$M_x =_{\mathrm{def}} \begin{pmatrix} 0 & -x \\ -x & 0 \end{pmatrix}.$$

The idea in the proof is to simulate a BPM($\mathbb{Z}$) program by a BPM($\mathbb{N}$) program where integers are represented as 2×2 matrices as above.

More specifically we define two mappings σ and ρ on matrices: First we expand $k \times m$ matrices over $\mathbb{Z}$ to $2k \times 2m$ matrices over $\mathbb{N}$ as follows: Let $M = (m_{i,j})_{\substack{1 \leq i \leq k \\ 1 \leq j \leq m}}$. Then define $\sigma(M)$ by substituting every entry x by the four entries given by the 2×2 matrix M_x. Formally, $\sigma(M) = (m'_{i,j})_{\substack{1 \leq i \leq 2k \\ 1 \leq j \leq 2m}}$, where $m'_{2i,2j} = m'_{2i-1,2j-1} = m_{i,j}$ and $m'_{2i-1,2j} = m'_{2i,2j-1} = 0$ if $m_{i,j} \geq 0$; and $m'_{2i,2j} = m'_{2i-1,2j-1} = 0$ and $m'_{2i-1,2j} = m'_{2i,2j-1} = -m_{i,j}$ if $m_{i,j} < 0$. Every $\sigma(M)$ is thus a matrix with entries from $\mathbb{N}$.

Second, we reduce $2k \times 2m$ matrices to $k \times m$ matrices as follows: Let $M = (m'_{i,j})_{\substack{1 \leq i \leq 2k \\ 1 \leq j \leq 2m}}$. Then define $\rho(M) = (m_{i,j})_{\substack{1 \leq i \leq k \\ 1 \leq j \leq m}}$ by $m_{i,j} = m'_{2i-1,2j-1} - m'_{2i-1,2j}$.

These mappings have the obvious property that $\rho(\sigma(M)) = M$ for every matrix M, and moreover, if M is a $2k \times 2l$ matrix and N is a $2l \times 2m$ matrix, then

$$\rho(M \cdot N) = \rho(M) \cdot \rho(N). \tag{5.2}$$

We now proceed as follows:

Let $\mathcal{P}$ be the given family of (Δ, v_0, v_f)-programs. Δ is a finite set of $k \times k$ matrices over $\mathbb{Z}$ (for some $k \in \mathbb{N}$). If the result of a matrix program P on an input x is now given by some product

$$z = v_0 \cdot \left(\prod_{i=1}^{l} M_i \right) \cdot v_f$$

(for $M_i \in \Delta$ for $1 \leq i \leq l$), then we conclude from the above that

$$z = \rho \left(\sigma(v_0) \cdot \left(\prod_{i=1}^{l} \sigma(M_i) \right) \cdot \sigma(v_f) \right).$$

Observe that $\sigma(v_0) \cdot \left(\prod_{i=1}^{l} \sigma(M_i) \right) \cdot \sigma(v_f)$ is a 2×2 matrix, hence the above product can be computed as

$$z = \begin{pmatrix} 1 & 0 \end{pmatrix} \cdot \sigma(v_0) \cdot \left(\prod_{i=1}^{l} \sigma(M_i) \right) \cdot \sigma(v_f) \cdot \begin{pmatrix} 1 \\ -1 \end{pmatrix}$$

$$= \begin{pmatrix} 1 & 0 \end{pmatrix} \cdot \sigma(v_0) \cdot \left(\prod_{i=1}^{l} \sigma(M_i) \right) \cdot \sigma(v_f) \cdot \begin{pmatrix} 1 \\ 0 \end{pmatrix}$$

$$- \begin{pmatrix} 1 & 0 \end{pmatrix} \cdot \sigma(v_0) \cdot \left(\prod_{i=1}^{l} \sigma(M_i) \right) \cdot \sigma(v_f) \cdot \begin{pmatrix} 0 \\ 1 \end{pmatrix}.$$

Finally we expand the matrices $\sigma(v_0)$ and $\sigma(v_f)$ to $2k \times 2k$ matrices and adjust the left and right vectors accordingly. Then we see that both parts of the above difference can be computed by BPM($\mathbb{N}$) programs, and we conclude that $\mathcal{P}$ can be simulated by the difference of two families of BPM($\mathbb{N}$) programs.

Gap-NC$^1 \subseteq$ BPM($\mathbb{Z}$): Let $f \in$ Gap-NC^1 be given by the family $\mathcal{C}$ of counting arithmetic circuits. In Theorem 5.15 we showed how such a family can be evaluated by a family $\mathcal{P}$ of straight-line programs using only three registers. Since in the circuits the only constant gates are of a type from $\{-1, 0, 1\}$, these are the only constants from $\mathcal{R}$ occurring in instructions in $\mathcal{P}$. We now simulate the computation of such a straight-line program by a corresponding matrix program. This works as follows:

We represent the contents of the three registers by a 1×3 vector over $\mathbb{Z}$, the so-called *register vector*: $\begin{pmatrix} r_1 & r_2 & r_3 \end{pmatrix}$ denotes the situation where register R_i

holds contents r_i for $i = 1, 2, 3$. Every instruction s of a straight-line program can then be simulated by multiplying the register vector by a suitable matrix M_s. More specifically, we define $M_s = (m_{ij})_{1 \leq i, j \leq 3}$ as:

- All entries in the diagonal are set to 1, i.e., $m_{11} = m_{22} = m_{33} = 1$.
- If s is the instruction $R_j \leftarrow R_j + c \cdot R_i$, then $m_{j,i} = c$.
- If s is the instruction $R_j \leftarrow R_j - c \cdot R_i$, then $m_{j,i} = -c$.
- If s is the instruction $R_j \leftarrow R_j + x_k \cdot R_i$, then $m_{j,i} = x_k$.
- If s is the instruction $R_j \leftarrow R_j - x_k \cdot R_i$, then $m_{j,i} = -x_k$.
- All other entries of M_s are set to 0.

The reader can check by a simple case inspection that the matrices perform as claimed.

Now let P_n be the straight-line program for input length n, $P_n = (s_i)_{1 \leq i \leq l}$, and let $x = x_1 \cdots x_n$ be an input. Then $\begin{pmatrix} 1 & 0 & 0 \end{pmatrix} \cdot \prod_{i=1}^{l} M_i$ yields the register vector after the execution of P_n on input x. Hence

$$f_P(x) = \begin{pmatrix} 1 & 0 & 0 \end{pmatrix} \cdot \left(\prod_{i=1}^{l} M_i \right) \cdot \begin{pmatrix} 1 \\ 0 \\ 0 \end{pmatrix}.$$

This shows how we can simulate the straight-line program family (and hence the given arithmetic circuit family) by a family of matrix programs, where the matrices have entries from $\{-1, 0, 1\}$.

Gap-BWBP $\subseteq$ Gap-NC^1: Follows directly from #BWBP $\subseteq$ #NC^1.   □

From this theorem, we immediately obtain the following corollary (note that it could also have been proved along the lines of the proof of Theorem 5.28).

Corollary 5.36. Gap-NC^1 = #NC^1 − #NC^1.

Proof. The inclusion from right to left is clear. For the other inclusion, we argue by Theorem 5.35, Def. 5.30, and Theorem 5.34, as follows: Gap-NC^1 $\subseteq$ Gap-BWBP = #BWBP − #BWBP $\subseteq$ #NC^1 − #NC^1.   □

All the above simulations can be made uniform. Recall that we introduced U_B-uniformity for M-programs in Sect. 4.5.3 (Def. 4.51).

Theorem 5.37. U_B-#BWBP = U_B-BPM($\mathbb{N}$) $\subseteq$ U_E^*-#NC^1.

Proof. We show that the constructions made in the proof of Theorem 5.34 preserve uniformity as claimed.

For the inclusion #BWBP $\subseteq$ BPM($\mathbb{N}$), observe that the matrix program constructed is essentially equal to the given Σ-program; we have just replaced letters in the instructions by constant-size matrices. Hence the matrix program is uniform if the Σ-program is.

For the inclusion BPM($\mathbb{N}$) $\subseteq$ #BWBP, the argument is essentially identical to the one above. This time, however, we have to replace every instruction in the matrix program by a constant number of instructions in the program over a free monoid, see the proof of Theorem 5.34. This does not affect uniformity.

Finally, uniformity of the NC^1 circuit family can be ensured, if we choose a gate encoding where the name of a gate g contains the path leading from the root to g in the binary tree constructed in the proof of Theorem 5.6. $\square$

Theorem 5.38. U_B-Gap-BWBP $= U_B$-BPM$(\mathbb{Z}) = U_E^*$-Gap-NC^1.

Proof. U_B-BPM$(\mathbb{Z}) \subseteq U_B$-Gap-BWBP: We first transform the BPM$(\mathbb{Z})$ program into two BPM$(\mathbb{N})$ programs by replacing single entries by 2×2 matrices. This obviously does not affect uniformity; neither does adjusting the left and right vectors. Finally these programs are simulated by uniform #BWBP programs, as argued in Theorem 5.37.

U_E^*-Gap-NC$^1 \subseteq U_B$-BPM$(\mathbb{Z})$: The main element of this proof is to show that Theorem 5.15 can be made uniform, i. e., if the given arithmetic circuit family is U_E^*-uniform then the resulting straight-line program family is uniform and thus the matrix program family constructed from this in Theorem 5.35 is U_B-uniform. We leave the proof as an exercise (Exercise 5.9).

U_B-Gap-BWBP $\subseteq U_E^*$-Gap-NC^1: Uniformity of this simulation immediately carries over from Theorem 5.37. $\square$

We remark that the above two theorems hold for U_B-uniformity as well as U_D-uniformity, see Exercise 5.9.

Let us finally turn to an upper bound for arithmetic NC^1 in terms of Boolean circuits.

Theorem 5.39. #NC$^1 \subseteq$ FDEPTH$(\log n \log^* n)$.

Proof. In Theorem 5.35 we showed that the evaluation of an arithmetic $\mathbb{Z}$-circuit can be simulated by evaluating a matrix program. In fact this proof shows how the evaluation of a $\mathbb{Z}$-circuit reduces under (constant-depth reductions) to the problem of determining the upper left entry of a product of polynomially many 3×3 matrices (see Exercise 5.10). Since #NC$^1 \subseteq$ Gap-NC^1 we immediately see, that the following modified iterated matrix product problem is hard for #NC^1 (under constant-depth reductions):

Problem:	MIMP
Input:	a sequence $M_1, \ldots, M_n$ of 3×3 matrices over $\mathbb{Z}$ and a number $m \in \mathbb{N}$, where m as well as all entries in the matrices have n bits in binary
Output:	$z \bmod m$, where z is the entry at row 1, column 1 of $\prod_{i=1}^n M_i \pmod{m}$

We design a circuit for MIMP as follows: For the first n primes p (observe that these primes have $O(\log n)$ bits) compute all matrices $M_i \pmod{p}$ and the product $\prod_{i=1}^n M_i \pmod{p}$. Then making use of the Chinese Remainder

Theorem (Appendix A7) recover $\prod_{i=1}^{n} M_i$ (mod m). In the proof of Theorem 1.40 we showed how the computation necessary for this latter step can be carried out in depth $O(\log n)$ by bounded fan-in circuits. Thus it remains to show how we can compute $\prod_{i=1}^{n} M_i$ (mod p).

Our approach is first to divide the sequence $M_1, \ldots, M_n$ into subsequences consisting of $\lfloor \log n \rfloor$ matrices, then compute the product (mod p) within all these subsequences (this is again an instance of MIMP) and combine the results in a $\lfloor \log n \rfloor$-ary tree of depth $\left\lceil \frac{\log n}{\log \lfloor \log n \rfloor} \right\rceil$.

Let $d(n)$ be the depth necessary to solve MIMP for n matrices. Then the above procedure leads to the following recurrence (where for brevity we leave out floors $\lfloor \cdots \rfloor$ and ceilings $\lceil \cdots \rceil$; the reader may check that this does not affect the result of our computation):

$$d(n) = c \log n + \frac{\log n}{\log \log n} \cdot d(\log n) \qquad (5.3)$$

for some constant $c \in \mathbb{N}$. Hence,

$$d(\log n) = c \log \log n + \frac{\log \log n}{\log \log \log n} \cdot d(\log \log n). \qquad (5.4)$$

Substituting $d(\log n)$ in (5.3) by the right-hand side of (5.4), we obtain:

$$d(n) = c \log n + \frac{\log n}{\log \log n} \cdot \left(c \log \log n + \frac{\log \log n}{\log \log \log n} \cdot d(\log \log n) \right)$$

$$= c \log n + c \log n + \frac{\log n}{\log \log \log n} \cdot d(\log \log n).$$

Continuing this process we see that $d(n) = O(\log n \log^* n)$. $\qquad \square$

The construction in this proof requires that we have a certain number of small prime numbers available. Non-uniformly and for ptime-uniformity this is no problem, but it is not known how this can be done by logspace-uniform circuits. This is the same problem as we already faced when designing circuits for iterated multiplication and division, see Theorem 1.40 and Exercises 1.19 and 2.4.

In terms of classes of the NC hierarchy, the best upper bound for $\#NC^1$ is the same as the one given for $\#SAC^1$ in Sect. 5.3.1, i.e., $\#NC^1 \subseteq FTC^1$. In particular, whether $\#NC^1 \subseteq FAC^1$ is open.

Both upper bounds given in this section for $\#NC^1$ of course also hold for Gap-NC^1 under a representation of the integers, e.g., using two's complement.

5.3.3 Counting in AC^0

As was the case for the class NC^1 in the preceding subsection, we will also see here that of the two classes $\#AC^0$ and Gap-AC^0, the gap class is actually the more natural one. It seems that closure under subtraction is important to obtain smooth classes. Remarkably we will obtain a characterization of TC^0 in terms of Gap-AC^0.

Recall that with the lower bound techniques developed in Chap. 3 we were only able to separate AC^0 and $AC^0[p]$ (for prime number p) from the classes above. Concerning TC^0, we currently do not know if, e. g., $TC^0 \stackrel{?}{=} NP$. Thus, one motivation for the result that we give below connecting TC^0 and AC^0 is the hope that one might obtain new lower bounds this way.

In the previous two subsections we proved Gap-$SAC^1 = \#SAC^1 - \#SAC^1$ (Theorem 5.28) and Gap-$NC^1 = \#NC^1 - \#NC^1$ (Corollary 5.36). An analogous result holds also for AC^0, but the technique which helped in the above cases no longer works here. We need a new, much more involved proof instead.

Theorem 5.40. Gap-$AC^0 = \#AC^0 - \#AC^0$.

Proof. We start as in the proof of Theorem 5.28. The direction from right to left is clear.

For the other direction, let $f \in$ Gap-AC^0 via the $\mathbb{Z}$-circuit family $\mathcal{C} = (C_n)_{n \in \mathbb{N}}$. Fix an input length n and consider C_n. We construct a circuit C'_n over the semi-ring $\mathbb{N}$ of the naturals as follows: For every gate g from C_n there will be two gates g_P and g_N in C'_n such that for all inputs x of length n, we have $\text{val}_g(x) = \text{val}_{g_P}(x) - \text{val}_{g_N}(x)$.

The construction is by induction on the depth of a gate g: If g is an input gate or a constant $(0, 1, -1)$, the gates g_N and g_P are clear. If g is a $+$ gate with predecessors $g^1, \ldots, g^s$, and $g_P^1, \ldots, g_P^s, g_N^1, \ldots, g_N^s$ are gates in C'_n such that for all x and all i, $1 \leq i \leq s$, $\text{val}_{g^i}(x) = \text{val}_{g_P^i}(x) - \text{val}_{g_N^i}(x)$, then we define g_P to be the sum of all gates $g_P^1, \ldots, g_P^s$ and g_N to be the sum of all gates $g_N^1, \ldots, g_N^s$.

Finally let g be a $\times$ gate with predecessors $g^1, \ldots, g^s$, and let $g_P^1, \ldots, g_P^s$, $g_N^1, \ldots, g_N^s$ be gates in C'_n such that for all x and all i, $1 \leq i \leq s$, $\text{val}_{g^i}(x) = \text{val}_{g_P^i}(x) - \text{val}_{g_N^i}(x)$. Hence,

$$\text{val}_g(x) = \prod_{i=1}^{s} \left(\text{val}_{g_P^i}(x) - \text{val}_{g_N^i}(x) \right). \tag{5.5}$$

We cannot simply multiply out this product as we did in Theorem 5.28, since this will not produce a polynomial-size circuit. In fact, multiplying out yields

$$\prod_{i=1}^{s} \left(\text{val}_{g_P^i}(x) - \text{val}_{g_N^i}(x) \right) =$$

$$\sum_{j=0}^{s} \sum_{\substack{l_1, \ldots, l_{s-j}, \\ r_1, \ldots, r_j}} (-1)^j \cdot \text{val}_{g_P^{l_1}}(x) \cdots \text{val}_{g_P^{l_{s-j}}}(x) \cdot \text{val}_{g_N^{r_1}}(x) \cdots \text{val}_{g_N^{r_j}}(x), \tag{5.6}$$

where the numbers $l_1, \ldots, l_{s-j}, r_1, \ldots, r_j$ are chosen such that $\{l_1, \ldots, l_{s-j}, r_1, \ldots, r_j\} = \{1, \ldots, s\}$. Obviously this is an exponential sum which we cannot realize in this naive way.

Instead we proceed as follows: Our aim is to express the above product (5.5) as a sum of the form

$$\sum_{k=1}^{s+1} c_k(s) \cdot \prod_{i=1}^{s} \left(\mathrm{val}_{g_P^i}(x) + k \cdot \mathrm{val}_{g_N^i}(x) \right). \tag{5.7}$$

We have to show that integers $c_k(s)$ for $1 \le k \le s+1$ exist, such that the sum (5.7) is identical to (5.5).

Let us multiply out (5.7):

$$\sum_{k=1}^{s+1} c_k(s) \cdot \prod_{i=1}^{s} \left(\mathrm{val}_{g_P^i}(x) + k \cdot \mathrm{val}_{g_N^i}(x) \right)$$

$$= \sum_{k=1}^{s+1} c_k(s) \cdot \sum_{j=0}^{s} \sum_{\substack{l_1, \ldots, l_{s-j}, \\ r_1, \ldots, r_j}} k^j \mathrm{val}_{g_P^{l_1}}(x) \cdots \mathrm{val}_{g_P^{l_{s-j}}}(x) \mathrm{val}_{g_N^{r_1}}(x) \cdots \mathrm{val}_{g_N^{r_j}}(x)$$

$$= \sum_{k=1}^{s+1} \sum_{j=0}^{s} \sum_{\substack{l_1, \ldots, l_{s-j}, \\ r_1, \ldots, r_j}} k^j c_k(s) \mathrm{val}_{g_P^{l_1}}(x) \cdots \mathrm{val}_{g_P^{l_{s-j}}}(x) \mathrm{val}_{g_N^{r_1}}(x) \cdots \mathrm{val}_{g_N^{r_j}}(x)$$

$$= \sum_{j=0}^{s} \sum_{\substack{l_1, \ldots, l_{s-j}, \\ r_1, \ldots, r_j}} \left(\sum_{k=1}^{s+1} k^j c_k(s) \right) \mathrm{val}_{g_P^{l_1}}(x) \cdots \mathrm{val}_{g_P^{l_{s-j}}}(x) \mathrm{val}_{g_N^{r_1}}(x) \cdots \mathrm{val}_{g_N^{r_j}}(x)$$

$$\tag{5.8}$$

Comparing the term (5.8) with (5.6) we see that we have to set

$$\sum_{k=1}^{s+1} k^j c_k(s) = (-1)^j$$

for $0 \le j \le s$, i.e., we can reach our aim if the following system of linear equations has integer solutions $c_k(s)$ ($1 \le k \le s+1$):

$$
\begin{array}{cccccc}
c_1(s) & + & c_2(s) & + & \cdots & + & c_{s+1}(s) & = & 1 \\
c_1(s) & + & 2c_2(s) & + & \cdots & + & (s+1)c_{s+1}(s) & = & -1 \\
c_1(s) & + & 4c_2(s) & + & \cdots & + & (s+1)^2 c_{s+1}(s) & = & 1 \\
\vdots & & \vdots & & \ddots & & \vdots & & \vdots \\
c_1(s) & + & 2^s c_2(s) & + & \cdots & + & (s+1)^s c_{s+1}(s) & = & (-1)^s.
\end{array}
$$

Let A be the coefficient matrix of this system, and let $A^{(k)}$ be the matrix that we get from A by replacing the kth column vector by $(1, -1, 1, \ldots, (-1)^s)^{\mathrm{T}}$.

All these matrices are Vandermonde matrices (see Appendix A9), in particular $A = V_\mathbb{Z}(1, 2, \ldots, s+1)$ and $A^{(k)} = V_\mathbb{Z}(1, 2, \ldots, k-1, -1, k+1, \ldots, s+1)$. Clearly $\det A \neq 0$, hence by Cramer's rule the unique solution of the above system is given by

$$c_k(s) = \frac{\det A^{(k)}}{\det A}.$$

Using the identity for Vandermonde determinants (p. 240), we get

$$c_k(s) = \frac{\displaystyle\prod_{\substack{i>j \\ i\neq k, j\neq k}} (i-j) \prod_{i<k}(-1-i) \prod_{j>k}(j-(-1))}{\displaystyle\prod_{i>j}(i-j)}$$

$$= \frac{\displaystyle\prod_{i=1}^{k-1}(-1-i) \prod_{j=k+1}^{s+1}(j-(-1))}{\displaystyle\prod_{i=1}^{k-1}(k-i) \prod_{j=k+1}^{s+1}(j-k)}$$

$$= \frac{\big((-1)^{k-1} 2 \cdot 3 \cdots k\big)\big((k+2)(k+3) \cdots (s+2)\big)}{\big((k-1)(k-2)\cdots 2 \cdot 1\big)\big(1 \cdot 2 \cdots (s+1-k)\big)}$$

$$= (-1)^{k-1} \frac{k(k+2)(k+3) \cdots (s+2)}{(s+1-k)!}$$

$$= (-1)^{k-1} \cdot k \cdot \binom{s+2}{k+1}$$

Hence there exist integers $c_1(s), \ldots, c_{s+1}(s)$ independent of the circuit input x, that identify (5.5) and (5.7). Therefore it is sufficient to define

$$g_P = \sum_{\substack{1 \leq k \leq s+1 \\ k \text{ odd}}} c_k(s) \times \prod_{i=1}^{s}(g_P^i + k \times g_N^i); \tag{5.9}$$

$$g_N = \sum_{\substack{1 \leq k \leq s+1 \\ k \text{ even}}} c_k(s) \times \prod_{i=1}^{s}(g_P^i + k \times g_N^i). \tag{5.10}$$

This finishes the simulation of a $\times$ gate and the induction.

As in the proof of Theorem 5.28, it is now obvious to construct from C_n' two N-circuits A_n and B_n such that $f_C(x) = f_A(x) - f_B(x)$. □

To make the above proof uniform, we have to show that the simulation of a $\times$ gate is uniform, i. e., that (5.9) and (5.10) can be computed by uniform circuits. We first prove a lemma.

Lemma 5.41. *The function* $f(x,y) =_{\text{def}} \binom{|x|}{|y|}$ *is in* $\mathrm{U_D}$-$\#\mathrm{AC}^0$, *in other words, the binomial coefficient* $\binom{m}{k}$ *can be computed by* $\mathrm{U_D}$-$\#\mathrm{AC}^0$ *circuits, given m and k in unary.*

Proof. To construct the required circuits we first consider some arithmetic operations restricted to numbers that reach in absolute value at most the input length n. We first claim that a logtime-uniform constant-depth circuit can compute the following terms, given that x and y are less than or equal to the input length n (see Exercise 5.11):

- $x \cdot y$
- $[x$ divides $y]$
- $[x$ is prime$]$
- $\lfloor \frac{x}{y} \rfloor$

By Exercise 5.14, $\mathrm{U_D}$-$\mathrm{AC}^0 \subseteq \mathrm{U_D}$-$\#\mathrm{AC}^0$, hence we now conclude that all the above functions are in $\mathrm{U_D}$-$\#\mathrm{AC}^0$.

Next observe that given two numbers x,y, among all numbers $1,2\ldots,y$ exactly $\lfloor \frac{y}{x} \rfloor$ are divisible by x.

Now let $p,m,k \leq n$ and p be prime. Define $d(p,m,k)$ to be the highest power of p that divides $\binom{m}{k} = \frac{\prod_{i=m-k+1}^m i}{\prod_{i=1}^k i}$, i. e.,

$$d(p,m,k) =_{\text{def}} \max\left\{ p^j \ \middle| \ j \geq 0 \text{ and } j \text{ divides } \binom{m}{k} \right\}.$$

By the above, in the product $\prod_{i=m-k+1}^m i$ exactly

$$a_1(p,m,k,j) =_{\text{def}} \left\lfloor \frac{m}{p^j} \right\rfloor - \left\lfloor \frac{m-k}{p^j} \right\rfloor$$

terms are divisible by p^j (for $j \in \mathbb{N}$), and in the product $\prod_{i=1}^k i$ exactly

$$a_2(p,m,k,j) =_{\text{def}} \left\lfloor \frac{k}{p^j} \right\rfloor$$

terms are divisible by p^j. Thus the total number of p's in the prime factorization of $\prod_{i=m-k+1}^m i$ is $\sum_{j=1}^n a_1(p,m,k,j)$, and the total number of p's in the prime factorization of $\prod_{i=1}^k i$ is $\sum_{j=1}^n a_2(p,m,k,j)$. Since in both products we have the same number of terms, the two numbers $a_1(p,m,k,j)$ and $a_2(p,m,k,j)$ can for a fixed j differ by at most 1. Hence $d(p,m,k) = p^l$, where $l = \left|\left\{ j \mid a_1(p,m,k,j) > a_2(p,m,k,j) \right\}\right|$. From this we conclude that

$$d(p,m,k) = \prod_{j=1}^n \begin{cases} p, & \text{if } a_1(p,m,k,j) > a_2(p,m,k,j) \\ 1, & \text{otherwise} \end{cases}$$

$$= \prod_{j=1}^n \left(1 + (p-1) \cdot [a_1(p,m,k,j) > a_2(p,m,k,j)]\right).$$

Finally we get:

$$\binom{m}{k} = \prod_{i=2}^{n} ([p \text{ is composite}] + [p \text{ is prime}] \cdot d(p, m, k)).$$

Using the arithmetic functions considered above, all computations necessary for this can be carried out by logtime-uniform circuits of constant depth. $\square$

Corollary 5.42. *For logtime-uniformity as well as for less strict uniformity conditions,* $\text{Gap-AC}^0 = \#\text{AC}^0 - \#\text{AC}^0$.

Proof. Using the circuits constructed in Lemma 5.41 we can compute the numbers $c_k(s)$ in the proof of Theorem 5.40. It is then clear that the terms (5.9) and (5.10) can be computed by logtime-uniform circuits. All other constructions in the proof of Theorem 5.40 clearly preserve uniformity. $\square$

Let us now turn to the promised characterization of TC^0 in terms of Gap-AC^0 functions. Our way to present the result is to introduce two complexity classes, defined using Gap-AC^0 functions, and then prove that both classes are equal to TC^0.

Definition 5.43.   1. The class $\text{C}_=\text{AC}^0$ consists of all sets A for which there is a function $f \in \text{Gap-AC}^0$ such that for all inputs x, we have: $x \in A$ if and only if $f(x) = 0$.
   2. The class CAC^0 consists of all sets A for which there is a function $f \in \text{Gap-AC}^0$ such that for all inputs x, we have: $x \in A$ if and only if $f(x) > 0$.

Lemma 5.44. $\text{C}_=\text{AC}^0 \subseteq \text{CAC}^0 \subseteq \text{TC}^0$.

Proof. Let $A \in \text{C}_=\text{AC}^0$ via f, i.e., for all x, $x \in A \iff f(x) = 0$. Define $f'(x) = 1 - (f(x) \times f(x))$. Clearly $f' \in \text{Gap-AC}^0$ and moreover, $f(x) = 0 \iff f'(x) > 0$. Hence f' witnesses that $A \in \text{CAC}^0$, proving the first inclusion of the lemma.

Since iterated addition and multiplication are in FTC^0 (Theorems 1.37 and 1.40), we immediately get $\#\text{AC}^0 \subseteq \text{FTC}^0$ and, under an appropriate encoding of the integers, $\text{Gap-AC}^0 \subseteq \text{FTC}^0$. Thus a TC^0 circuit for $A \in \text{CAC}^0$ computes the underlying arithmetic function and accepts if and only if the result is positive. $\square$

Lemma 5.45. $\text{TC}^0 \subseteq \text{C}_=\text{AC}^0$.

Proof. The *exact threshold functions* are given by

$$ET_m^n(x_1, \ldots, x_n) = 1 \iff \sum_{i=1}^{n} x_i = m.$$

Let $m \in \mathbb{N}$ and let C be a circuit with n inputs over basis $\{ET_{\frac{n}{2}}^m\}$. We first show how C can be simulated by an arithmetic circuit.

Define $\mu =_{\text{def}} \prod_{\substack{0 \le j \le m \\ j \ne \frac{m}{2}}} \left(\frac{m}{2} - j \right)$, and

$$\Delta(x_1, \ldots, x_m) =_{\text{def}} \frac{\prod_{\substack{0 \le j \le m \\ j \ne \frac{m}{2}}} \left(\sum_{i=1}^{m} x_i - j \right)}{\mu}.$$

For all $x_1, \ldots, x_m \in \{0,1\}$, clearly $\Delta(x_1, \ldots, x_m) = 1$ if $\sum_{i=1}^{m} x_i = \frac{m}{2}$, and $\Delta(x_1, \ldots, x_m) = 0$ otherwise. Hence $\Delta(x_1, \ldots, x_m) = ET_{\frac{m}{2}}^m(x_1, \ldots, x_m)$.

Let g be a gate in C. We prove by induction on the depth of g that there is a Gap-AC^0 function f and a number $t \in \mathbb{N}$ such that for all circuit inputs $x = x_1 \cdots x_n$ the following holds: If g has value 1 then $f(x) = \mu^t$, and if g has value 0 then $f(x) = 0$.

For the base case where we consider a constant or a circuit input, the above claim is clear.

Now let g be an $ET_{\frac{m}{2}}^m$ gate whose predecessors are $g_1, \ldots, g_m$, and let (by induction hypothesis) $h_1, \ldots, h_m \in$ Gap-AC^0 and $t_1, \ldots, t_m \in \mathbb{N}$ be such that for $1 \le i \le m$, $h_i(x) = \mu^{t_i}$ if g_i outputs 1, and $h_i(x) = 0$ otherwise.

Let $t = \max\{t_1, \ldots, t_m\}$ and $f_i(x) = h_i(x) \cdot \mu^{t - t_i}$ $(1 \le i \le m)$. Consider the following term:

$$\Delta\left(\frac{f_1}{\mu^t}, \ldots, \frac{f_m}{\mu^t} \right) = \frac{\prod_{\substack{0 \le j \le m \\ j \ne \frac{m}{2}}} \left(\frac{\sum_{i=1}^{m} f_i}{\mu^t} - j \right)}{\mu}$$

$$= \frac{\prod_{\substack{0 \le j \le m \\ j \ne \frac{m}{2}}} \left(\sum_{i=1}^{m} f_i - j \cdot \mu^t \right)}{\mu^{mt+1}}$$

Now if g outputs 1 then exactly $\frac{m}{2}$ of its predecessors output 1, i.e., exactly as many terms $\frac{f_i}{\mu^t}$ have value 1. Hence $\Delta\left(\frac{f_1}{\mu^t}, \ldots, \frac{f_m}{\mu^t} \right) = 1$. If g outputs 0 then by analogous reasoning, $\Delta\left(\frac{f_1}{\mu^t}, \ldots, \frac{f_m}{\mu^t} \right) = 0$.

If we now define

$$f(x) =_{\text{def}} \prod_{\substack{0 \le j \le m \\ j \ne \frac{m}{2}}} \left(\sum_{i=1}^{m} f_i(x) - j \cdot \mu^t \right),$$

then $f(x) = \mu^{mt+1}$ if g (on circuit input x) outputs 1, and $f(x) = 0$ otherwise. This finishes the induction.

Now it is easy to see that a language A is in TC^0 if and only if there is a family $C = (C_n)_{n \in \mathbb{N}}$ of polynomial-size constant-depth circuits that accepts A, where for every n there is a number m_n such that C_n has only $ET_{\frac{m_n}{2}}^{m_n}$

gates (see Exercise 5.13). Let $\mathcal{C} = (C_n)_{n \in \mathbb{N}}$ be such a family. Define $\mu_n = \prod_{\substack{0 \le j \le m_n \\ j \ne \frac{m_n}{2}}} \left(\frac{m_n}{2} - j \right)$, and let f^n and t_n be the function and natural number constructed in the above induction for the output gate of C_n, that is, for all circuit inputs x of length n, if C_n accepts x then $f^n(x) = \mu_n^{t_n}$, and if C_n does not accept x then $f^n(x) = 0$. Let $f = (f^n)_{n \in \mathbb{N}}$. By the above construction, it follows immediately that $f \in \text{Gap-AC}^0$, and if we now set $f'(x) = f(x) - \mu_n^{t_{|x|}}$, then f' witnesses that the language accepted by $\mathcal{C}$ is in $\text{C}_=\text{AC}^0$.   □

Theorem 5.46. $\text{TC}^0 = \text{C}_=\text{AC}^0 = \text{CAC}^0$.

Proof. Immediate from the preceding two lemmas.   □

We remark that the proof just given shows that $\text{TC}^0 = \text{C}_=\text{AC}^0 = \text{CAC}^0$ holds also ptime-uniformly. In fact, Lemma 5.45 holds even for logtime-uniformity, but to prove $\text{CAC}^0 \subseteq \text{TC}^0$ we used the TC^0 circuits for iterated addition and multiplication from Chap. 1, which are only ptime-uniform. We come back to this point in the Bibliographic Remarks for the present chapter, see Question 3 on p. 209.

How can Gap-AC^0 be related to classes of Boolean circuits? Since iterated addition and multiplication are in TC^0, we immediately conclude:

Theorem 5.47. $\text{Gap-AC}^0 \subseteq \text{FTC}^0$.

Can this upper bound be improved? We close this section by observing that $\text{Gap-AC}^0 \not\subseteq \text{FAC}^0$.

Theorem 5.48. $\oplus \in \text{Gap-AC}^0$.

Proof. Let $x_1, \ldots, x_n \in \{0, 1\}$. First observe that $(-1)^{\sum_{i=1}^n x_i} = \prod_{i=1}^n (1 - 2x_i)$. The parity of $x_1, \ldots, x_n$ is now given by

$$\sum_{i=1}^n \left(\prod_{j=1}^{i-1} (1 - 2x_j) \right) x_i,$$

a function which is in Gap-AC^0.   □

Corollary 5.49. $\text{FAC}^0 \subsetneq \text{Gap-AC}^0$.

5.4 Relations among Arithmetic and Boolean Circuit Classes

Figure 5.6 summarizes the relations established in this chapter among classes defined by arithmetic circuits and classes defined by Boolean circuits. Moreover, relations to counting classes (on the right) defined via Turing machines are given.

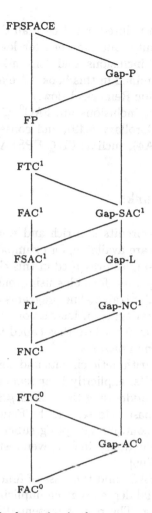

Fig. 5.6. Arithmetic and Boolean circuit classes

The class #L is the class of all functions $f: \{0,1\}^* \to N$, for which there exists a nondeterministic logarithmic space-bounded Turing machine such that for all x, $f(x)$ is equal to the number of accepting paths of M on input x [AJ93]; and #P is defined analogously for nondeterministic polynomially time-bounded machines [Val79]. The corresponding gap classes are Gap-L $=_{\text{def}}$ #L $-$ #L [AO96] and Gap-P $=_{\text{def}}$ #P $-$ #P [FFK94]. The inclusions Gap-NC^1 $\subseteq$ Gap-L $\subseteq$ Gap-SAC^1 are given in Exercise 5.17.

A few remarks on uniformity: The inclusion FSAC^1 $\subseteq$ Gap-SAC^1 (given in Exercise 5.16) is only known non-uniformly. The inclusions of the counting classes Gap-SAC^1 and Gap-NC^1 in FTC^1, and similarly of Gap-AC^0 in FTC^0 rely on the fact that iterated multiplication is in FTC^0; however this is only known for ptime-uniformity and not for stricter conditions examined in this book, as has already been pointed out. Hence all these inclusions are known to hold for ptime-uniformity, but are open for logspace- or stricter uniformity conditions. All other inclusions hold non-uniformly as well as under all uniformity conditions examined in this book. (See also the open questions at the end of the Bibliographic Remarks below.)

The only known strict inclusions are $FAC^0 \subsetneq FTC^0$ (see Sect. 4.5.2), $FAC^0 \subsetneq Gap\text{-}AC^0$ (see Corollary 5.49), and consequences of the hierarchy theorems (see Appendix A4), such as $FL \subsetneq FPSPACE$.

Bibliographic Remarks

The theory of arithmetic circuits is a rich and well-established field, and a number of survey articles are available, for example [Bor82, vzG93], see also [Sav98, Chap. 6]. The examples presented in this chapter are from [vzG93].

The evaluation of algebraic formulas using only three registers (Theorem 5.15) is from [BOC92]. Straight-line programs are usually defined allowing somewhat more general types of instructions than we did in Def. 5.14, see, e. g., [AHU74]. Ben-Or and Cleve thus called these particular programs *linear bijection straight-line programs.*

The relation between arithmetic circuits and counting accepting subtrees has been known for a while; explicitly it appears in [Vin91, Ven92, Jia92]. A recent survey, not only addressing the classes $\#SAC^1$, $\#NC^1$, and $\#AC^0$, but also Turing machine based classes, can be found in [All98].

The problem ASC of counting accepting subcircuits is complete for $\#P$ (under so called metric reductions), in fact, even restricted to circuits of depth 4, as was shown in [MVW99].

Correspondence of $\#SAC^1$ and the class of functions computed by arithmetic circuits of polynomial degree is given implicitly in [VSBR83], but only in the non-uniform setting. The results presented in Sect. 4.3 of this book, due to [Ruz80, Ven91], show that this correspondence also holds uniformly in the case of the Boolean semi-ring. This was extended in [Vin91] to the case of the ring of the integers. Vinay's proof relies on a correspondence between SAC^1 circuits and auxiliary pushdown automata, see Exercises 4.8–4.10 and the Bibliographic Remarks given there. Finally a direct proof for general commutative semi-rings appeared in [AJMV98]. Our presentation follows this paper.

The characterization of $Gap\text{-}NC^1$ in terms of counting paths in finite automata presented in Sect. 5.3.2 appeared in [CMTV98]. Matrix programs were also defined in this paper, and the equivalence of polynomial-size matrix programs with programs over free monoids was given (see also [Cau96], where the above notions were additionally related to a type of nondeterministic branching programs). The construction of the graphs in Lemma 5.33 relies on a similar construction from [Tod92] (see also [Tan94, 41–42]).

The upper bound for $\#NC^1$ on Boolean circuits (Theorem 5.39) was given in [Jun85]; our presentation follows a proof which appeared in [All98]. Jung's paper proves the more general result that every bounded fan-in arithmetic circuit of depth $d(n)$ can be simulated by a Boolean circuit of depth $O(d(n) \cdot \log^* n)$.

In Corollary 2.52 we characterized $U_E^*\text{-}NC^1$ via alternating Turing machines. This connection can also be used to obtain a machine-based way of describing the class $U_E^*\text{-}\#NC^1$, see Exercise 5.18.

The equality $\text{Gap-}AC^0 = \#AC^0 - \#AC^0$ (Theorem 5.40) is from [AML98]. The characterization of TC^0 via $\text{Gap-}AC^0$ (Theorem 5.46) appeared in [AAD97].

Although, as mentioned at the beginning of Sect. 5.3.3, one of the motivations for the study of the class $\text{Gap-}AC^0$ is the hope of obtaining lower bounds for threshold circuits, this has not led to success so far. The question whether this avenue might prove fruitful at all is discussed in [All98].

It should be remarked that some of the notions of this chapter are defined in direct analogy to corresponding notions from the polynomial-time context. The idea to consider, in addition to the counting classes $\#\mathcal{K}$, classes $\text{Gap-}\mathcal{K}$ is from [FFK94], where the class Gap-P was defined. The operators C and $C_=$ were introduced in [Wag86]. Applied to Gap-P these yield the classes PP and $C_=P$.

The study of arithmetic circuits is a very active field of research, and a lot of exciting developments should be expected in the future. As the reader has noticed during the study of this chapter, a number of very important issues are still unsettled. Therefore let us finish by pointing out some central open questions.

1. Can the FTC^1 upper bound for $\#SAC^1$ and $\#NC^1$ be improved?
   In particular, are these classes maybe subclasses of FAC^1? In Theorem 5.39 we showed that $\#NC^1 \subseteq FDEPTH(\log n \log^* n)$. It is intriguing to conjecture that the $\log^* n$ factor could be removed here, thus yielding $\#NC^1 \subseteq FNC^1$.
2. Is $\#BWBP = \#NC^1$?
   In Sect. 5.3.2, where we compared counting paths in finite automata with counting accepting subtrees in NC^1 circuits, we proved that the corresponding gap classes coincide (Gap-BWBP = Gap-NC^1, Theorem 5.35), but concerning the above question, only $\#BWBP \subseteq \#NC^1$ could be shown. The proof of Theorem 5.35 relies heavily on the techniques used to prove Barrington's Theorem (Theorem 4.46); these seem to be of no help here.
   However we want to remark that if the $\log^* n$ factor in Theorem 5.39 can be removed, i. e., if $\#NC^1 \subseteq FNC^1$, then $\#BWBP = \#NC^1$. This is simply because $FNC^1 \subseteq \#BWBP$ (see Exercise 5.19).
3. How much can the simulations of this chapter be made uniform?

The proof given for Theorem 5.39 works only for P-uniformity. The characterization of TC^0 in terms of Gap-AC^0 in Sect. 5.3.3 (Theorem 5.46), as presented here, works only for ptime-uniformity, though it is known that the result is also valid for logspace-uniformity, as shown in [AAD97]. Can these results be improved? Observe that the inclusions $TC^0 \subseteq C_=AC^0 \subseteq CAC^0$ in Lemmas 5.44–5.45 are easily seen to hold for logtime-uniformity (see Exercise 5.12). Hence the only problematic case in the identity $TC^0 = C_=AC^0 = CAC^0$ is the inclusion $CAC^0 \subseteq TC^0$.

Exercises

5.1. Let $\mathcal{R}$ be a semi-ring. Develop a reasonable definition of reducibility between functions $f, g \colon \mathcal{R}^* \to \mathcal{R}^*$.

5.2. Explain how the **if-then-else** statement in Fig. 5.1 is implemented in an arithmetic circuit.

5.3.* Consider the following problem:

> *Problem:* MATINV
> *Input:* an $n \times n$ matrix A over field $\mathcal{F} = (F, +, -, \times, /, 0, 1)$
> *Output:* the inverse of A

(1) Show that POLYPROD $\leq^{\mathcal{R}}_{cd}$ MATPROD.
(2) Show that POLYINV $\leq^{\mathcal{R}}_{cd}$ MATINV.
(3) Show that POLYINV $\leq^{\mathcal{R}}_{cd}$ ITMATPROD.

Hint: Let $f \in \mathcal{F}[x]$, $f(x) = a_0 + a_1 x + a_2 x^2 + \cdots + a_n x^n$. Associate with f the following matrix:

$$\begin{pmatrix} a_0 & a_1 & a_2 & \cdots & a_n \\ 0 & a_0 & a_1 & \cdots & a_{n-1} \\ 0 & 0 & a_0 & \cdots & a_{n-2} \\ \vdots & \vdots & \vdots & \ddots & \vdots \\ 0 & 0 & 0 & \cdots & a_0 \end{pmatrix}$$

5.4. (1) Show that AST is in FP.
(2) Show that the following problem ASTD is P-complete:
> *Problem:* ASTD
> *Input:* circuit C over $\{\wedge, \vee\}$, an input $x \in \{0, 1\}^*$, a number v
> *Question:* Is $\#C(x) = v$?

(3) Show that ASC is in $\#P$.
(4) Show that the following problem ASC' is in $\#P$:
> *Problem:* ASC'

>   *Input:*   circuit C over $\{\wedge, \vee\}$, an input $x \in \{0,1\}^*$
>   *Output:* $\#_c C(x)$

5.5. Show that the circuit family C' constructed in the proof of Theorem 5.25 simulates the given family C.
Hint: Show by induction that each gate $\langle g, i \rangle$ in C'_n computes the sum of all monomials of degree i in the formal polynomial corresponding to gate g in C^*_n.

5.6. Prove: $\#SAC^1 \subseteq FTC^1$.
Hint: Use the fact that iterated addition and multiplication are in FTC^0, as proved in Chap. 1.

5.7. Show that the matrix M_x constructed in the proof of Theorem 5.34 has the property that for all $i, j \leq s$, $(m_x)_{ij}$ is the number of paths on which q_j can be reached starting from q_i and reading the input x.
Hint: Use the same reasoning as in Exercise 1.6.

5.8. Prove (5.2) on p. 196.

5.9.* Give a detailed account of the uniformity considerations at the end of Sect. 5.3.2; i. e., in particular, fill in the details in the proof of Theorem 5.37, prove that Lemma 5.15 can be made uniform, and prove that Theorems 5.37–5.38 also hold for U_D-uniformity.

5.10. Let $k \in \mathbb{N}$. Consider the following problem:

>   *Problem:*   IT-k-MATPROD
>   *Input:* $k \times k$ square matrices $A_1, A_2 \ldots, A_n$ with entries from
> $\{-1, 0, 1\}$ over $\mathcal{R} = (R, +, \times, 0, 1)$
>   *Output:*   their product

Show that for $k \geq 3$, the above problem is complete for Gap-NC^1 under $\leq_{cd}$.

5.11. (1) Show that each of the following functions can be computed by logtime-uniform unbounded fan-in circuits of constant depth and exponential size:
- $f_1(x, y) =_{\text{def}} x \cdot y$
- $f_2(x, y) =_{\text{def}} [x \text{ divides } y]$
- $f_3(x) =_{\text{def}} [x \text{ is prime}]$
- $f_4(x, y) =_{\text{def}} x \div y$ (integer division of x by y)

(2) Show that the above functions can be computed by logtime-uniform unbounded fan-in circuits of constant depth and polynomial size, if the inputs are given in unary.

Hint: For f_1, reduce multiplication to iterated addition as in Theorem 1.23. For iterated addition, a circuit of exponential size, completely analogous to the circuit in Sect. 1.1, can be given. For f_2, use the fact that x divides y if and only if there is a number $z \leq n$ such that $x \cdot z = y$. The existential quantifier can be realized by an $\vee$ gate of appropriate fan-in n. For f_3, use the fact that x is prime if and only if for all z, $1 < z < n$, z does not divide n. The universal quantifier can be realized by an $\wedge$ gate of fan-in n.

5.12. Prove: $U_D\text{-}TC^0 \subseteq U_D\text{-}C_=AC^0 \subseteq U_D\text{-}CAC^0$.
Hint: Show that the simulation in the proof of Lemma 5.45 is uniform.

5.13. Show that a language A is in TC^0 if and only if there is a family $C = (C_n)_{n \in \mathbb{N}}$ of polynomial-size constant-depth circuits that accepts A where for every n there is a number m such that C_n has only $ET^m_{\frac{m}{2}}$ gates.

5.14.* A circuit C is said to be *unambiguous* if, for every input x that is accepted by C, there is exactly one accepting subtree.

(1) Show that for every $A \in AC^0$ ($A \in NC^1$) there is a circuit family of polynomial size and constant depth over the standard unbounded fan-in basis (circuit family of logarithmic depth over the standard bounded fan-in basis, respectively), which accepts A and contains only unambiguous circuits.
(2) Let $\mathcal{K}$ be either AC^0 or NC^1. Let $A \in \mathcal{K}$. Using the above statement, show that $c_A \in \#\mathcal{K}$.
(3) Extending (2), prove that $FAC^0 \subseteq \#AC^0$ and $FNC^1 \subseteq \#NC^1$.
(4) Show that the above results also hold under logtime-uniformity.

Hint: For (1), use the fact that $\alpha \vee \beta$ can be expressed unambiguously as $\alpha \vee (\beta \wedge \neg\alpha)$. Propagate to the inputs the negations so introduced.

5.15.* Prove:

(1) $\text{Gap-}NC^1 = FNC^1 - \#NC^1 = \#NC^1 - FNC^1$.
(2) $\text{Gap-}AC^0 = FAC^0 - \#AC^0 = \#AC^0 - FAC^0$.

5.16.** (1) Show that SAC^1-circuits can be made unambiguous, that is, show that, for every $A \in SAC^1$, there is an SAC^1 circuit family which accepts A and contains only unambiguous circuits.
(2) Prove: $FSAC^1 \subseteq \#SAC^1$.

5.17.* Show that $\#NC^1 \subseteq \#L \subseteq \#SAC^1$.
Hint: Use the techniques of Chap. 2, which lead to the inclusions $NC^1 \subseteq L \subseteq SAC^1$ (Theorems 2.31–2.32).

5.18. Prove: A function $f : \{0,1\}^* \to \mathbb{N}$ is in $U^*_E\text{-}\#NC^1$ if and only if there is an alternating Turing machine running in logarithmic time such that, for all x, $f(x)$ is equal to the number of accepting computation subtrees of M on input x.
Hint: Modify the proof leading to Corollary 2.52.

5.19.* Prove: $FNC^1 \subseteq \#BWBP$.

5.20.* Show that a language A is in $AC^0[2]$ if and only if its characteristic function is in Gap-AC^0.
Hint: Use Theorem 5.48.

5.21.** In analogy to Def. 5.43, define the class $C_=NC^1$ to consist of all sets A for which there is a function $f \in$ Gap-NC^1 such that, for all inputs x, we have $x \in A$ if and only if $f(x) = 0$; and define CNC^1 to consist of all sets A for which there is a function $f \in$ Gap-NC^1 such that, for all inputs x, we have $x \in A$ if and only if $f(x) > 0$.
    Prove:

(1) $C_=NC^1 \subseteq L$.
(2) $CNC^1 \subseteq L$.

5.22.* (1) Let $f \in$ Gap-AC^0 and $k \in \mathbb{N}$. Prove that the following functions are in Gap-AC^0:

a) $h_1(x) =_{\text{def}} \sum_{i=1}^{n^k} f(x, i)$.

b) $h_2(x) =_{\text{def}} \prod_{i=1}^{n^k} f(x, i)$.

c) $h_3(x) =_{\text{def}} \binom{f(x)}{k}$.

(For binomial coefficients $\binom{n}{k}$ where k is a natural number but n may be an integer, see Appendix A7.)
(2) Show that (1) also holds with Gap-AC^0 replaced by Gap-NC^1.

Notes on the Exercises

5.3. Solutions can be found in [vzG93].

5.5. See [AJMV98].

5.6. This was observed in [AJMV98]. In fact, in that paper it was shown that TC^1 is *exactly* the class of functions that can be computed by N-circuits of polynomial size and logarithmic depth with unbounded fan-in $+$ gates and fan-in two $\times$ and integer division gates.

5.10. See [BOC92] and [CMTV98].

5.13. This appeared first in [PS88].

5.14 and 5.15. A solution for the case of the class AC^0 can be found in [AAD97], and a solution for class NC^1 can be found in [CMTV98], see also [Jia92].

5.16. See [AR97].

5.17. See [CMTV98].

5.19. A solution is given in [All98].

5.20. See [AAD97].

5.21. In [LZ77] it is shown that a logspace machine can check an entry of an iterated matrix product for equality with zero, carrying out the product modulo enough small primes. In [Mac98] it is shown how two numbers given by their residues modulo enough small primes can be compared, implying that a logspace machine can determine whether an entry in an iterated matrix product is positive. As observed in [CMTV98], these results imply $C_=NC^1 \subseteq L$ and $CNC^1 \subseteq L$, respectively. We remark that if $\#NC^1 \subseteq FNC^1$ (see Question 1 on p. 209), then $C_=NC^1 = CNC^1 = NC^1$.

5.22. These results are due to [AAD97, CMTV98]. Both papers also consider other closure properties of Gap-AC^0 and Gap-NC^1.

6. Polynomial Time and Beyond

In the previous chapters we studied mostly subclasses of the class NC. As we have stressed already, NC is meant to stand for problems with feasible highly parallel solutions (meaning very efficient runtime using a reasonable number of processors). In the theory of sequential algorithms, the class $P = DTIME(n^{O(1)})$ is meant to capture the notion of problems with feasible sequential algorithms (meaning reasonable runtime on a sequential machine). Every feasible highly parallel solution can be used to obtain a feasible sequential solution (NC $\subseteq$ P). The question whether every problem with a feasible sequential algorithm also has a feasible parallel algorithm was discussed in detail in Sect. 4.6.

What does "feasible sequential algorithm" mean in terms of circuits? In Chap. 2 we showed that $P = U_E^*\text{-SIZE}(n^{O(1)}) = U_P\text{-SIZE}(n^{O(1)})$. In this chapter we want to look at the class $SIZE(n^{O(1)})$, both non-uniformly and under different uniformity conditions. Informally, we will say that a language has small (uniform) circuits if it is in $SIZE(n^{O(1)})$ ($U_P\text{-SIZE}(n^{O(1)})$, respectively).

We will introduce all the complexity-theoretic concepts that we need in this chapter. However from time to time we will have to refer, without proof, to results that are usually presented in a course on Complexity Theory. As general textbooks we recommend [WW86, BDG95, BDG90, BC94, Pap94]. An quick overview is given in David Johnson's catalog of complexity classes [Joh90]. We will give detailed references in the Bibliographic Remarks for this chapter.

6.1 The class P/Poly

In this chapter, we will often require the notion of an *oracle Turing machine*. Such a machine M has (besides input and regular work tapes) an additional *oracle tape* and three distinguished states $q_?, q_Y, q_N$. During its computation M is allowed to write a word w on the oracle tape and then enter the so-called *query state* $q_?$. Then in one time step M is transferred into q_Y if the word w is in the oracle language, and into q_N if w is not in the oracle language; moreover, the contents of the oracle tape are deleted.

Let $B \subseteq \{0,1\}^*$. Then P^B is the class of all languages A for which there exists a polynomially time-bounded oracle Turing machine M such that M with oracle language B accepts A. We also say that A is *polynomial time Turing-reducible* to B and write $A \leq_T^p B$.

Analogously, NP^B is the class of all languages A for which there exists a nondeterministic polynomially time-bounded oracle Turing machine M such that M with oracle language B accepts A. (Here, a word is accepted by M if there is at least one computation path of M with input x, that accepts using oracle B.) If $\mathcal{K}$ is a class of sets then $\mathrm{P}^{\mathcal{K}} =_{\mathrm{def}} \bigcup_{B \in \mathcal{K}} \mathrm{P}^B$ and $\mathrm{NP}^{\mathcal{K}} =_{\mathrm{def}} \bigcup_{B \in \mathcal{K}} \mathrm{NP}^B$.

It turns out that the class $\mathrm{SIZE}(n^{O(1)})$ has a nice characterization in terms of polynomially time-bounded oracle Turing machines.

Definition 6.1. A set $A \subseteq \{0,1\}^*$ is *sparse* if there is a polynomial p such that, for every n, the number of words in A of length n is bounded above by $p(n)$.

Theorem 6.2. *A set A is in* P/Poly *if and only if there is a sparse set S such that $A \in \mathrm{P}^S$.*

Proof. ($\Rightarrow$): Let $A \in$ P/Poly. Let $B \in$ P and p be a polynomial such that $f \in \mathrm{F}(p)$ and for all x we have: $x \in A \iff \langle x, f(1^{|x|}) \rangle \in B$. Let $f'(1^n) =_{\mathrm{def}} 1f(1^n)$, and define S as follows: For every n and i, the ith string of length n is in S if and only if the ith symbol of $f'(1^n)$ is a 1. Since f is polynomially length-bounded, S is sparse. A polynomial-time oracle Turing machine on input x, $|x| = n$, can compute $p(n)$, use the oracle language S to reconstruct $f(1^n)$, and then simulate a machine for B to decide if $x \in A$. Hence $A \in \mathrm{P}^S$.

($\Leftarrow$): Let $A \in \mathrm{P}^S$ via oracle machine M. Let p be a polynomial such that $|S \cap \{0,1\}^n| \leq p(n)$. Let q be a polynomial bounding the runtime of M. If for $n \in \mathbb{N}$, $s_1, s_1, \ldots, s_k$ are the words of length at most $q(n)$ in S, then we define

$$f(1^n) =_{\mathrm{def}} \langle s_1, s_1, \ldots, s_k \rangle.$$

Since $k \leq p(q(n))$, we have $f \in$ Poly. Observe that due to the runtime restriction, M on an input of length n can never ask the oracle for words longer than $q(n)$. Thus all words in the oracle that are relevant for a computation of M on such an input can be found in $f(1^n)$. Thus, a set $B \in \Gamma$ can be defined such that for all x, $\langle x, f(1^{|x|}) \rangle \in B$ if and only if M with oracle S accepts x. Hence $A \in$ P/Poly. $\qquad \square$

6.1.1 BPP and P/Poly

In Sect. 3.3.1 we defined probabilistic circuits and proved that they can be turned into regular, i.e., deterministic, circuits with a slight increase of size and depth. From these results it follows that allowing probabilism in polynomial-size circuits does not lead us out of the class P/Poly. Another consequence is that randomized polynomial-time computations can be simulated by polynomial-size circuits as we show next.

We now define the class BPP, a class of immense importance in cryptography and the theory of efficient algorithms. For example, the problem of deciding whether a number is a prime number is in BPP, but no deterministic polynomial-time algorithm is known. It should be remarked that our definition, though formulated quite differently, is equivalent to the original definition (see Bibliographic Remarks, p. 228).

Definition 6.3. A set $A \subseteq \{0,1\}^*$ belongs to the class BPP if, for every $k \in \mathbb{N}$, there is a set $B \in P$ and a polynomial p such that for every $x \in \{0,1\}^*$, $|x| = n$,

$$\mathrm{prob}_{y \in \{0,1\}^{p(n)}} \left[c_A(x) = c_B(\langle x, y \rangle) \right] \geq 1 - n^{-k}.$$

Here, the strings y are chosen uniformly at random.

Theorem 6.4. BPP $\subseteq$ P/Poly.

Proof. Let $A \in$ BPP. For $k \in \mathbb{N}$, let $B_k \in P$ and $p_k \in n^{O(1)}$ be such that for every x, $|x| = n$, we have $\mathrm{prob}_{y \in \{0,1\}^{p_k(n)}} \left[c_A(x) = c_{B_k}(\langle x, y \rangle) \right] \geq 1 - n^{-k}$. Let C_k be a polynomial-size circuit family that accepts B_k. If we now interpret C_k as a family of probabilistic circuits by making the input gates corresponding to the second input component y probabilistic inputs, we see that for every k there is a probabilistic circuit family of polynomial size (over B_0) that accepts A with error probability n^{-k} for inputs of length n. Hence by Theorem 3.23 we conclude that $A \in \mathrm{SIZE}(n^{O(1)})$. $\qquad\Box$

While $\mathrm{U_P\text{-}SIZE}(n^{O(1)}) = P$, we have shown that $\mathrm{SIZE}(n^{O(1)}) \supseteq$ BPP. Although we do not know at present whether there are problems in BPP which do not have polynomial-time deterministic algorithms, this result is interesting since it shows how the power of non-uniformity can be helpful.

6.1.2 NP and P/Poly

The famous $P \overset{?}{=} NP$-problem is, in terms of circuits, the problem of whether every language in NP has small uniform circuits.

As we showed earlier (see Sect. 1.5.2 and the discussion at the beginning of Chap. 2), generally non-uniformity helps, i.e., yields provably more computational power. Therefore, relaxing the above problem, the question whether every language in NP might have non-uniform small circuits is also interesting. We show in this section that under some reasonable assumptions the answer is no.

For this we define a sequence of complexity classes known as the *polynomial-time hierarchy*.

Definition 6.5. Let $A \subseteq \{0,1\}^*$. The *relativized polynomial-time hierarchy* with respect to A (or, polynomial-time hierarchy relative to A) is the following sequence of classes: $\Sigma_0^{p,A} = \Pi_0^{p,A} = P^A$, and

$$\Sigma_k^{p,A} = NP^{\Sigma_{k-1}^{p,A}} \quad \text{and} \quad \Pi_k^{p,A} = coNP^{\Sigma_{k-1}^{p,A}}$$

for $k \geq 1$. Let $PH^A = \bigcup_{k \geq 0} \Sigma_k^{p,A}$. The (unrelativized) polynomial-time hierarchy consists of the classes $\Sigma_k^p = \Sigma_k^{p,\emptyset}$ and $\Pi_k^p = \Pi_k^{p,\emptyset}$ for $k \geq 0$, and we set $PH = PH^{\emptyset}$.

It is clear that for every A and every $i \geq 0$, $\Sigma_i^{p,A} = co\Pi_i^{p,A}$ and $\Sigma_i^{p,A} \cup \Pi_i^{p,A} \subseteq \Sigma_{i+1}^{p,A} \cap \Pi_{i+1}^{p,A}$.

The classes of the polynomial-time hierarchy can be characterized in terms of polynomially length-bounded quantifiers, as stated in the following proposition.

Proposition 6.6. *Let $A, B \subseteq \{0,1\}^*$ and $k \geq 0$.*

1. *$A \in \Sigma_k^{p,B}$ if and only if there exist a set $R \in P^B$ and polynomials $p_1, \ldots, p_k$ such that*

$$x \in A \iff (\exists y_1, |y_1| \leq p_1(|x|)) (\forall y_2, |y_2| \leq p_2(|x|)) \cdots$$
$$(Q_k y_k, |y_k| \leq p_k(|x|)) \langle x, y_1, y_2, \ldots, y_k \rangle \in R$$

2. *$A \in \Pi_k^{p,B}$ if and only if there exist a set $R \in P^B$ and polynomials $p_1, \ldots, p_k$ such that*

$$x \in A \iff (\forall y_1, |y_1| \leq p_1(|x|)) (\exists y_2, |y_2| \leq p_2(|x|)) \cdots$$
$$(\overline{Q}_k y_k, |y_k| \leq p_k(|x|)) \langle x, y_1, y_2, \ldots, y_k \rangle \in R$$

Here $Q_k = \exists$ and $\overline{Q}_k = \forall$ if k is odd, and $Q_k = \forall$ and $\overline{Q}_k = \exists$ if k is even.

Yet another characterization of the classes of the polynomial-time hierarchy can be given in terms of alternating Turing machines:

Proposition 6.7. *A language A is in Σ_k^p (Π_k^p) if and only if there is a polynomially time-bounded alternating Turing machine M that for every input makes at most $k - 1$ alternations, starting in an existential state (universal state, respectively), and accepts A.*

It is generally assumed that the classes of the polynomial hierarchy form a strict hierarchy. While this question is still open, the following is known: If for some $i \geq 1$, $\Sigma_i^p = \Pi_i^p$ then $PH = \Sigma_i^p$. On the other hand, if for all $i \geq 0$, $\Sigma_i^p \neq \Pi_i^p$ then $PH \neq PSPACE$.

Hence we see that if we can prove that any language in Σ_k^p for some $k \geq 0$ does not have small circuits then we have proved $P \neq NP$.

Now we come back to the question whether $NP \subseteq SIZE(n^{O(1)})$. We show that if the answer is yes, then the polynomial-time hierarchy collapses, contradicting widespread belief.

Theorem 6.8. *If* NP $\subseteq$ P/Poly *then the polynomial hierarchy collapses to its second level* (PH = Σ_2^p).

Proof. Let SAT be the set of all (encodings of) satisfiable propositional formulas (see Appendix A5). If NP $\subseteq$ P/Poly, then in particular SAT $\in$ P/Poly. From the definition of P/Poly, it now follows that there is a set $B \in$ P and a function $h \in$ Poly such that for all n the following holds: For all x, $|x| \le n$: $x \in$ SAT $\iff$ $\langle x, h(n) \rangle \in B$. Let us say that w is n-*valid*, if for all x, $|x| \le n$: $x \in$ SAT $\iff$ $\langle x, w \rangle \in B$. From the above we see that there is a polynomial p such that, for every n, there is a word $w \in \{0,1\}^{p(n)}$ which is n-valid. Define $B_w = \{ x \mid \langle x, w \rangle \in B \}$.

We now claim that if we set

$$V =_{\text{def}} \{ \langle x, w \rangle \mid w \text{ is } |x|\text{-valid} \},$$

then $V \in$ coNP.

Let M be a deterministic Turing machine which operates as follows: The input to M is a formula ϕ and a string $w \in \{0,1\}^*$ where $|w| \ge p(|\phi|)$. If ϕ has no propositional variables then M simply evaluates ϕ and accepts if and only if $\phi \in$ SAT. Otherwise, M produces the two formulas ϕ_0 and ϕ_1 which are obtained from ϕ by substituting the truth value 0 (1, respectively) for the lexicographically first propositional variable in ϕ. Now M accepts $\langle \phi, w \rangle$, if and only if $\phi_0 \in B_w$ or $\phi_1 \in B_w$.

Let C be the language accepted by M. Clearly, M is polynomially time-bounded, hence $C \in$ P.

We now prove the following:

$$\langle x, w \rangle \in V \text{ if and only if } (\forall y, |y| \le |x|)(y \in B_w \iff \langle y, w \rangle \in C). \quad (6.1)$$

For the direction from left to right, if $\langle x, w \rangle \in V$ then w is $|x|$-valid. Hence for all y, $|y| \le |x|$, we have $y \in$ SAT $\iff$ $y \in B_w$; hence M will accept $\langle \phi, w \rangle$ (for any ϕ such that $|\phi| \le |x|$) if and only if $\phi_0 \in B_w$ or $\phi_1 \in B_w$ if and only if $\phi_0 \in$ SAT or $\phi_1 \in$ SAT if and only if $\phi \in$ SAT if and only if $\phi \in B_w$.

For the direction from right to left, we prove by induction on the number of free variables in ϕ (where $|\phi| \le |x|$): $\phi \in$ SAT $\iff$ $\phi \in B_w$. If ϕ has no free variables then the definition of M ensures our claim. If ϕ has m free variables, then ϕ_0 and ϕ_1 have $m - 1$ free variables, and since (under a reasonable encoding of formulas) $|\phi_0|, |\phi_1| \le |\phi| \le |x|$ we conclude from the induction hypothesis that $\phi_0 \in$ SAT $\iff$ $\phi_0 \in B_w$ and $\phi_1 \in$ SAT $\iff$ $\phi_1 \in B_w$. Hence $\phi \in$ SAT if and only if $\phi_0 \in$ SAT or $\phi_1 \in$ SAT if and only if $\phi_0 \in B_w$ or $\phi_1 \in B_w$ if and only if $\langle \phi, w \rangle \in C$ (by definition of M) if and only if $\phi \in B_w$.

This completes the induction, and (6.1) is proved; hence $V \in$ coNP by Proposition 6.6.

We are now ready to show that under the above, $\Sigma_3^p \subseteq \Sigma_2^p$. Let $A \in \Sigma_3^p$. Since SAT is NP-complete, we get $A \in$ NP$^{\text{NP}^{\text{SAT}}}$. From Proposition 6.6 we conclude that there are a set $R \in$ P^{SAT} and polynomials p_1, p_2 such that

$$x \in A \iff (\exists y_1, |y_1| \le p_1(|x|))(\forall y_2, |y_2| \le p_2(|x|))\langle x, y_1, y_2 \rangle \in R.$$

Let R be accepted by the polynomial-time oracle machine M. Let the runtime of M on inputs of the form $\langle x, y_1, y_2 \rangle$, $|x| = n$, be $q(n)$. Hence on such inputs, the maximal length of an oracle query that M can ask is also bounded by $q(n)$. Thus, if we replace the oracle SAT by B_w for any w that is $q(n)$-valid, M accepts the same set of inputs. Define a set R' by $\langle x, y_1, y_2, w \rangle \in R'$ if and only if M with oracle B_w accepts $\langle x, y_1, y_2 \rangle$.

Thus we conclude:

$$x \in A \iff (\exists w, |w| \le p(q(|x|)))$$
$$\left[(\forall y, |y| \le q(|x|))(y \in B_w \iff \langle y, w \rangle \in C) \right.$$
$$\left. \wedge (\exists y_1, |y_1| \le p_1(|x|))(\forall y_2, |y_2| \le p_2(|x|))\langle x, y_1, y_2, w \rangle \in R' \right]$$
$$\iff (\exists w, |w| \le p(q(|x|)))(\exists y_1, |y_1| \le p_1(|x|))$$
$$(\forall y, |y| \le q(|x|))(\forall y_2, |y_2| \le p_2(|x|))$$
$$\left[(y \in B_w \iff \langle y, w \rangle \in C) \wedge \langle x, y_1, y_2, w \rangle \in R' \right]$$

Using pairing functions we can merge adjacent quantifiers of the same type (of course, we also have to adjust the predicate in brackets accordingly), hence by Proposition 6.6, $A \in \Sigma_2^p$. ☐

The preceding theorem gives a structural (or, conditional) lower bound: If $\Sigma_2^p \ne \Pi_2^p$, then SAT or any other NP-complete set does not have polynomial-size circuits.

6.1.3 Lower Bounds for P/Poly

In this section, we want to prove lower bounds for polynomial-size circuits, that is, exhibit problems outside P/Poly. From the results in Sect. 1.5.1, it follows that such sets exist, but nothing is stated about their complexity. It will turn out below that we can bound the complexity of these problems in a number of interesting ways.

We start with lower bounds for the class $\text{SIZE}(n^k)$ for fixed k. For convenience, let us fix the basis $\mathbb{B}^2$ for the rest of this section.

Theorem 6.9. *For every $k \in \mathbb{N}$, there is a language $L_k \in \Sigma_2^p \cap \Pi_2^p$ that does not have circuits of size $O(n^k)$.*

Proof. Fix $k \in \mathbb{N}$. From Sect. 1.5.1 we know that the number of functions from $\{0,1\}^n \to \{0,1\}$ that can be computed by circuits over $\mathbb{B}^2$ of size n^{k+1} is bounded above by $O(2^{n^{k+2}})$. (Use the bound for $N_{q,n}$ derived in the proof of Theorem 1.47, for $q = n^{k+1}$.) Let $\{0,1\}^n = \{s_1^n, s_2^n, \ldots, s_{2^n}^n\}$. Then there are $2^{n^{2k}}$ subsets of $\{s_1^n, s_2^n, \ldots, s_{n^{2k}}^n\}$. Hence we conclude that for large enough

n and k, there is at least one such subset, S_0, that cannot be accepted by any circuit of size n^{k+1}. On the other hand, clearly S_0 can be accepted by a circuit C_0 over $\mathbb{B}^2$ of size $(n+1) \cdot n^{2k} + n^{2k} \leq 2n^{2k+1}$. C_0 simply realizes the characteristic function of B_0 in disjunctive normal form: this needs n^{2k} conjuncts of size at most $(n+1)$ each, and an $\vee$ gate of fan-in n^{2k} which can be realized by n^{2k} $\vee$ gates of fan-in 2.

As an intermediate step, we now prove: For every $k \in \mathbb{N}$, there is a language $L'_k \in \Sigma^p_4 \cap \Pi^p_4$ that does not have circuits of size $O(n^k)$.

Define an order $\prec$ on all circuits of size at most $2n^{2k+1}$ as follows: $C \prec C'$ if the encoding of C (under any reasonable encoding scheme) is lexicographically smaller than the encoding of C'. We define $L'_k \cap \{0,1\}^n$ to be a subset of $\{s^n_1, s^n_2, \ldots, s^n_{2n}\}$ which cannot be accepted by a circuit of size n^{k+1}; more specifically, among all those subsets, the set $L'_k \cap \{0,1\}^n$ will be the one which is accepted by the minimum (with respect to $\prec$) circuit $\widehat{C}$ of size $2n^{2k+1}$, which cannot be simulated by a circuit of size n^{k+1}.

We have to show that L'_k is in $\Sigma^p_4 \cap \Pi^p_4$. Let $|x| = n$. Then

$$x \in L'_k \iff (\exists \text{ a circuit } \widehat{C} \text{ of size } 2n^{2k+1})$$
$$[\quad (\forall \text{ circuits } C \text{ of size } n^{k+1})$$
$$(\exists y, |y| = n)$$
$$f_{\widehat{C}}(y) \neq f_C(y)$$
$$\wedge (\forall \text{ circuits } C' \text{ of size } 2n^{2k+1} \text{ such that } C' \prec \widehat{C})$$
$$(\exists \text{ a circuit } C \text{ of size } n^{k+1})$$
$$(\forall y, |y| = n)$$
$$f_{C'}(y) = f_C(y)]$$
$$\wedge f_{\widehat{C}}(x) = 1.$$

The quantifiers that range over circuits can be replaced by quantifiers that range over encodings of circuits. Then all quantifiers are polynomially length-bounded and we conclude that $L'_k \in \Sigma^p_4$. Moreover, just by replacing the final line in the above by "$f_{\widehat{C}}(x) = 0$", we see that $L'_k \in \Pi^p_4$.

We now prove the claim of the theorem. We distinguish two cases.

$NP \subseteq P/Poly$: In this case we conclude from Theorem 6.8 that $\Sigma^p_4 = \Pi^p_4 = \Sigma^p_2 = \Pi^p_2$, hence $L_k =_{\text{def}} L'_k$ from the above fulfills the claim.

$NP \not\subseteq P/Poly$: In this case, $SAT \notin P/Poly$, hence we may choose $L_k =_{\text{def}} SAT$.

$\square$

We now exhibit a language which does not have circuits of size n^k for any k. This language will be located in the exponential-time hierarchy which we define next.

Definition 6.10. A language A is in $\Sigma_k^{\exp}$ ($\Pi_k^{\exp}$) if and only if there is an alternating Turing machine M that is time-bounded by $2^{n^{O(1)}}$, makes for every input at most $k-1$ alternations, starting in an existential state (universal state, respectively), and accepts A.

Theorem 6.11. *There is a language $L \in \Sigma_2^{\exp} \cap \Pi_2^{\exp}$ that does not have polynomial-size circuits.*

Proof. The proof is similar to the proof of Theorem 6.9. Again we distinguish two cases: If NP $\not\subseteq$ P/Poly then $L =_{\text{def}}$ SAT fulfills the theorem. In the other case we conclude from Theorem 6.8 that PH $= \Sigma_2^p$ and continue as follows:

As in the proof of Theorem 6.9 we obtain a language L' which cannot be computed by circuits of size $O\left(n^{\log n}\right)$ (hence not by polynomial-size circuits), but by circuits of size $2^{n^{O(1)}}$. L' can be defined using the same quantifier structure as in the proof of Theorem 6.9, but now the quantifiers range over exponentially long strings. This defining formula can clearly be checked by an exponentially time-bounded alternating machine with four alternations, hence we conclude that $L' \in \Sigma_4^{\exp}$.

Let the machine M witnessing $L' \in \Sigma_4^{\exp}$ be time-bounded by the function 2^{n^k}, where p is a polynomial. Then the set $L'' =_{\text{def}} \left\{ x10^{2^{|x|^k}-|x|-1} \mid x \in L' \right\}$ is in Σ_4^p. A polynomial-time alternating machine M' for L'' works as follows: On input $x10^{2^{|x|^k}-|x|-1}$, M' strips away the tailing $10^{2^{|x|^k}-|x|-1}$ and simulates M on the remaining input. M will need time $2^{|x|^k}$, but measured in the length of M'''s input $x10^{2^{|x|^k}-|x|-1}$, this is only polynomial.

However, since PH $= \Sigma_2^p$ we conclude that $L'' \in \Sigma_2^p \cap \Pi_2^p$. Let M'' be a machine witnessing $L'' \in \Sigma_2^p$. Define a machine M''' working as follows: On input x, M''' simulates M'' on input $x10^{2^{|x|^k}-|x|-1}$. The simulation takes time polynomial in $\left|x10^{2^{|x|^k}-|x|-1}\right|$, hence exponential in $|x|$. Thus M'' is exponentially time-bounded, and we conclude that $L' \in \Sigma_2^{\exp}$. Analogously we obtain $L' \in \Pi_2^{\exp}$, and the proof is completed. $\qquad\Box$

Corollary 6.12. *There is a language in* EXPSPACE *that does not have small circuits.*

Proof. Follows from the theorem above together with the fact that $\Sigma_2^{\exp} \subseteq$ EXPSPACE, see Exercise 6.4. $\qquad\Box$

6.1.4 Lower Bounds for Uniform Circuits

In this subsection and the next section, we want to turn to lower bounds for specific problems, but this time we will consider *uniform* circuit families. All lower bounds given in this book so far hold for non-uniform families (and therefore, of course, for uniform families as well). Below we will present examples for problems which are difficult for uniform circuits, as we show, but no corresponding lower bound for non-uniform circuits is known.

Our examples in this subsection address the power of uniform circuit size n^k for fixed k. We consider different uniformity conditions. Let us start with logtime-uniformity.

Theorem 6.13. *For every $k \in \mathbb{N}$, there is a language $L_k \in \mathrm{P}$ that does not have logtime-uniform circuits of size $O(n^k)$.*

Proof. For every $k \in \mathbb{N}$, there is a $k' > k$ such that $U_D\text{-SIZE}(n^k) \subseteq \mathrm{DTIME}(n^{k'})$, see Exercise 6.5. However it is well known (see, e.g., [HU79]) that there are languages in $\mathrm{P} \setminus \mathrm{DTIME}(n^{k'})$ for every k'. $\qquad\qquad\square$

The proof above relies heavily on logtime-uniformity; inclusion of uniform $\text{SIZE}(n^k)$ in $\mathrm{DTIME}(n^{k'})$ for a fixed k' is only possible since the time needed for construction of the circuit can be bounded by a particular polynomial in the size of the circuit. This is not possible for the class $U_P\text{-SIZE}(n^k)$. Here we do not have one fixed polynomial that bounds the time needed to construct the circuit for the appropriate input length. Thus a result as above cannot be shown. However the following holds:

Theorem 6.14. *For every $k \in \mathbb{N}$, there is a language $L_k \in \mathrm{NP}$ that does not have ptime-uniform circuits of size $O(n^k)$.*

Proof. We distinguish two cases:

Let us first assume that $\mathrm{NP} \not\subseteq U_P\text{-SIZE}(n^{O(1)})$. Then the claim of the theorem holds trivially; take, e.g., $L_k = \mathrm{SAT}$.

Now assume $\mathrm{NP} \subseteq U_P\text{-SIZE}(n^{O(1)}) = \mathrm{P}$, hence $\mathrm{PH} = \mathrm{P}$. Now the language L_k constructed in Theorem 6.9 does not have (even non-uniform) circuits of size $O(n^k)$, but $L_k \in \Sigma_2^p \cap \Pi_2^p = \mathrm{NP}$. $\qquad\qquad\square$

6.2 A Lower Bound for Uniform Threshold Circuits

In this final section we want to prove a lower bound for the circuit complexity of a particular problem, the so-called *permanent*, but again we will consider uniform circuits.

The frontiers of our knowledge about circuit lower bounds, as given in Chap. 3 and the previous subsections, can be summarized as follows: Although we know that most problems have exponential circuit size (Theorem 1.47), we do not currently know of any problem in NP which has a non-linear lower bound with respect to the size of *non-uniform* circuits (we only know that such problems exist in $\Sigma_2^p \cap \Pi_2^p$, see Theorem 6.9). Concerning polynomial circuit size, it is consistent with our current knowledge that all problems in NEXPTIME are in $\text{SIZE}(n^{O(1)})$ (but we know that $\Sigma_2^{\exp} \cap \Pi_2^{\exp} \setminus \text{SIZE}(n^{O(1)}) \neq \emptyset$ by Theorem 6.11). On the other hand, concerning *uniform* circuit size, we proved in Sect. 6.1.4 that there are problems which do not have uniform circuits of size $O(n^k)$ (for fixed $k \in \mathbb{N}$), and these problems can be located in NP or even P, depending on the uniformity condition we consider. Of course, any problem outside the sequential class P cannot have uniform polynomial-size circuits, thus $\text{EXPTIME} - \text{U}_P\text{-SIZE}(n^{O(1)}) \neq \emptyset$ (see Appendix A4), but we do not even know if P = PSPACE.

As argued in Chap. 3, the difficulty of obtaining lower bounds in the unrestricted circuit model led researchers to the examination of *constant-depth circuits*, and here remarkable lower bounds are known; probably the most impressive is given by Smolensky's Theorem (Theorem 3.31). But again, the techniques developed for constant-depth circuits are limited: we do not know whether $\text{ACC}^0 \supseteq \text{NEXPTIME}$.

Concerning uniform circuits, the following can be said: if A is complete for PSPACE (under logspace-reductions, say), then A cannot be in $\text{U}_L\text{-NC}^1$. This holds since $\text{U}_L\text{-NC}^1 \subseteq L$, and it is well known that $L \subsetneq \text{PSPACE}$ (see Appendix A4). Hence we know that for all the uniform constant-depth classes considered in this book, there are problems in PSPACE that do not fall into one of these classes. However, the results presented so far leave it open whether such problems can be found in a (possibly strict) subclass of PSPACE such as PH or even NP.

We now prove the existence of a problem in PP (defined below), which does not have uniform TC^0 circuits. A number of definitions are in order.

Definition 6.15. The class PP is defined as follows: A language A belongs to PP if there is a nondeterministic polynomially time-bounded Turing machine M such that for all $x \in \{0,1\}^*$ the following holds: $x \in A$ if and only if M on input x has more accepting than rejecting computations paths.

M is sometimes referred to as a *counting Turing machine*, and A is said to be accepted by M.

When we allow M access to an oracle language B, then we get in a natural way the relativized class PP^B. For a class of languages $\mathcal{K}$ we set as before $\text{PP}^{\mathcal{K}} = \bigcup_{B \in \mathcal{K}} \text{PP}^B$.

Definition 6.16. The *counting polynomial-time hierarchy* consists of the following classes:

$$C^0P =_{\text{def}} P;$$
$$C^iP =_{\text{def}} PP^{C^{i-1}P} \quad \text{for } i \geq 1.$$

Let $CH = \bigcup_{i \geq 0} C^iP$.

Observe that $CP = C^1P = PP$. Moreover, it can be seen (Exercise 6.6) that $NP \subseteq PP$ and $coNP \subseteq PP$, which implies $PH \subseteq CH$.

In the *leaf language* approach to the characterization of complexity classes, the acceptance of a word input to a nondeterministic machine depends only on the values printed at the leaves of the computation tree. To be more precise, let M be a nondeterministic Turing machine which, on every path, (a) halts, and (b) prints a symbol from the alphabet $\{0,1\}$. Let the nondeterministic choices of M be ordered arbitrarily, but fixed (e.g., based on the textual presentation of M's Turing program). Then, leafstring$^M(x)$ is the concatenation of the symbols printed at the leaves of the computation tree of M on input x (according to the order of M's paths induced by the order of M's nondeterministic choices). Given now a language $B \subseteq \{0,1\}^*$, we define Leaf$^M(B) = \{ x \mid \text{leafstring}^M(x) \in B \}$.

Let us call a computation tree of a machine M *balanced*, if all of its configurations are either leaves or have exactly two successor configurations, all computation paths have the same length, and moreover, if we identify every path with the string over $\{0,1\}$ describing the sequence of nondeterministic choices on this path, then there is some string z such that all paths y with $|y| = |z|$ and $y \preceq z$ (in lexicographic ordering) exist, but no path y with $y \succ z$ exists. Informally, a tree is balanced if it is a full binary tree with a missing right part.

Definition 6.17. Let $B \subseteq \{0,1\}^*$. The class Leaf$^P(B)$ consists of all languages A for which there exists a nondeterministic polynomial time machine M whose computation tree is always balanced, such that $A = \text{Leaf}^M(B)$. B is called *leaf language* defining the class Leaf$^P(B)$.

Let $\mathcal{K}$ be a class of languages. Then the class Leaf$^P(\mathcal{K})$ consists of the union over all $B \in \mathcal{K}$ of the classes Leaf$^P(B)$.

Example 6.18. Consider the language L_1 defined by the regular expression $0^*1(0+1)^*$. Then, obviously, Leaf$^P(L_1) = \text{NP}$. If we define L_2 to consist of all words over the alphabet $\{0,1\}$ with more 1's than 0's, then Leaf$^P(L_2) = \text{PP}$.

Interesting cases arise when we take as leaf languages elements from a constant-depth circuit class. In the following lemma, we consider the class TC^0. More examples can be found in Exercises 6.8, 6.9(1), and 6.11.

Lemma 6.19. *1.* Leaf$^P(\text{U}_\text{D}\text{-TC}^0) = \text{CH}$.
2. Leaf$^P(\text{P}) = \text{EXPTIME}$.

Proof. Leaf$^P(\text{U}_\text{D}\text{-TC}^0) \subseteq \text{CH}$: Let $A \in \text{Leaf}^M(B)$ for $B \in \text{U}_\text{D}\text{-TC}^0$ and a polynomially time-bounded Turing machine M. Let $\mathcal{C} = (C_n)_{n \in \mathbb{N}}$ be a uniform circuit family accepting B. We assume without loss of generality that all circuits C_n only consist of majority gates. Let $d \in \mathbb{N}$ be the depth of $\mathcal{C}$. We now prove by induction on d that $A \in \text{CH}$.

Let $d = 1$. Then $\mathcal{C}$ computes the majority of some of its inputs, hence $A \in \text{PP}$ by the following counting Turing machine $\widehat{M}$: On input x, $\widehat{M}$ first determines $m = |\text{leafstring}^M(x)|$. (Observe that m can be exponential in $|x|$.) Then $\widehat{M}$ uses the uniformity machine for $\mathcal{C}$ to obtain information about the structure of C_m. Since $\mathcal{C}$ is logtime-uniform, this machine runs in time polynomial in $|x|$. More exactly, $\widehat{M}$ branches nondeterministically on all predecessors of the output gate g of C_m. On each such branch, $\widehat{M}$ computes the corresponding circuit input by simulating machine M. More precisely, if the current predecessor gate of g is found to be the input gate x_i, then $\widehat{M}$ simulates the computation on the ith path of M. This simulation is deterministic, since the nondeterminism of M is given now by the bits of the binary representation of i. It is clear that $\widehat{M}$ accepts its input (in the PP sense) if and only if the majority of g's predecessors are 1.

Now let $d > 1$. Consider the following counting Turing machine $\widehat{M}$: On input x, $\widehat{M}$ again first determines $m = |\text{leafstring}^M(x)|$. $\widehat{M}$ then (using the uniformity machine for C) computes the number of the output gate g of C_m and, as above, branches on all possible predecessors of g. These predecessors are output gates of shallower threshold circuits, hence we see that, by induction hypothesis, we can compute the values of these gates asking queries to an oracle $B \in \text{CH}$. Now $x \in A$ if and only if $\widehat{M}$ with oracle B accepts x, and we conclude that $A \in \text{CH}$.

$\text{CH} \subseteq \text{Leaf}^P(\text{U}_D\text{-TC}^0)$: We prove the inclusion $\text{C}^i\text{P} \subseteq \text{Leaf}^P(\text{U}_D\text{-TC}^0)$ by induction on i. In the base case $i = 1$, we have a language $A \in \text{PP}$, accepted by counting Turing machine M. Let n^k be the runtime of M. We define a nondeterministic polynomial time machine M_k as follows: Given input $x = a_1 \cdots a_n$, M_k branches on 2^{n^k} paths. On the ith path, M_k outputs a_i for $1 \le i \le n$, 1 for $i = n + 1$, and 0 for $i > n + 1$. In other words, $\text{leafstring}^{M_k}(x) = x10^{2^{|x|^k}-1-|x|}$.

Now consider the following circuit $C = C_{2^{n^k}}$: The input of C is a word $z = \text{leafstring}^{M_k}(x)$. The output g of C_n is a threshold gate. g has predecessor gates g_p for every possible path p (on *any* input of length 2^{n^k}) of M. Every path p prints an output symbol a_p from $\{0, 1\}$, depending on the machine's input x. Gate g_p now determines a_p, depending on the input z. This is achieved by simulating path p of M, using the circuit input $x10^{2^{|x|^k}-1-|x|}$ instead of M's input. The symbol a_p is the symbol that is printed by M on input x on path x. If p does not correspond to a legal path of M on the actual input x, we set $a_p = 0$. Choosing an appropriate threshold value, we see that C_n can simulate the behavior of M. It remains to consider uniformity of the circuit family. We choose a gate number for g_p which allows us to reconstruct from this number the corresponding path p of the machine in polynomial time (given the gate number in binary). To compute a connection between gates in C_n we have to simulate paths of M. This may take polynomial time, which is, however, logarithmic in the size of C_n. Hence C is logtime-uniform, and we have proved $A \in \text{Leaf}^P(\text{U}_D\text{-TC}^0)$.

Now let $i > 1$, and $A \in \text{C}^i\text{P}$. Let A be accepted by counting Turing machine M with oracle $B \in \text{C}^{i-1}\text{P}$. Given $x = a_1 \cdots a_n$, basically we proceed as above to construct $C_{2^{n^k}}$, but now the values a_p depend not only on M's input but also on the oracle B. Thus we obtain a circuit with oracle gates, where the oracle gates compute a language which (by induction hypothesis) is in $\text{U}_D\text{-TC}^0$. Thus these gates may be replaced by threshold circuits, and we obtain a TC^0 family. Choosing an appropriate numbering of gates, logtime-uniformity is easily verified as in the case $i = 1$ we considered above.

$\text{Leaf}^P(P) \subseteq \text{EXPTIME}$: Let $A \in \text{Leaf}^M(B)$ for $B \in P$ and a polynomially time-bounded Turing machine M. The exponential time algorithm for A works as follows: Given input x, compute $\text{leafstring}^M(x)$ by traversing the computation tree of M on x. Then just check whether $\text{leafstring}^M(x) \in B$. Observe that the length of $\text{leafstring}^M(x)$ is exponential in $|x|$, hence this procedure takes exponential time establishing $A \in \text{EXPTIME}$.

$\text{EXPTIME} \subseteq \text{Leaf}^P(P)$: Let $A \in \text{EXPTIME}$ via the deterministic Turing machine M. Let the runtime of M be bounded by 2^{n^k} for some $k \in \mathbb{N}$. Define machine M_k as above, i.e., $\text{leafstring}^{M_k}(x) = x10^{2^{|x|^k}-1-|x|}$. Now define a deterministic machine M'' as follows: Given a word $x10^{2^{|x|^k}-1-|x|}$, M'' simulates M on input x. The time needed for this simulation is $2^{|x|^k}$. Hence M'' is a polynomial-time machine, and we have proved that $A \in \text{Leaf}^P(P)$. $\qquad\Box$

Theorem 6.20. $U_D\text{-}TC^0 \subsetneq CH$.

Proof. Assume $U_D\text{-}TC^0 = CH$. Then $\text{Leaf}^P(U_D\text{-}TC^0) = \text{Leaf}^P(CH)$. Since by Lemma 6.19, $\text{Leaf}^P(U_D\text{-}TC^0) = CH$ and $\text{Leaf}^P(CH) \supseteq \text{Leaf}^P(P) = \text{EXPTIME}$, we obtain $\text{EXPTIME} \subseteq \text{Leaf}^P(CH) \subseteq \text{Leaf}^P(U_D\text{-}TC^0) \subseteq CH \subseteq U_D\text{-}TC^0 \subseteq P$, which contradicts the time hierarchy theorem (see Appendix A4). $\qquad\Box$

It is interesting to note that the preceding proof only depends on the inclusions $\text{EXPTIME} \subseteq \text{Leaf}^P(P)$ and $\text{Leaf}^P(U_D\text{-}TC^0) \subseteq CH$ from Lemma 6.19.

Corollary 6.21. $U_D\text{-}TC^0 \neq PP$.

Proof. Assume $U_D\text{-}TC^0 = PP$. We will show that then $CH \subseteq U_D\text{-}TC^0$, contradicting Theorem 6.20.

We prove by induction on i, that $C^iP \subseteq U_D\text{-}TC^0$ for all i. The case $i = 1$ is the assumption. Now assume $A \in C^{i+1}P$. Let A be accepted (in the PP sense) by counting Turing machine M with oracle B, $B \in C^iP$. By induction hypothesis, $B \in U_D\text{-}TC^0$. Hence there is a $U_D\text{-}TC^0$ circuit family C that accepts B. Hence we may replace the oracle queries to B by calls to a subroutine which evaluates a circuit from C. This shows that $A \in PP$, hence $A \in U_D\text{-}TC^0$ and $C^{i+1}P \subseteq U_D\text{-}TC^0$. $\qquad\Box$

Corollary 6.22. *Let A be hard for PP under $\leq_{cd}$. Then $A \notin U_D\text{-}TC^0$.*

Proof. If A is hard for PP under $\leq_{cd}$ and $A \in U_D\text{-}TC^0$, then clearly $PP \subseteq U_D\text{-}TC^0$. This contradicts Corollary 6.21. $\qquad\Box$

We can relax the reducibility notion here and allow somewhat more powerful circuits to compute the reduction function, see Exercise 6.12.

One prominent example of a PP-hard language is the following:

Definition 6.23. Let M be an $n \times n$ matrix, $M = (m_{i,j})_{1 \leq i,j \leq n}$. Then the *permanent* of M is defined to be

$$\text{perm}(M) =_{\text{def}} \sum_{\sigma \in S_n} \prod_{i=1}^{n} m_{i,\sigma(i)}.$$

We also define $\text{PERM} =_{\text{def}} \{ \langle M, k \rangle \mid \text{perm}(M) \geq k \}$.

It is known that the permanent is complete for #P under logtime-uniform AC^0 reductions, and that PERM is hard for PP under the same type of reductions.

From Corollary 6.22 we now obtain immediately:

Corollary 6.24. *The permanent cannot be computed by U_D-FTC^0 circuits.*

Proof. If the permanent could be computed by U_D-FTC^0 circuits, then clearly $\text{PERM} \in U_D$-TC^0, which contradicts Corollary 6.22. $\square$

Bibliographic Remarks

The result that a language has small circuits if and only if it is recognizable in polynomial time relative to a sparse oracle (Theorem 6.2) is credited to A. R. Meyer in [BH77].

The class BPP was defined in [Gil77] via so-called polynomial-time bounded-error probabilistic Turing machines. It has been shown that these machines can be made to run with "arbitrarily small error" (see [Sch86, pp. 35ff.]), and from this it can be seen immediately that Gill's original definition is equivalent to Def. 6.3; see also [BDG95, Sect. 6.4]. That primality testing is in BPP is shown in [Rab76]; see also [Pap94, Sect. 11.1]. That every set in BPP has small circuits (Theorem 6.4) appeared as a remark in [BG81, 105]. Explicitly it was published in [Sch86].

The polynomial-time hierarchy was defined in [MS72]. The characterization of the classes of this hierarchy in terms of polynomially length-bounded quantifiers (Proposition 6.6) is due to [Sto77, Wra77]; see also [BDG95, Sect. 8.3]. The characterization of the classes of the polynomial-time hierarchy using alternating Turing machines (Proposition 6.7) is from [CKS81]; see also [BDG90, Chap. 3]. Theorem 6.8 relating NP to P/Poly is from [KL82]. Our presentation follows a generalization of the result which appeared in [BBS86], see also [Sch86, 61–62].

All results in Sects. 6.1.3–6.1.4 are due to [Kan82].

The line of research that led to the result presented in Sect. 6.2 was started with [AG94], where it was shown that the permanent is not in uniform ACC^0. In [CMTV98] the existence of some problem in the counting hierarchy that does not have uniform TC^0 circuits (Theorem 6.20) was proved. The final step (Corollary 6.21) was obtained in [All99], showing that the permanent and any problem complete for PP cannot be in uniform TC^0. The latter paper also obtained improved size lower bounds.

The class PP was introduced in [Sim75, Gil77]. The counting hierarchy was defined in [Wag86]. Completeness of the permanent for #P under so called metric reductions is from [Val79]; see also [Pap94, Sect. 18.1]. That these reductions can actually be computed by logtime-uniform AC^0 circuits has been observed, e. g., in [Imm87, Zan91, AG94].

Leaf languages, the main technical tool in [CMTV98], were introduced in another context in [BCS92, Ver93]. The use of leaf languages (and leaf language classes) to characterize complexity classes goes back to [HLS$^+$93, JMT96], see also [Pap94, 504-505]. For a survey see [Vol98a, Vol99]. It should be remarked that Theorem 6.20 can also be obtained without using leaf languages at all. In [All99] a proof by direct diagonalization was given. Diagonalization is of course also present in the proof given in this chapter, because it makes use of the classical time-hierarchy theorem (see Appendix A4).

An important connection between circuit classes within NC^1 and Turing machine based classes within PSPACE is that lower bounds for the former can often be used to obtain *oracle separations* for the latter. We do not develop this topic here, but refer the reader to [FSS84, Yao85, Hås86]; for a survey, see [AW90]. The tool of leaf languages mentioned above was originally introduced in [BCS92, Ver93], exactly for the purpose of simplifying the construction of separating oracles. In [Vol98b] it is shown how the leaf language model can be used to obtain oracle separations based on circuit lower bounds in a unified way.

Exercises

6.1. A set A is a *tally set*, if $A \subseteq \{0\}^*$.

Show that a set A is in P/Poly if and only if there is a tally set T such that $A \in P^T$.

6.2. Show that PH $\subseteq$ PSPACE.

6.3.* Let $f: \mathbb{N} \to \mathbb{N}$. A language A is in Σ_k^f (Π_k^f) if and only if there is an alternating Turing machine time-bounded by f that, for every input, makes at most $k - 1$ alternations, starting in an existential state (universal state, respectively).

Now let f be time-constructible such that $n^k \in o(f)$ for all k. Prove: There is a language $L \in \Sigma_2^f \cap \Pi_2^f$ that does not have polynomial-size circuits.

6.4. Show that for every $k \in \mathbb{N}$, $\Sigma_k^{\exp} \subseteq$ EXPSPACE.

6.5.* Show that for every $k \in \mathbb{N}$ there is a $k' > k$ such that $U_D\text{-SIZE}(n^k) \subseteq$ DTIME$(n^{k'})$.

6.6. Prove:

(1) $NP \subseteq PP$.
(2) $coNP \subseteq PP$.
(3) $PH \subseteq CH$.

6.7. Prove:

(1) A language A belongs to $C_=P$ if and only if there is a function $f \in Gap\text{-}P$ such that for all x, we have: $x \in A \iff f(x) = 0$.
(2) A language A belongs to PP if and only if there is a function $f \in Gap\text{-}P$ such that for all x, we have: $x \in A \iff f(x) > 0$.

6.8. Prove:

(1) $Leaf^P(U_D\text{-}AC^0) = PH$.
(2) $Leaf^P(U_D\text{-}NC^1) = PSPACE$.

6.9.* Let $k > 1$. Define the class $Mod_k P$ as follows: A language A belongs to $Mod_k P$ if and only if there is a function $f \in \#P$ (see p. 207) such that for all $x \in \{0,1\}^*$, $x \in A \iff f(x) \not\equiv 0 \mod k$. The relativized class $Mod_k P^{\mathcal{K}}$ for a class $\mathcal{K}$ of oracle languages is defined in the obvious way. The *modular polynomial-time hierarchy* consists of the following classes:

- P is a class of the modular polynomial-time hierarchy.
- If $\mathcal{K}$ is a class of the modular polynomial-time hierarchy, then $NP^{\mathcal{K}}$ and $coNP^{\mathcal{K}}$ are classes of the modular polynomial-time hierarchy.
- If $\mathcal{K}$ is a class of the modular polynomial-time hierarchy and $k > 1$ then $Mod_k P^{\mathcal{K}}$ is a class of the modular polynomial-time hierarchy.

Let MOD-PH be the union of all classes of the modular polynomial-time hierarchy.
   Prove:

(1) $Leaf^P(U_D\text{-}ACC^0) = MOD\text{-}PH$.
(2) $U_D\text{-}ACC^0 \subsetneq MOD\text{-}PH$.

6.10.** Show that there is a regular language B such that $PSPACE = Leaf^P(B)$.
Hint: Use the characterization $PSPACE = ATIME(n^{O(1)})$ (see Sect. 2.5.2). Interpret the computation tree of an alternating machine as a circuit and use the ideas of Barrington's Theorem (Theorem 4.46) to evaluate this circuit.

6.11.** Prove: $Leaf^P(U_L\text{-}AC^0) = PSPACE$.

6.12. Define a sensible notion of TC^0-reductions and show that Corollary 6.22 holds for these reductions as well.

Notes on the Exercises

6.1. See [BBS86, Sch86].

6.3. This result is due to [Kan82].

6.7. These results appear in different terminology in [Sim75, Wag86].

6.8. See [HLS⁺93].

6.9. This exercise is due to [CMTV98]. The classes Mod_kP were introduced in [CH89, BGH90]; the class MOD-PH is from [HLS⁺93].

6.10. See [HLS⁺93].

6.11. See [CMTV98].

6.12. This result is due to [All99].

Appendix: Mathematical Preliminaries

A1 Alphabets, Words, Languages

We will work with the alphabet $\{0,1\}$, the *binary alphabet*, and we will treat the two values 0 and 1 both as letters from this alphabet and as representations of the truth values "false" and "true". The empty word will be denoted by ϵ. The length of a word $w \in \{0,1\}^*$ is denoted by $|w|$; $|\epsilon| = 0$.

The cardinality of a set M will be denoted by $|M|$; in contexts where confusion with previous notation might arise we also use $\#M$.

Normally we do not differentiate between k tuples over $\{0,1\}$ and words of length k over $\{0,1\}$. For the set of all such k tuples (or length k words) we write $\{0,1\}^k$.

If A is an arbitrary set, then the characteristic function of A is defined as follows:

$$c_A(x) =_{\text{def}} \begin{cases} 0 & \text{if } x \notin A, \\ 1 & \text{if } x \in A. \end{cases}$$

If we have a function $f \colon \{0,1\}^k \to \{0,1\}$, this is a function on k tuples over $\{0,1\}$ as well as a characteristic function of some subset of $\{0,1\}^k$.

The *marked union* of two sets A and B is $A \uplus B =_{\text{def}} \{\, 0x \mid x \in A \,\} \cup \{\, 1x \mid x \in B \,\}$. The *power-set* of a set A is $\mathcal{P}(A) =_{\text{def}} \{\, B \mid B \subseteq A \,\}$.

We need to be able to work on *pairs of words*, so we define an encoding of pairs of words in words as follows: Let $w = w_1 w_2 \cdots w_n$, $w_i \in \{0,1\}$ for $1 \leq i \leq n$. Set $d(w) =_{\text{def}} w_1 w_1 w_2 w_2 \cdots w_n w_n$. Define the *pairing function* $\langle \cdot, \cdot \rangle \colon \{0,1\}^* \times \{0,1\}^* \to \{0,1\}^*$ as follows: $\langle v, w \rangle =_{\text{def}} v01d(w)$. Furthermore, set $\langle u, v, w \rangle =_{\text{def}} \langle \langle u, v \rangle, w \rangle$ and so on for higher arities.

We will also use the notation $\langle \cdots \rangle$ to denote encodings of sequences with a non-constant number of elements. If $w_0, w_1, w_2 \cdots$ is such a sequence, then we simply let $\langle w_0, w_1, w_2 \cdots \rangle =_{\text{def}} w_0 01d(w_1)01d(w_2) \cdots$.

A2 Binary Encoding

Circuits work with the two *binary values* $\{0,1\}$; Turing machines operate on strings. However we want to discuss number-theoretic functions computed by circuits or machines. Thus we have to encode numbers in such a way that these devices can work on them. For this we use *binary encoding*. Let $a_n, \ldots, a_0 \in \{0, 1\}$, then $a_n \cdots a_0$ is the encoding of the number $A = \sum_{i=0}^{n} a_i \cdot 2^i$. Note that in this way every natural number receives a code, but on the other hand there are infinitely many codes for one number; $a_n \cdots a_0, 0a_n \cdots a_0, 00a_n \cdots a_0, \ldots$ all encode the same number A. We write: $\text{bin}(A) =_{\text{def}} a_n \cdots a_0$ if $A = \sum_{i=0}^{n} a_i \cdot 2^i$ and $a_n = 1$. Additionally, we set $\text{bin}(0) =_{\text{def}} 0$. For $n \in \mathbb{N}$, $n > 0$, define $\ell(n) =_{\text{def}} \lfloor \log_2 n + 1 \rfloor$. Additionally, let $\ell(0) =_{\text{def}} 1$. Note that $\ell(n)$ is the length of the word $\text{bin}(n)$ over the alphabet $\{0, 1\}$. For $k \geq \ell(n)$, we define the *length k binary representation of* n by $\text{bin}_k(n) =_{\text{def}} 0^{k-\ell(n)}\text{bin}(n)$. When k is clear from the context, we will often abbreviate $\text{bin}_k(n)$ by $\text{bin}(n)$.

If it is clear that in a given context a string has to be interpreted as a natural number, we will even omit the "bin" and write simply i instead of $\text{bin}(i)$.

A3 Asymptotic Behavior of Functions

In computational complexity we often do not want to determine exactly the number of steps needed to compute a function or the number of gates in a circuit. Most of the time we are satisfied with a rough estimate. For this the following notations have proved to be very useful. Let $f, g \colon \mathbb{N} \to \mathbb{N}$. Then we write:

1. $f = O(g)$ if there are numbers $c, d \in \mathbb{N}$ such that $f(n) \leq c \cdot g(n)$ for all $n \geq d$.
2. $f = \Omega(g)$ if $g = O(f)$.
3. $f = \Theta(g)$ if $f = O(g)$ and $f = \Omega(g)$.
4. $f = o(g)$ if $\lim_{n \to \infty} f(n)/g(n) = 0$.
5. $f = \omega(g)$ if $g = o(f)$.

Formally one should consider $O(g)$ as a class of functions, and instead of "$f = O(g)$" one should write "$f \in O(g)$." However the notation above has become standard, and we will follow it in this book. (For some of the subtleties of using the "$=$"-sign in this definition, see [Knu97, pp. 107ff.].)

In particular we consider the following classes of functions:

- $(\log n)^{O(1)} = \{\, f \mid \text{for some } k, k' \text{ and all } n \geq k', f(n) \leq (\log n)^k \,\}$.
- $n^{O(1)} = \{\, f \mid \text{for some } k, k' \text{ and all } n \geq k', f(n) \leq n^k \,\}$.
- $2^{(\log n)^{O(1)}} = \{\, f \mid \text{for some } k, k' \text{ and all } n \geq k', f(n) \leq 2^{(\log n)^k} \,\}$.

A4 Turing Machines

The Turing machines we work with are standard multi-tape Turing machines, see, e.g., [HU79, 161–162]. We assume our machines to have a read-only input tape whose head never moves to the left. For computation of a function, our machines will have an additional write-only output tape whose head again never moves to the left. Space on input and output tapes is not considered when discussing space-bounds. Work tapes are two-way infinite, input and output tapes are one-way infinite. A k tape machine is a machine with k work tapes plus input (and maybe output) tape. Where we discuss sub-linear time-bounds, we use *random access* to the input. This means that the Turing machine has an *index tape* on which it writes a number in binary. When the machine enters the special *input query state*, it reads the input symbol at the position given by the index tape and switches to the pre-specified state corresponding to that symbol. If the number on the index tape is larger than the input length, a particular blank symbol is returned. The index tape is not erased by this operation. (This model and variants are discussed in [RV97], see also [BDG90, 77].)

Following the standard literature [HU79] we denote time- and space-bounded classes on deterministic and nondeterministic multi-tape Turing machines by DTIME(f), DSPACE(f), NTIME(f), and NSPACE(f) (if the bound on time or space is given by $f\colon \mathbb{N} \to \mathbb{N}$). We assume that for these resource-bounded classes all machines involved always halt on all paths for all of their inputs. We use the following abbreviations:

$$L =_{\text{def}} \text{DSPACE}(\log n)$$
$$NL =_{\text{def}} \text{NSPACE}(\log n)$$
$$P =_{\text{def}} \text{DTIME}(n^{O(1)})$$
$$NP =_{\text{def}} \text{NTIME}(n^{O(1)})$$
$$PSPACE =_{\text{def}} \text{DSPACE}(n^{O(1)})$$
$$EXPTIME =_{\text{def}} \text{DTIME}(2^{n^{O(1)}})$$
$$NEXPTIME =_{\text{def}} \text{NTIME}(2^{n^{O(1)}})$$
$$EXPSPACE =_{\text{def}} \text{DSPACE}(2^{n^{O(1)}})$$

For a complexity class $\mathcal{K}$, we denote by $\text{co}\mathcal{K}$ the class of all problems whose complements are in $\mathcal{K}$, i.e., $\text{co}\mathcal{K} = \{ A \subseteq \{0,1\}^* \mid \{0,1\}^* \setminus A \in \mathcal{K} \}$. In this sense, we write, e.g., coNL, coNP, and so on.

While the above classes form the obvious inclusion chain $L \subseteq NL \subseteq P \subseteq NP \subseteq PSPACE \subseteq EXPTIME \subseteq NEXPTIME \subseteq EXPSPACE$, most of the converse inclusions are open. However, as consequences of the space hierarchy and time hierarchy theorems [HS65, HLS65] (see also [HU79, Sect. 12.3]), some separations are known: $NL \neq PSPACE \neq EXPSPACE$, $P \neq EXPTIME$, and $NP \neq NEXPTIME$.

For deterministic Turing machines M, it is clear how to define the function computed by M—the function value is just the value found on the output tape at the end of the computation. We use the notations FDTIME(f) and FDSPACE(f) to denote the corresponding complexity classes; i.e., prefixing the name of a decision class yields the corresponding function class. In this vein we also use the notations FL and FP respectively for the class of all functions computable in logarithmic space and polynomial time.

Our classes are defined via *multi-tape* Turing machines, but it is well known that the space classes do not change if we choose single-tape machines instead. Regarding time, a t time-bounded machine with k tapes can be simulated by a single-tape machine running in time t^2 [HU79, Chap. 12.2].

A function $f: \mathbb{N} \rightarrow \mathbb{N}$ is *space constructible* if there is a deterministic Turing machine M which on inputs of length n uses exactly $f(n)$ tape cells. A function $f: \mathbb{N} \rightarrow \mathbb{N}$ is *time constructible* if there is a deterministic Turing machine M which on inputs of length n runs for exactly $f(n)$ time steps.

A5 Logic

We use the following Boolean connectives $\wedge, \vee, \oplus, \equiv, \neg$ from propositional logic (which we identify with the corresponding Boolean functions):

x	y	$\neg x$	$x \wedge y$	$x \vee y$	$x \oplus y$	$x \equiv y$
0	0	1	0	0	0	1
0	1	1	0	1	1	0
1	0	0	0	1	1	0
1	1	0	1	1	0	1

A *literal* is either a propositional variable or the negation of a propositional variable. A clause is a disjunction of literals. A formula in *conjunctive normal form* is a conjunction of clauses, i.e., a formula of the form

$$\bigwedge_{i=1}^{k} \bigvee_{j=1}^{m_i} X_{i,j},$$

where the $X_{i,j}$ are literals. A formula in *disjunctive normal form* is a disjunction of conjuncts, i.e., a formula of the form

$$\bigvee_{i=1}^{k} \bigwedge_{j=1}^{m_i} X_{i,j},$$

where the $X_{i,j}$ are literals.

A Horn clause is a clause with at most one unnegated literal. A Horn formula is a formula in conjunctive normal form, all of whose clauses are Horn clauses.

If ϕ is any Boolean expression then

$$[\phi] =_{\text{def}} \begin{cases} 0, & \text{if } \phi \text{ is false,} \\ 1, & \text{if } \phi \text{ is true.} \end{cases}$$

For $x \in \{0,1\}$, $\bar{x}$ is the complement of x, i.e., $\bar{x} = 1 \iff x = 0$.

A propositional formula with constants is either one of the truth values 0 or 1, a literal, or one of the Boolean connectives above applied to shorter propositional formulas with constants. SAT is the set of all (encodings of) satisfiable propositional formulas with constants. It is well known that SAT is NP-complete [Coo71].

For general background on logic, we recommend [Sho67, EFT94].

A6 Graphs

A *directed graph* is a pair $G = (V, E)$ where V is the set of nodes (or vertices) and $E \subseteq V \times V$ the set of edges. G is acyclic if G contains no *directed cycles*, i.e., no sequence $v_0, \ldots, v_k \in V$ such that $(v_i, v_{i+1}) \in E$ (for $0 \le i < k$) and $(v_k, v_0) \in E$.

A sequence $v_0, \ldots, v_k \in V$ such that $(v_i, v_{i+1}) \in E$ (for $0 \le i < k$) is called a (directed) *path* in G from v_0 to v_k. The *length* of this path is k.

If $r \in V$ is such that every node $v \in V$ is reachable from r, i.e., there is a directed path that starts in r and ends in v, then r is called a *root* of G. G is a *directed tree* if it has a root and every $v \in V$ is reachable from r by a unique directed path. The depth of a tree G is the length of a longest directed path in G. We say that G is *finite* if V is finite. For $v \in V$, we say that $\left| \{ u \mid (u, v) \in E \} \right|$ is the *in-degree* of v, and that $\left| \{ u \mid (v, u) \in E \} \right|$ is the *out-degree* of v. For graphs $G = (V, E)$ and $G' = (V', E')$, $G \subseteq G'$ denotes that $V \subseteq V'$ and $E \subseteq E'$.

An undirected graph is a directed graph $G = (V, E)$ with a symmetric edge relation E, i.e., for all nodes u, v, $(u, v) \in E \iff (v, u) \in E$.

A *subgraph* G' of $G = (V, E)$ is a graph $G' = (V', E')$ where $V' \subseteq V$ and $E' \subseteq E$. A subtree T' of a tree T is a subgraph of T which is a tree.

For background on graph theory, consult [Eve79, Tar83].

A7 Numbers and Functions

The set of natural numbers is $\mathbb{N} = \{0, 1, 2, \ldots\}$. A *sequence* a is a function from $\mathbb{N}$ into an arbitrary set M, i.e., $a\colon \mathbb{N} \to M$. Instead of $a(k)$ we often write a_k, and we use the notation $a = (a_k)_{k \in \mathbb{N}}$.

We use the notation $\log n$ for the logarithm to the base 2. However, since most of the time we are only interested in asymptotic behavior, the actual base of the logarithm is not important, i.e., for every $b > 1$ we have $\log_b n = \Theta(\log n)$.

We use $f^{(k)}(n)$ to denote the k-fold application of the function f to the input n, i.e., $f^{(k)}(n) =_{\text{def}} \underbrace{f(f(\cdots f(n) \cdots))}_{k \text{ times}}$. For $n \in \mathbb{N}$ define $\log^* n =_{\text{def}}$ $\min\{ k \mid \log^{(k)} n \le 1 \}$.

Let $\mathbb{Z}$ denote the set of the integers. We want to generalize *binomial coefficients* $\binom{n}{k}$ to the case where $k \in \mathbb{N}$, $n \in \mathbb{Z}$. This is achieved by defining

$$\binom{n}{k} =_{\text{def}} \frac{n \cdot (n-1) \cdots (n-k+1)}{k!},$$

if $k > 0$, and $\binom{n}{k} =_{\text{def}} 1$ if $k = 0$. Let now $m, n, k \in \mathbb{N}$. Then the following holds (see [FFK94]):

$$\binom{m-n}{k} = \sum_{i=0}^{k} (-1)^i \binom{m+1}{k-i} \binom{n+i}{i}.$$

We use mod in two senses, as an infix operator ($a \bmod b$ is the remainder when a is divided by b) and as a notation for the number theoretical concept of *congruence*: $a \equiv b \pmod{m}$ denotes that $a - b$ is an integral multiple of m, in other words: $a \bmod m = b \bmod m$ (see [Knu97, 40–41]).

We need the following results from number theory. Proofs can be found, e. g., in [HW54].

Prime Number Theorem. Let $\Pi(n)$ denote the number of prime numbers not greater than n. Then

$$\Pi(n) = \Theta(\frac{n}{\log n}).$$

Let p_k for $k \in \mathbb{N}$ denote the k-th prime number.

Corollary. $p_n = O(n \log n)$.

Fermat's Theorem. Let p be prime. Then $a^{p-1} \equiv 1 \pmod{p}$ if a is not a multiple of p.

Chinese Remainder Theorem. Let $k \in \mathbb{N}$, $P = p_1 \cdot p_2 \cdots p_k$. Let $a_1, \ldots, a_k$ be given. Then there is a unique number $a \in \{0, \ldots, P-1\}$ such that $a \equiv a_i \pmod{p_i}$ for $1 \leq i \leq k$. More specifically, we have

$$a = \left[\sum_{i=1}^{k} (a_i \bmod p_i) \cdot r_i \cdot s_i \right] \bmod P,$$

where $r_i = \frac{P}{p_i}$ and $s_i = \left(\frac{P}{p_i} \right)^{-1} \bmod p_i$ (i.e., s_i is the multiplicative inverse of r_i modulo p_i).

A8 Algebraic Structures

A *semi-ring* $\mathcal{R} = (R, +, \times, 0, 1)$ is given by a set R with two binary operations $+$ and $\times$ over R and two elements $0, 1 \in R$ such that $(R, +, 0)$ is a commutative monoid, $(R, \times, 1)$ is a monoid, the distributive laws $a \cdot (b+c) = a \times b + a \times c$ and $(a + b) \times c = a \times c + b \times c$ hold for all $a, b, c \in R$, and $0 \times a = a \times 0 = 0$ for every a. If $\times$ is commutative, then $\mathcal{R}$ is called a commutative semi-ring. If $(R, +, 0)$ is a group, then $\mathcal{R}$ is called a *ring*. If additionally, $(R \setminus \{0\}, \cdot, 1)$ is an Abelian group, then $\mathcal{R}$ is called a *field*.

Important examples in this book are the Boolean semi-ring given by $(\{0,1\}, \vee, \wedge, 0, 1)$, the semi-ring of the natural numbers $(\mathbb{N}, +, \cdot, 0, 1)$ (both are in fact commutative semi-rings), and the ring of the integers $(\mathbb{Z}, +, \cdot, 0, 1)$. We will use $\mathbb{N}$ throughout this book as a shorthand for $(\mathbb{N}, +, \cdot, 0, 1)$, and analogously $\mathbb{Z}$.

For a prime number p and $n \geq 0$, let $\mathrm{GF}(p^n)$ be the (uniquely determined) field with p^n elements, see [Jac85, Chap. 4]. It is known that $\mathrm{GF}(p)$ is a subfield of $\mathrm{GF}(p^n)$ for every $n \geq 1$, and that $\mathrm{GF}(p^n)$ is a subfield of $\mathrm{GF}(p^{in})$ for every $i \geq 1$.

For the special case $n = 1$, $\mathrm{GF}(p)$ is isomorphic to the field $\mathbb{Z}_p$, defined to consist of the set $\{0, 1, \ldots, p-1\}$ with operations addition and multiplication in $\mathbb{Z}$ modulo p. The corresponding multiplicative group is cyclic, i.e., there is an element g such that $\{ g^i \mid 0 \leq i < p - 2 \} = \{1, \ldots, p-1\}$. g is called a *generator* of $\mathbb{Z}_p$.

A *polynomial* p in variables $x_1, \ldots, x_n$ over some ring $\mathcal{R} = (R, +, \times, 0, 1)$ is a function given by $p(x_1, \ldots, x_n) = \sum_{i=0}^{k} c_i \prod_{j=1}^{n} x_j^{a_{i,j}}$ for some $c_i \in R$, $a_{i,j} \in \mathbb{N}$ for $0 \leq i \leq k, 1 \leq j \leq n$. The set of all such polynomials is denoted by $\mathcal{R}[x_1, \ldots, x_n]$. The terms $\prod_{j=1}^{n} x_j^{a_{i,j}}$ are called *monomials* of p. If all the $a_{i,j}$ are at most 1, then this is called a *multi-linear* monomial. If all monomials of p are multi-linear, then p is called a *multi-linear polynomial*. The *degree* of p is $\max_{0 \leq i \leq k} \sum_{j=1}^{n} a_{i,j}$.

The *symmetric group* of all permutations on n elements is denoted by S_n.

Let G be a group, and $g, h \in G$. The *commutator* of g and h is defined to be $(g, h) =_{\mathrm{def}} ghg^{-1}h^{-1}$. The *commutator subgroup* (or, derived subgroup) G' of G is the subgroup generated by all elements of the form (g, h) where $g, h \in G$. The second derived subgroup is $G'' =_{\mathrm{def}} (G')'$, and so on, i.e., $G^{(k)} =_{\mathrm{def}} (G^{(k-1)})'$. A finite group G is *solvable* if there exists a $k \geq 1$ such that $G^{(k)}$ is the trivial group consisting of only the identity element, see [Jac85, pp. 244ff.]. Thus a finite group G is non-solvable if and only if it has a non-trivial subgroup that coincides with its commutator subgroup.

A9 Linear Algebra

Let $\mathcal{R} = (R, +, \times, 0, 1)$ be a ring. An $n \times n$ matrix of the form

$$V_{\mathcal{R}}(x_1, \ldots, x_n) =_{\text{def}} \begin{pmatrix} 1 & 1 & 1 & \cdots & 1 \\ x_1 & x_2 & x_3 & \cdots & x_n \\ x_1^2 & x_2^2 & x_3^2 & \cdots & x_n^2 \\ x_1^3 & x_2^3 & x_3^3 & \cdots & x_n^3 \\ \vdots & \vdots & \vdots & \ddots & \vdots \\ x_1^{n-1} & x_2^{n-1} & x_3^{n-1} & \cdots & x_n^{n-1} \end{pmatrix}$$

where all entries are drawn from $\mathcal{R}$, is called *Vandermonde matrix* given by $x_1, \ldots, x_n$ [Lan93, 257]. Its determinant is

$$\det V_{\mathcal{R}}(x_1, \ldots, x_n) = \prod_{j > i} (x_j - x_i).$$

Bibliography

[Adl78]   L. Adleman. Two theorems on random polynomial time. In *Proceedings 19th Symposium on Foundations of Computer Science*, p. 75–83. IEEE Computer Society Press, 1978.

[AAD97]   M. Agrawal, E. Allender, and S. Datta. On TC^0, AC^0, and arithmetic circuits. In *Proceedings 12th Computational Complexity*, p. 134–148. IEEE Computer Society Press, 1997.

[AHU74]   A. Aho, J. E. Hopcroft, and J. D. Ullman. *The Design and Analysis of Computer Algorithms*. Series in Computer Science and Information Processing. Addison-Wesley, Reading, MA, 1974.

[Ajt83]   M. Ajtai. Σ_1^1 formulae on finite structures. *Annals of Pure and Applied Logic*, 24:1–48, 1983.

[All89]   E. Allender. P-uniform circuit complexity. *Journal of the ACM*, 36:912–928, 1989.

[All95]   E. Allender. Combinatorial methods in complexity theory. Lecture Notes, Rutgers University, New Brunswick, 1995.

[All96]   E. Allender. Circuit complexity before the dawn of the new millennium. In *Proceedings 16th Foundations of Software Technology and Theoretical Computer Science*, Lecture Notes in Computer Science 1180, p. 1–18, Springer-Verlag, Berlin, 1996.

[All98]   E. Allender. Making computation count: arithmetic circuits in the nineties. *SIGACT News*, 28(4):2–15, 1998.

[All99]   E. Allender. The permanent requires large uniform circuits. *Chicago Journal of Theoretical Computer Science*, 1999. To appear. A preliminary version appeared as: A note on uniform circuit lower bounds for the counting hierarchy, in *Proceedings 2nd Computing and Combinatorics Conference*, Lecture Notes in Computer Science 1090, p. 127–135, Springer-Verlag, Berlin, 1996.

[AG91]   E. Allender and V. Gore. Rudimentary reductions revisited. *Information Processing Letters*, 40:89–95, 1991.

[AG94]   E. Allender and V. Gore. A uniform circuit lower bound for the permanent. *SIAM Journal on Computing*, 23:1026–1049, 1994.

[AH94]   E. Allender and U. Hertrampf. Depth reduction for circuits of
            unbounded fan-in. *Information and Computation*, 112:217–238,
            1994.

[AJMV98]   E. Allender, J. Jiao, M. Mahajan, and V. Vinay. Non-commuta-
            tive arithmetic circuits: depth reduction and size lower bounds.
            Theoretical Computer Science, 209:47–86, 1998.

[AO96]   E. Allender and M. Ogihara. Relationships among PL, #L, and
            the determinant. *RAIRO – Theoretical Informatics and Appli-
            cations*, 30:1–21, 1996.

[AR97]   E. Allender and K. Reinhardt. Making nondeterminism un-
            ambiguous. In *Proceedings 37th Symposium on Foundations of
            Computer Science*, p. 244–253. IEEE Computer Society Press,
            1997.

[AW90]   E. Allender and K. W. Wagner. Counting hierarchies: polynomial
            time and constant depth circuits. *Bulletin of the EATCS*, 40:182–
            194, 1990.

[AJ93]   C. Álvarez and B. Jenner. A very hard log space counting class.
            Theoretical Computer Science, 107:3–30, 1993.

[AML98]   A. Ambainis, D. A. Mix Barrington, and H. LêThanh. On count-
            ing AC^0 circuits with negative constants. In *Proceedings 23rd
            Mathematical Foundations of Computer Science*, Lecture Notes
            in Computer Science 1450, p. 409–417, Springer-Verlag, Berlin,
            1998.

[BBS86]   J. L. Balcázar, R. V. Book, and U. Schöning. The polynomial-
            time hierarchy and sparse oracles. *Journal of the ACM*, 33:603–
            617, 1986.

[BDG95]   J. L. Balcázar, J. Díaz, and J. Gabarró. *Structural Complexity
            I*. Texts in Theoretical Computer Science – An EATCS series.
            Springer-Verlag, Berlin, 2nd edition, 1995.

[BDG90]   J. L. Balcázar, J. Díaz, and J. Gabarró. *Structural Complex-
            ity II*. EATCS Monographs on Theoretical Computer Science.
            Springer-Verlag, Berlin, 1990.

[Bar86]   D. A. Barrington. A note on a theorem of Razborov. Technical
            report, University of Massachusetts, 1986.

[Bar89]   D. A. Mix Barrington. Bounded-width polynomial size branching
            programs recognize exactly those languages in NC^1. *Journal of
            Computer and System Sciences*, 38:150–164, 1989.

[BI94]   D. A. Mix Barrington and N. Immerman. Time, hardware, and
            uniformity. In *Proceedings 9th Structure in Complexity Theory*,
            p. 176–185. IEEE Computer Society Press, 1994.

[BIS90]   D. A. Mix Barrington, N. Immerman, and H. Straubing. On uni-
            formity within NC^1. *Journal of Computer and System Sciences*,
            41:274–306, 1990.

[BT88]   D. A. Mix Barrington and D. Thérien. Finite monoids and the fine structure of NC^1. *Journal of the ACM*, 35:941–952, 1988.

[BCH86] P. W. Beame, S. A. Cook, and H. J. Hoover. Log depth circuits for division and related problems. *SIAM Journal on Computing*, 15:994–1003, 1986.

[BLM93] F. Bédard, F. Lemieux, and P. McKenzie. Extensions to Barrington's M-program model. *Theoretical Computer Science*, 107:31–61, 1993.

[Bei93] R. Beigel. The polynomial method in circuit complexity. In *Proceedings 8th Structure in Complexity Theory*, p. 82–95. IEEE Computer Society Press, 1993. Revised version, 1995.

[BGH90] R. Beigel, J. Gill, and U. Hertrampf. Counting classes: thresholds, parity, mods, and fewness. In *Proceedings 7th Symposium on Theoretical Aspects of Computer Science*, Lecture Notes in Computer Science 415, p. 49–57, Springer-Verlag, Berlin, 1990.

[BRS91] R. Beigel, N. Reingold, and D. Spielman. The perceptron strikes back. In *Proceedings 6th Structure in Complexity Theory*, p. 286–291. IEEE Computer Society Press, 1991.

[BT94]   R. Beigel and J. Tarui. On ACC. *Computational Complexity*, 4:350–366, 1994. Special issue on circuit complexity.

[BG81]   C. Bennett and J. Gill. Relative to a random oracle $P^A \neq NP^A \neq coNP^A$ with probability 1. *SIAM Journal on Computing*, 10:96–113, 1981.

[BOC92] M. Ben-Or and R. Cleve. Computing algebraic formulas using a constant number of registers. *SIAM Journal on Computing*, 21:54–58, 1992.

[Ber84] S. J. Berkowitz. On computing the determinant in small parallel time using a small number of processors. *Information Processing Letters*, 18:147–150, 1984.

[BH77]   L. Berman and J. Hartmanis. On isomorphism and density of NP and other complete sets. *SIAM Journal on Computing*, 6:305–322, 1977.

[Blu84] N. Blum. A Boolean function requiring $3n$ network size. *Theoretical Computer Science*, 36:59–70, 1984.

[BS90]   R. B. Boppana and M. Sipser. The complexity of finite functions. In J. van Leeuwen (ed.), *Handbook of Theoretical Computer Science*, vol. A, p. 757–804. Elsevier, Amsterdam, 1990.

[Bor77] A. B. Borodin. On relating time and space to size and depth. *SIAM Journal on Computing*, 6:733–744, 1977.

[Bor82] A. B. Borodin. Structured versus general models in computational complexity. *L'Enseignement Mathématique*, 30:47–65, 1982. Special Issue *Logic and Algorithms*, Symposium in honour of Ernst Specker.

[BCD+89] A. B. Borodin, S. A. Cook, P. W. Dymond, W. L. Ruzzo, and M. Tompa. Two applications of inductive counting for complementation problems. *SIAM Journal on Computing*, 18:559–578, 1989. See Erratum in *SIAM Journal on Computing* 18:1283.

[BC94]   D. P. Bovet and P. Crescenzi. *Introduction to the Theory of Complexity*. International Series in Computer Science. Prentice Hall, London, 1994.

[BCS92]   D. P. Bovet, P. Crescenzi, and R. Silvestri. A uniform approach to define complexity classes. *Theoretical Computer Science*, 104:263–283, 1992.

[Büc62]   J. R. Büchi. On a decision method in restricted second-order arithmetic. In *Proceedings Logic, Methodology and Philosophy of Sciences 1960*, Stanford University Press, Stanford, CA, 1962.

[Bus87]   S. R. Buss. The Boolean formula value problem is in ALOGTIME. In *Proceedings 19th Symposium on Theory of Computing*, p. 123–131. ACM Press, 1987.

[Cai89]   J.-Y. Cai. With probability one, a random oracle separates PSPACE from the polynomial-time hierarchy. *Journal of Computer and System Sciences*, 38:68–85, 1989.

[CH89]   J.-Y. Cai and L. Hemachandra. On the power of parity polynomial time. In *Proceedings 6th Symposium on Theoretical Aspects of Computer Science*, Lecture Notes in Computer Science 349, p. 229–239, Springer-Verlag, Berlin, 1989.

[CC95]   L. Cai and J. Chen. On input read-modes of alternating Turing machines. *Theoretical Computer Science*, 148:33–55, 1995.

[CCH97]   L. Cai, J. Chen, and J. Håstad. Circuit bottom fan-in and computational power. In *Proceedings 12th Computational Complexity*, p. 158–164. IEEE Computer Society Press, 1997.

[Cau96]   H. Caussinus. *Contributions à l'étude du non-déterminisme restreint*. PhD thesis, Dép. d'informatique et recherche opérationnelle, Université de Montréal, 1996.

[CMTV98] H. Caussinus, P. McKenzie, D. Thérien, and H. Vollmer. Nondeterministic NC^1 computation. *Journal of Computer and System Sciences*, 57:200–212, 1998.

[CKS81]   A. K. Chandra, D. Kozen, and L. J. Stockmeyer. Alternation. *Journal of the ACM*, 28:114–133, 1981.

[CSV84]   A. K. Chandra, L. Stockmeyer, and U. Vishkin. Constant depth reducibility. *SIAM Journal on Computing*, 13:423–439, 1984.

[CK99]   P. Clote and E. Kranakis. *Boolean Functions and Computation Models*. Springer-Verlag, Berlin, 1999. To appear.

[Col88]   R. Cole. Parallel merge sort. *SIAM Journal on Computing*, 17:770–785, 1988.

[Coo71]   S. A. Cook. The complexity of theorem proving procedures. In *Proceedings 3rd Symposium on Theory of Computing*, p. 151–158. ACM Press, 1971.

[Coo79]   S. A. Cook. Deterministic CFL's are accepted simultaneously in polynomial time and log squared space. In *Proceedings 11th Symposium on Theory of Computing*, p. 338–345. ACM Press, 1979.

[Coo85]   S. A. Cook. A taxonomy of problems with fast parallel algorithms. *Information and Control*, 64:2–22, 1985.

[Dev93]   K. Devlin. *The Joy of Sets, Fundamentals of Contemporary Set Theory*. Undergraduate Texts in Mathematics. Springer-Verlag, New York, 2nd edition, 1993.

[DLR79]   D. P. Dobkin, R. J. Lipton, and S. P. Reiss. Linear programming is log-space hard for P. *Information Processing Letters*, 8:96–97, 1979.

[Dun88]   P. E. Dunne. *The Complexity of Boolean Networks*, vol. 29 of *A. P. I. C. Studies in Data Processing*. Academic Press, London, 1988.

[EF95]   H.-D. Ebbinghaus and J. Flum. *Finite Model Theory*. Perspectives in Mathematical Logic. Springer-Verlag, Berlin, 1995.

[EFT94]   H.-D. Ebbinghaus, J. Flum, and W. Thomas. *Mathematical Logic*. Springer-Verlag, Berlin, 2nd edition, 1994.

[Eil76]   S. Eilenberg. *Automata, Languages, and Machines*, vol. B. Academic Press, New York, 1976.

[Eve79]   S. Even. *Graph Algorithms*. Computer Science Press, Rockville, MD, 1979.

[Fag74]   R. Fagin. Generalized first-order spectra and polynomial time recognizable sets. In R. Karp (ed.), *Complexity of Computations*, vol. 7 of *SIAM-AMS Proceedings*, p. 43–73. American Mathematical Society, Providence, RI, 1974.

[FFK94]   S. Fenner, L. Fortnow, and S. Kurtz. Gap-definable counting classes. *Journal of Computer and System Sciences*, 48:116–148, 1994.

[Fic93]   F. Fich. The complexity of computation on the parallel random access machine. In J. H. Reif (ed.), *Synthesis of Parallel Algorithms*, p. 843–899. Morgan Kaufmann, San Mateo, CA, 1993.

[FRW88]   F. Fich, R. Ragde, and A. Wigderson. Relations between concurrent-write models of parallel computation. *SIAM Journal on Computing*, 17:606–627, 1988.

[Fis74]   M. J. Fischer. Lectures on network complexity. Universität Frankfurt/Main, 1974.

[FW78]   S. Fortune and J. Wyllie. Parallelism in random access machines. In *Proceedings 10th Symposium on Theory of Computing*, p. 114–118. ACM Press, 1978.

246   Bibliography

[FSS84]   M. Furst, J. B. Saxe, and M. Sipser. Parity, circuits, and
           the polynomial-time hierarchy. *Mathematical Systems Theory*,
           17:13–27, 1984.
[GJ79]   M. R. Garey and D. S. Johnson. *Computers and Intractability, A
           Guide to the Theory of NP-Completeness*. Freeman, New York,
           1979.
[Gil77]   J. Gill. Computational complexity of probabilistic Turing ma-
           chines. *SIAM Journal on Computing*, 6:675–695, 1977.
[Gol82]   L. M. Goldschlager. A universal interconnection pattern for par-
           allel computers. *Journal of the ACM*, 29:1073–1086, 1982.
[Got97]   G. Gottlob. Relativized logspace and generalized quantifiers over
           ordered finite structures. *Journal of Symbolic Logic*, 62:545–574,
           1997.
[GHR95]   R. Greenlaw, H. J. Hoover, and W. L. Ruzzo. *Limits to Parallel
           Computation: P-Completeness Theory*. Oxford University Press,
           New York, 1995.
[GL84]   Y. Gurevich and H. Lewis. A logic for constant-depth circuits.
           Information and Control, 61:65–74, 1984.
[HP93]   P. Hájek and P. Pudlák. *Metamathematics of First-Order Arith-
           metic*. Perspectives in Mathematical Logic. Springer-Verlag,
           Berlin, 1993.
[HMP+87]   A. Hajnal, W. Maass, P. Pudlák, M. Szegedy, and G. Turán.
           Threshold circuits of bounded depth. In *Proceedings 18th Sym-
           posium on Foundations of Computer Science*, p. 99–110. IEEE
           Computer Society Press, 1987.
[HW54]   G. H. Hardy and E. M. Wright. *An Introduction to the Theory
           of Numbers*. Oxford University Press, Oxford, 3rd edition, 1954.
[HLS65]   J. Hartmanis, P. M. Lewis II, and R. E. Stearns. Hierarchies of
           memory limited computations. In *Proceedings 6th Symposium on
           Switching Circuit Theory and Logical Design*, p. 179–190. IEEE
           Computer Society Press, 1965.
[HS65]   J. Hartmanis and R. E. Stearns. On the computational complex-
           ity of algorithms. *Transactions of the American Mathematical
           Society*, 117:285–306, 1965.
[Hås86]   J. Håstad. Almost optimal lower bounds for small depth circuits.
           In *Procedings 18th Symposium on Theory of Computing*, p. 6–20.
           IEEE Computer Society Press, 1986.
[Hås88]   J. Håstad. *Computational Limitations of Small Depth Circuits*.
           MIT Press, Cambridge, MA, 1988.
[HS66]   F. C. Hennie and R. E. Stearns. Two-tape simulation of multi-
           tape Turing machines. *Journal of the ACM*, 13:533–546, 1966.
[HLS+93]   U. Hertrampf, C. Lautemann, T. Schwentick, H. Vollmer, and
           K. W. Wagner. On the power of polynomial time bit-reductions.

In *Proceedings 8th Structure in Complexity Theory*, p. 200–207. IEEE Computer Society Press, 1993.

[HU79]   J. E. Hopcroft and J. D. Ullman. *Introduction to Automata Theory, Languages, and Computation*. Series in Computer Science. Addison-Wesley, Reading, MA, 1979.

[Hro97]   J. Hromkovič. *Communication Complexity and Parallel Computing*. Texts in Theoretical Computer Science – An EATCS Series. Springer-Verlag, Berlin, 1997.

[Imm87]   N. Immerman. Languages that capture complexity classes. *SIAM Journal on Computing*, 16:760–778, 1987.

[Imm88]   N. Immerman. Nondeterministic space is closed under complementation. *SIAM Journal on Computing*, 17:935–938, 1988.

[Imm89a]   N. Immerman. Descriptive and computational complexity. In J. Hartmanis (ed.), *Computational Complexity Theory*, vol. 38 of *Proceedings of Symposia in Applied Mathematics*, p. 75–91. American Mathematical Society, Providence, RI, 1989.

[Imm89b]   N. Immerman. Expressibility and parallel complexity. *SIAM Journal on Computing*, 18:625–638, 1989.

[Imm99]   N. Immerman. *Descriptive Complexity*. Graduate Texts in Computer Science. Springer-Verlag, New York, 1999.

[Jac85]   N. Jacobson. *Basic Algebra*, vol. I. Freeman & Co., New York, 1985.

[JMT96]   B. Jenner, P. McKenzie, and D. Thérien. Logspace and logtime leaf languages. *Information and Computation*, 129:21–33, 1996.

[Jia92]   J. Jiao. Some questions concerning circuit counting classes and other low-level complexity classes. Master's essay, Department of Computer Science, Rutgers University, New Brunswick, 1992.

[Joh90]   D. S. Johnson. A catalog of complexity classes. In J. van Leeuwen (ed.), *Handbook of Theoretical Computer Science*, vol. A, p. 67–161. Elsevier, Amsterdam, 1990.

[Jon75]   N. D. Jones. Space-bounded reducibility among combinatorial problems. *Journal of Computer and System Sciences*, 15:68–85, 1975.

[JL76]   N. D. Jones and W. T. Laaser. Complete problems for deterministic polynomial time. *Theoretical Computer Science*, 3:105–117, 1976.

[Jun85]   H. Jung. Depth efficient transformations of arithmetic circuits into Boolean circuits. In *Proceedings Fundamentals of Computation Theory*, Lecture Notes in Computer Science 199, p. 167–173, Springer-Verlag, Berlin, 1985.

[Kan82]   R. Kannan. Circuit-size lower bounds and non-reducibility to sparse sets. *Information and Control*, 55:40–56, 1982.

[KL82]   R. Karp and R. J. Lipton. Turing machines that take advice. *L'enseignement mathématique*, 28:191–209, 1982.

[KR90]   R. Karp and V. Ramachandran. Parallel algorithms for shared-memory machines. In J. van Leeuwen (éd.), *Handbook of Theoretical Computer Science*, vol. A, p. 869–941. Elsevier, Amsterdam, 1990.

[KW85]   R. Karp and A. Wigderson. A fast parallel algorithm for the maximal independent set problem. *Journal of the ACM*, 32:762–773, 1985.

[Kha79]   L. G. Khachian. A polynomial time algorithm for linear programming. *Doklady Akademii Nauk SSSR*, 244:1093–1096, 1979. In Russian.

[Knu97]   D. E. Knuth. *The Art of Computer Programming Vol. I: Fundamental Algorithms*. Addison-Wesley, Reading, MA, 3rd edition, 1997.

[Kuč82]   L. Kučera. Parallel computation and conflicts in memory access. *Information Processing Letters*, 14:93–96, 1982.

[Lad75]   R. E. Ladner. The circuit value problem is log space complete for P. *SIGACT News*, 7(1):12–20, 1975.

[LF80]   R. E. Ladner and M. J. Fischer. Parallel prefix computation. *Journal of the ACM*, 27:831–838, 1980.

[Lan93]   S. Lang. *Algebra*. Addison-Wesley, Reading, MA, 3rd edition, 1993.

[LMSV99]   C. Lautemann, P. McKenzie, T. Schwentick, and H. Vollmer. The descriptive complexity approach to LOGCFL. In *Proceedings 16th Symposium on Theoretical Aspects of Computer Science*, Lecture Notes in Computer Science 1563, p. 444–454, Springer-Verlag, Berlin, 1999.

[Lee59]   C. Y. Lee. Representation of switching functions by binary decision programs. *Bell Systems Technical Journal*, 38:985–999, 1959.

[LW90]   T. Lengauer and K. W. Wagner. The binary network flow problem is logspace complete for P. *Theoretical Computer Science*, 75:357–363, 1990.

[Lev80]   G. Lev. *Size Bounds and Parallel Algorithms for Networks*. PhD thesis, Department of Computer Science, University of Edinburgh, 1980.

[Lin92]   S. Lindell. A purely logical characterization of circuit uniformity. In *Proceedings 7th Structure in Complexity Theory*, p. 185–192. IEEE Computer Society Press, 1992.

[Lin66]   P. Lindström. First order predicate logic with generalized quantifiers. *Theoria*, 32:186–195, 1966.

[LZ77]   R. J. Lipton and Y. Zalcstein. Word problems solvable in logspace. *Journal of the ACM*, 24:522–526, 1977.

[Lup58]   O. B. Lupanov. A method of circuit synthesis. *Izvestia VUZ Radiofizika*, 1:120–140, 1958.

[Mac98]   I. Macarie. Space-efficient deterministic simulation of probabilistic automata. *SIAM Journal on Computing*, 27:448–465, 1998.

[May90]   E. W. Mayr. Basic parallel graph algorithms. *Computing Supplementum*, 7:69–91, 1990.

[MS92]   E. W. Mayr and A. Subramanian. The complexity of circuit value and network stability. *Journal of Computer and System Sciences*, 44:302–323, 1992.

[MT94]   P. McKenzie and D. Thérien (eds.). *Special Issue on Circuit Complexity*, vol. 4(4) of *Computational Complexity*. Birkhäuser, Basel, 1994.

[MVW99]   P. McKenzie, H. Vollmer, and K. W. Wagner. Arithmetic circuits and polynomial replacement systems. Technical report, Fachbereich Mathematik und Informatik, Universität Würzburg, 1999.

[MP71]   R. McNaughton and S. A. Papert. *Counter-Free Automata*. MIT Press, Cambridge, MA, 1971.

[MC81]   C. Mead and L. Conway. *Introduction to VLSI Systems*. Addison-Wesley, Reading, MA, 1981.

[Mei86]   C. Meinel. *Modified Branching Programs and Their Computational Power*. Lecture Notes in Computer Science 370, Springer-Verlag, Berlin, 1986.

[MS72]   A. R. Meyer and L. J. Stockmeyer. The equivalence problem for regular expressions with squaring requires exponential time. In *Proceedings 13th Symposium on Switching and Automata Theory*, p. 125–129. IEEE Computer Society Press, 1972.

[MP88]   M. L. Minsky and S. A. Papert. *Perceptrons*. MIT Press, Cambridge, MA, 1988. Expanded edition. Originally published in 1969.

[Pap94]   C. H. Papadimitriou. *Computational Complexity*. Addison-Wesley, Reading, MA, 1994.

[Par94]   I. Parberry. *Circuit Complexity and Neural Networks*. Foundations of Computing. MIT Press, Cambridge, MA, 1994.

[PS88]   I. Parberry and G. Schnitger. Parallel computation with threshold functions. *Journal of Computer and System Sciences*, 36:287–302, 1988.

[Pin86]   J. E. Pin. *Varieties of Formal Languages*. Plenum Press, New York, 1986.

[Pip79]   N. J. Pippenger. On simultaneous resource bounds. In *Proceedings 20th Symposium on Foundations of Computer Science*, p. 307–311. IEEE Computer Society Press, 1979.

[PF79]   N. J. Pippenger and M. J. Fischer. Relations among complexity measures. *Journal of the ACM*, 26:361–381, 1979.

[Rab76]   M. O. Rabin. Probabilistic algorithms. In J. Traub (ed.), *Algorithms and Complexity: New Directions and Results*, p. 21–39. Academic Press, London, 1976.

[Raz85]   A. A. Razborov. Lower bounds on the monotone complexity of some boolean functions. *Doklady Akademii Nauk SSSR*, 281:798–801, 1985. In Russian. English translation in *Soviet Mathematics Doklady*, 31:354–357, 1985.

[Raz87]   A. A. Razborov. Lower bounds on the size of bounded depth networks over a complete basis with logical addition. *Matematicheskie Zametki*, 41:598–607, 1987. In Russian. English translation in *Mathematical Notes of the Academy of Sciences of the USSR* 41:333–338, 1987.

[Reg93]   K. Regan. Log-time ATMs and first-order logic. Unpublished manuscript, 1993.

[Reg97]   K. Regan. Polynomials and combinatorial definitions of languages. In L. A. Hemaspaandra and A. L. Selman (eds.), *Complexity Theory Retrospective II*, p. 261–293. Springer-Verlag, New York, 1997.

[RV97]   K. Regan and H. Vollmer. Gap-languages and log-time complexity classes. *Theoretical Computer Science*, 188:101–116, 1997.

[RS42]   J. Riordan and C. Shannon. The number of two-terminal series-parallel networks. *Journal of Mathematics and Physics*, 21:83–93, 1942.

[Ruz80]   W. L. Ruzzo. Tree-size bounded alternation. *Journal of Computer and System Sciences*, 21:218–235, 1980.

[Ruz81]   W. L. Ruzzo. On uniform circuit complexity. *Journal of Computer and System Sciences*, 21:365–383, 1981.

[Sav76]   J. E. Savage. *The Complexity of Computing*. John Wiley & Sons, New York, 1976.

[Sav98]   J. E. Savage. *Models of Computation – Exploring the Power of Computing*. Addison-Wesley, Reading, MA, 1998.

[Sav70]   W. J. Savitch. Relationships between nondeterministic and deterministic tape complexities. *Journal of Computer and Systems Sciences*, 4:177–192, 1970.

[Sch74]   C. P. Schnorr. Zwei lineare untere Schranken für die Komplexität Boolescher Funktionen. *Computing*, 13:155–171, 1974.

[Sch76]   C. P. Schnorr. The network complexity and the Turing machine complexity of finite functions. *Acta Informatica*, 7:95–107, 1976.

[Sch86]   U. Schöning. *Complexity and Structure*. Lecture Notes in Computer Science 211, Springer-Verlag, Berlin, 1986.

[Sch89]   U. Schöning. *Logic for Computer Scientists*, vol. 8 of *Progress in Computer Science and Applied Logic*. Birkhäuser, Boston, MA, 1989.

[SP98]   U. Schöning and R. Pruim. *Gems of Theoretical Computer Science*. Springer-Verlag, Berlin, 1998.

[Ser90]   M. J. Serna. *The parallel approximability of P-complete problems.* PhD thesis, Universitat Politècnica de Catalunya, 1990.

[Sha38]   C. Shannon. A symbolic analysis of relay and switching circuits. *Transactions AIEE*, 57:59–98, 1938.

[Sha49]   C. Shannon. The synthesis of two-terminal switching circuits. *Bell Systems Technical Journal*, 28:59–98, 1949.

[SV81]   Y. Shiloach and U. Vishkin. Finding the maximum, merging and sorting in a parallel computation model. *Journal of Algorithms*, 2:88–102, 1981.

[Sho67]   J. R. Shoenfield. *Mathematical Logic*. Series in Logic. Addison-Wesley, Reading, MA, 1967.

[Sim75]   J. Simon. *On Some Central Problems in Computational Complexity*. PhD thesis, Cornell University, 1975.

[Smo87]   R. Smolensky. Algebraic methods in the theory of lower bounds for Boolean circuit complexity. In *Proceedings 19th Symposium on Theory of Computing*, p. 77–82. ACM Press, 1987.

[Smo91]   C. Smoryński. *Logical Number Theory I*. Springer-Verlag, Berlin, 1991.

[Ste92]   I. A. Stewart. Using the Hamilton path operator to capture NP. *Journal of Computer and System Sciences*, 45:127–151, 1992.

[Sto77]   L. J. Stockmeyer. The polynomial-time hierarchy. *Theoretical Computer Science*, 3:1–22, 1977.

[SV84]   L. J. Stockmeyer and U. Vishkin. Simulation of parallel random access machines by circuits. *SIAM Journal on Computing*, 13:409–422, 1984.

[Str94]   H. Straubing. *Finite Automata, Formal Logic, and Circuit Complexity*. Birkhäuser, Boston, MA, 1994.

[Sud78]   I. H. Sudborough. On the tape complexity of deterministic context-free languages. *Journal of the ACM*, 25:405–414, 1978.

[Sze87]   R. Szelepcsényi. The method of forcing for nondeterministic automata. *Bulletin of the EATCS*, 33:96–100, 1987.

[Tan94]   S. Tan. *Calcul et vérification parallèles des problèmes d'algèbre linéaire*. PhD thesis, Université de Paris-Sud, U.F.R. Scientifique d'Orsay, 1994.

[Tar83]   R. E. Tarjan. *Data Structures and Network Algorithms*, vol. 44 of *CBMS-NSF Regional Conference Series in Applied Mathematics*. Society for Industrial and Applied Mathematics, 1983.

[Tar93]   J. Tarui. Probabilistic polynomials, AC^0 functions, and the polynomial hierarchy. *Theoretical Computer Science*, 113:167–183, 1993.

[Tod91]   S. Toda. PP is as hard as the polynomial-time hierarchy. *SIAM Journal on Computing*, 20:865–877, 1991.

[Tod92]   S. Toda. Classes of arithmetic circuits capturing the complexity of computing the determinant. *IEICE Transactions on Communications/Electronics/Information and Systems*, E75-D:116–124, 1992.

[Tor93]   J. Torán. P-completeness. In A. Gibbons and P. Spirakis (eds.), *Lectures on Parallel Computation*, vol. 4 of *Cambridge International Series on Parallel Computation*, p. 177–196. Cambridge University Press, Cambridge, 1993.

[Tra61]   B. A. Trakhtenbrot. Finite automata and logic of monadic predicates. *Doklady Akademii Nauk SSSR*, 140:326–329, 1961. In Russian.

[Val79]   L. G. Valiant. The complexity of computing the permanent. *Theoretical Computer Science*, 8:189–201, 1979.

[VSBR83]   L. Valiant, S. Skyum, S. Berkowitz, and C. Rackoff. Fast parallel computation of polynomials using few processors. *SIAM Journal on Computing*, 12:641–644, 1983.

[VV86]   L. G. Valiant and V. V. Vazirani. NP is as easy as detecting unique solutions. *Theoretical Computer Science*, 47:85–93, 1986.

[Ven86]   H. Venkateswaran. *Characterizations of Parallel Complexity Classes*. PhD thesis, Department of Computer Science, University of Washington, 1986.

[Ven91]   H. Venkateswaran. Properties that characterize LOGCFL. *Journal of Computer and System Sciences*, 43:380–404, 1991.

[Ven92]   H. Venkateswaran. Circuit definitions of non-deterministic complexity classes. *SIAM Journal on Computing*, 21:655–670, 1992.

[Ver93]   N. K. Vereshchagin. Relativizable and non-relativizable theorems in the polynomial theory of algorithms. *Izvestija Rossijskoj Akademii Nauk*, 57:51–90, 1993. In Russian.

[Vin91]   V. Vinay. *Semi-Unboundedness and Complexity Classes*. PhD thesis, Department of Computer Science and Automation, Indian Institute of Science, Bangalore, 1991.

[Vis83a]   U. Vishkin. Implementation of simultaneous memory access in models that forbid it. *Journal of Algorithms*, 4:45–50, 1983.

[Vis83b]   U. Vishkin. Synchronous parallel computation – a survey. Technical report, Courant Institute, New York University, 1983.

[Vol91]   H. Vollmer. The gap-language technique revisited. In *4th Computer Science Logic, Selected Papers*, Lecture Notes in Computer Science 533, p. 389–399, Springer-Verlag, Berlin, 1991.

[Vol98a]   H. Vollmer. A generalized quantifier concept in computational complexity theory. Technical report, Institut für Informatik, Universität Würzburg, 1998.

[Vol98b]   H. Vollmer. Relating polynomial time to constant depth. *Theoretical Computer Science*, 207:159–170, 1998.

[Vol99]   H. Vollmer. Uniform characterizations of complexity classes. *SIGACT News*, 30(1):17–27, 1999.

[vzG93]   J. von zur Gathen. Parallel linear algebra. In J. H. Reif (ed.), *Synthesis of Parallel Algorithms*, p. 573–617. Morgan Kaufmann, San Mateo, CA, 1993.

[Wag86]   K. W. Wagner. The complexity of combinatorial problems with succinct input representation. *Acta Informatica*, 23:325–356, 1986.

[WW86]   K. W. Wagner and G. Wechsung. *Computational Complexity*. VEB Verlag der Wissenschaften, Berlin, 1986.

[Weg87]   I. Wegener. *The Complexity of Boolean Functions*. Wiley-Teubner series in computer science. B. G. Teubner & John Wiley, Stuttgart, 1987.

[Weg89]   I. Wegener. *Effiziente Algorithmen für grundlegende Funktionen*. Leitfäden und Monographien der Informatik. B. G. Teubner, Stuttgart, 1989.

[Wil87]   C. Wilson. Relativized NC. *Mathematical Systems Theory*, 20:13–29, 1987.

[Wil90]   C. Wilson. On the decomposability of NC and AC. *SIAM Journal on Computing*, 19:384–396, 1990.

[Wra77]   C. Wrathall. Complete sets and the polynomial-time hierarchy. *Theoretical Computer Science*, 3:23–33, 1977.

[Yao85]   A. C. Yao. Separating the polynomial-time hierarchy by oracles. In *Proceedings 26th Foundations of Computer Science*, p. 1–10. IEEE Computer Society Press, 1985.

[Yao90]   A. C. Yao. On ACC and threshold circuits. In *Proceedings 31st Foundations of Computer Science*, p. 619–627. IEEE Computer Society Press, 1990.

[Zan91]   V. Zankó. #P-completeness via many-one reductions. *International Journal of Foundations of Computer Science*, 2:77–82, 1991.

List of Figures

Author Index

Subject Index

Monographs in Theoretical Computer Science · An EATCS Series

C. Calude
Information and Randomness
An Algorithmic Perspective

K. Jensen
Coloured Petri Nets
Basic Concepts, Analysis Methods
and Practical Use, Vol. 1
2nd ed.

K. Jensen
Coloured Petri Nets
Basic Concepts, *Analysis Methods*
and Practical Use, Vol. 2

K. Jensen
Coloured Petri Nets
Basic Concepts, Analysis Methods
and *Practical Use,* Vol. 3

A. Nait Abdallah
The Logic of Partial Information

Z. Fülöp, H. Vogler
Syntax-Directed Semantics
Formal Models
Based on Tree Transducers

A. de Luca, S. Varricchio
**Finiteness and Regularity
in Semigroups
and Formal Languages**

Texts in Theoretical Computer Science · An EATCS Series

J. L. Balcázar, J. Díaz, J. Gabarró
Structural Complexity I
2nd ed. (see also overleaf, Vol. 22)

M. Garzon
Models of Massive Parallelism
Analysis of Cellular Automata
and Neural Networks

J. Hromkovič
**Communication Complexity
and Parallel Computing**

A. Leitsch
The Resolution Calculus

G. Păun, G. Rozenberg, A. Salomaa
DNA Computing
New Computing Paradigms

A. Salomaa
Public-Key Cryptography
2nd ed.

K. Sikkel
Parsing Schemata
A Framework for Specification
and Analysis of Parsing Algorithms

H. Vollmer
Introduction to Circuit Complexity
A Uniform Approach

Former volumes appeared as
EATCS Monographs on Theoretical Computer Science

Vol. 5: W. Kuich, A. Salomaa
Semirings, Automata, Languages

Vol. 6: H. Ehrig, B. Mahr
Fundamentals of Algebraic Specification 1
Equations and Initial Semantics

Vol. 7: F. Gécseg
Products of Automata

Vol. 8: F. Kröger
Temporal Logic of Programs